INTELLIGENT CONTROL SYSTEMS WITH AN INTRODUCTION TO SYSTEM OF SYSTEMS ENGINEERING

SYSTEM OF SYSTEMS ENGINEERING SERIES

Series Editor

Mo Jamshidi

*University of Texas
Electrical and Computer Engineering Department
San Antonio, Texas, U.S.A.*

Intelligent Control Systems with an Introduction to System of Systems Engineering, *Thrishantha Nanayakkara and Mo Jamshidi*

Systems of Systems Engineering: Principles and Applications, *Mo Jamshidi*

SYSTEM OF SYSTEMS ENGINEERING SERIES

INTELLIGENT CONTROL SYSTEMS WITH AN INTRODUCTION TO SYSTEM OF SYSTEMS ENGINEERING

THRISHANTHA NANAYAKKARA
FERAT SAHIN
MO JAMSHIDI

CRC Press
Taylor & Francis Group
Boca Raton London New York

CRC Press is an imprint of the
Taylor & Francis Group, an **informa** business

MATLAB® is a trademark of The MathWorks, Inc. and is used with permission. The MathWorks does not warrant the accuracy of the text or exercises in this book. This book's use or discussion of MATLAB® software or related products does not constitute endorsement or sponsorship by The MathWorks of a particular pedagogical approach or particular use of the MATLAB® software.

MATLAB® code is available for download at: http://www.crcpress.com/product/isbn/9781420079241

CRC Press
Taylor & Francis Group
6000 Broken Sound Parkway NW, Suite 300
Boca Raton, FL 33487-2742

© 2010 by Taylor and Francis Group, LLC
CRC Press is an imprint of Taylor & Francis Group, an Informa business

No claim to original U.S. Government works

Printed in the United States of America on acid-free paper
10 9 8 7 6 5 4 3 2 1

International Standard Book Number: 978-1-4200-7924-1 (Hardback)

This book contains information obtained from authentic and highly regarded sources. Reasonable efforts have been made to publish reliable data and information, but the author and publisher cannot assume responsibility for the validity of all materials or the consequences of their use. The authors and publishers have attempted to trace the copyright holders of all material reproduced in this publication and apologize to copyright holders if permission to publish in this form has not been obtained. If any copyright material has not been acknowledged please write and let us know so we may rectify in any future reprint.

Except as permitted under U.S. Copyright Law, no part of this book may be reprinted, reproduced, transmitted, or utilized in any form by any electronic, mechanical, or other means, now known or hereafter invented, including photocopying, microfilming, and recording, or in any information storage or retrieval system, without written permission from the publishers.

For permission to photocopy or use material electronically from this work, please access www.copyright.com (http://www.copyright.com/) or contact the Copyright Clearance Center, Inc. (CCC), 222 Rosewood Drive, Danvers, MA 01923, 978-750-8400. CCC is a not-for-profit organization that provides licenses and registration for a variety of users. For organizations that have been granted a photocopy license by the CCC, a separate system of payment has been arranged.

Trademark Notice: Product or corporate names may be trademarks or registered trademarks, and are used only for identification and explanation without intent to infringe.

Library of Congress Cataloging-in-Publication Data

Nanayakkara, Thrishantha.
   Intelligent control systems with an introduction to system of systems engineering / Thrishantha Nanayakkara, Mo Jamshidi, Ferat Sahin.
      p. cm. -- (System of systems engineering series)
   Includes bibliographical references and index.
   ISBN 978-1-4200-7924-1 (alk. paper)
   1. Intelligent control systems. 2. Systems engineering. 3. Large scale systems. I. Jamshidi, Mohammad. II. Sahin, Ferat. III. Title.

TJ217.5.N35 2010
629.8--dc22   2009039083

Visit the Taylor & Francis Web site at
http://www.taylorandfrancis.com

and the CRC Press Web site at
http://www.crcpress.com

Contents

Preface xiii
About the Authors xv

1 Introduction 1
   References 7

2 Elements of a Classical Control System 9
   2.1 Introduction 9
      2.1.1 Plant 9
      2.1.2 Variables 10
      2.1.3 System 10
      2.1.4 Disturbances 11
   2.2 How the Model of a Dynamic System Can Help to Control It 11
      2.2.1 Newtonian Dynamics 11
      2.2.2 Control Experiments on a Point Mass 12
      2.2.3 Projects 15
   2.3 Control of Robot Manipulators 16
      2.3.1 Derivation of Dynamics of a Robot Manipulator 16
      2.3.2 Worked Out Project 1: Manipulator Dynamics 21
      2.3.3 Worked Out Project 2: Manipulator Dynamics 23
      2.3.4 Effect of Parameter Uncertainty on Model-Based Control Performance 24
      2.3.5 Worked Out Project 3: Hybrid Controllers 27
   2.4 Stability 31
      2.4.1 Equilibrium Points and Their Stability 31
      2.4.2 Lyapunov Stability 34
      2.4.3 Worked Out Example 35
   References 36

3 Introduction to System of Systems 37
   3.1 Introduction 37
   3.2 Definitions of SoS 38
   3.3 Challenging Problems in SoS 39
      3.3.1 Theoretical Problems 39
         3.3.1.1 Open Systems Approach to SoSE 39
         3.3.1.2 Engineering of SoS 40
         3.3.1.3 Standards of SoS 40
         3.3.1.4 SoS Architecting 41
         3.3.1.5 SoS Simulation 41
         3.3.1.6 SoS Integration 42

		3.3.1.7	Emergence in SoS	43
		3.3.1.8	SoS Management: Governance of Paradox	44
	3.3.2	Implementation Problems for SoS		45
		3.3.2.1	SE for the Department of Defense SoS	45
		3.3.2.2	E-enabling and SoS Aircraft Design via SoSE	46
		3.3.2.3	An SoS Perspective on Infrastructures	47
		3.3.2.4	Sensor Networks	48
		3.3.2.5	An SoS View of Services	48
		3.3.2.6	SoSE in Space Exploration	50
		3.3.2.7	Communication and Navigation in Space SoS	50
		3.3.2.8	Electric Power Systems Grids as SoS	52
		3.3.2.9	SoS Approach for Renewable Energy	52
		3.3.2.10	Sustainable Environmental Management from an SoSE Perspective	53
		3.3.2.11	Robotic Swarms as an SoS	54
		3.3.2.12	Transportation Systems	55
		3.3.2.13	Health Care Systems	57
		3.3.2.14	Global Earth Observation SoS	58
		3.3.2.15	Deepwater Coastguard Program	59
		3.3.2.16	Future Combat Missions	60
		3.3.2.17	National Security	61
	3.4	Conclusions		62
	References			62
4	**Observer Design and Kalman Filtering**			**67**
	4.1	State Space Methods for Model-Based Control		67
		4.1.1	Derivation of Dynamics for the Cart-Pole Balancing Problem	67
		4.1.2	Introduction to State Space Representation of Dynamics	70
		4.1.3	Regulator Control of State Vectors	72
	4.2	Observing and Filtering Based on Dynamic Models		73
		4.2.1	Why Observers?	74
		4.2.2	Elements of an Observer	74
		4.2.3	Simulations on an Inverted Pendulum Control	77
	4.3	Derivation of the Discrete Kalman Filter		83
	4.4	Worked Out Project on the Inverted Pendulum		87
	4.5	Particle Filters		94
	References			97
5	**Fuzzy Systems—Sets, Logic, and Control**			**99**
	5.1	Introduction		99
	5.2	Classical Sets		101

	5.3		Classical Set Operations 102
		5.3.1	Union 102
		5.3.2	Intersection 102
		5.3.3	Complement 103
	5.4	Properties of Classical Set 103
	5.5	Fuzzy Sets 104
	5.6	Fuzzy Set Operations 106
		5.6.1	Union 106
		5.6.2	Intersection 108
		5.6.3	Complement 108
	5.7	Properties of Fuzzy Sets 108
		5.7.1	Alpha-Cut Fuzzy Sets 109
			5.7.1.1 Alpha-Cut Sets 110
		5.7.2	Extension Principle 110
	5.8	Classical Relations versus Fuzzy Relations 112
	5.9	Predicate Logic 114
		5.9.1	Tautologies 118
		5.9.2	Contradictions 118
		5.9.3	Deductive Inferences 119
	5.10	Fuzzy Logic 121
	5.11	Approximate Reasoning 123
	5.12	Fuzzy Control 124
		5.12.1	Inference Engine 126
		5.12.2	Defuzzification 129
		5.12.3	Fuzzy Control Design 129
		5.12.4	Analysis of Fuzzy Control Systems 132
		5.12.5	Stability of Fuzzy Control Systems 135
			5.12.5.1 Time-Domain Methods 136
			5.12.5.2 Frequency-Domain Methods 138
		5.12.6	Lyapunov Stability 138
		5.12.7	Stability via Interval Matrix Method 143
	5.13	Conclusions 147
	References 147
6	**Neural Network-Based Control** 151
	6.1	Introduction to Function Approximation 151
		6.1.1	Biological Inspirations 151
		6.1.2	Construction of Complex Functions by Adaptive Combination of Primitives 160
		6.1.3	Concept of Radial Basis Functions 161
	6.2	NN-Based Identification of Dynamics of a Robot Manipulator 165
		6.2.1	An Approach to Estimate the Elements of the Mass, Centrifugal, Coriolis, and Gravity Matrices 168
		6.2.2	Optimum Parameter Estimation 169

6.3	Structure of NNs		174
6.4	Generating Training Data for an NN		178
6.5	Dynamic Neurons		179
6.6	Attractors, Strange Attractors, and Chaotic Neurons		181
	6.6.1	Attractors and Recurrent Neurons	181
	6.6.2	A Chaotic Neuron	183
6.7	Cerebellar Networks and Exposition of Neural Organization to Adaptively Enact Behavior		184

References ... 188

7 System of Systems Simulation ... 189

- 7.1 Introduction ... 189
- 7.2 SoS in a Nutshell ... 190
- 7.3 An SoS Simulation Framework ... 195
  - 7.3.1 DEVS Modeling and Simulation ... 195
  - 7.3.2 XML and DEVS ... 196
- 7.4 SoS Simulation Framework Examples ... 197
  - 7.4.1 Case 1: Data Aggregation Simulation ... 198
    - 7.4.1.1 DEVS-XML Format ... 198
    - 7.4.1.2 Programming Environment ... 199
    - 7.4.1.3 Simulation Results ... 200
  - 7.4.2 Case Study 2: A Robust Threat Detection System Simulation ... 202
    - 7.4.2.1 DEVS-XML Format ... 202
    - 7.4.2.2 Simulation Setup ... 203
    - 7.4.2.3 Robust Threat Detection Simulation ... 205
  - 7.4.3 Case 3: Threat Detection Scenario with Several Swarm Robots ... 208
    - 7.4.3.1 XML Messages ... 209
    - 7.4.3.2 DEVS Components of the Scenario ... 210
    - 7.4.3.3 DEVS-XML SoS Simulation ... 215
- 7.5 Agent-in-the-Loop Simulation of an SoS ... 222
  - 7.5.1 XML SoS Real-Time Simulation Framework ... 223
    - 7.5.1.1 Threat Detection Scenario ... 223
    - 7.5.1.2 Synchronization ... 223
  - 7.5.2 DEVS Modeling ... 224
    - 7.5.2.1 DEVS Components ... 224
    - 7.5.2.2 Mobile Swarm Agent ... 224
    - 7.5.2.3 Base Station ... 226
  - 7.5.3 Agent-in-the-Loop Simulation ... 227
- 7.6 Conclusion ... 228

Acknowledgment ... 228
References ... 229

8 Control of System of Systems ... 233
- 8.1 Introduction ... 233
- 8.2 Hierarchical Control of SoS ... 233
- 8.3 Decentralized Control of SoS ... 240
  - 8.3.1 Decentralized Navigation Control ... 243
  - 8.3.2 The Decentralized Control Law ... 244
    - 8.3.2.1 Motion Coordination ... 245
- 8.4 Other Control Approaches ... 252
  - 8.4.1 Consensus-Based Control ... 252
  - 8.4.2 Cooperative Control ... 255
  - 8.4.3 Networked Control ... 255
- 8.5 Conclusions ... 258

References ... 259

9 Reward-Based Behavior Adaptation ... 261
- 9.1 Introduction ... 261
  - 9.1.1 Embodiment ... 263
  - 9.1.2 Situatedness ... 263
  - 9.1.3 Internal Models ... 264
  - 9.1.4 Policy ... 264
  - 9.1.5 Reward ... 264
  - 9.1.6 Emergence ... 264
- 9.2 Markov Decision Process ... 265
  - 9.2.1 A Markov State ... 265
  - 9.2.2 Value Function ... 266
- 9.3 Temporal Difference-Based Learning ... 267
- 9.4 Extension to Q Learning ... 270
- 9.5 Exploration versus Exploitation ... 271
- 9.6 Vector Q Learning ... 272
  - 9.6.1 Summary of Results ... 276

References ... 279

10 An Automated System to Induce and Innovate Advanced Skills in a Group of Networked Machine Operators ... 281
- 10.1 Introduction ... 281
- 10.2 Visual Inspection and Acquisition of Novel Motor Skills ... 282
- 10.3 Experimental Setup ... 283
- 10.4 Dynamics of Successive Improvement of Individual Skills ... 285
- 10.5 Proposed Model of Internal Model Construction and Learning ... 289
- 10.6 Discussion and Conclusion ... 294

References ... 294

11 A System of Intelligent Robots–Trained Animals–Humans in a Humanitarian Demining Application 297
11.1 Introduction 297
11.1.1 Mine Detection Technology 298
11.1.2 Metal Detectors (Electromagnetic Induction Devices) 299
11.1.3 Ground-Penetrating Radar 299
11.1.4 Multisensor Systems Using GPR and Metal Detectors 300
11.1.5 Trace Explosive Detection Systems 301
11.1.6 Biosensors 301
11.1.7 Magnetic Quadrupole Resonance 301
11.1.8 Seismoacoustic Methods 301
11.2 A Novel Legged Field Robot for Landmine Detection 302
11.2.1 Key Concerns Addressed by the Moratuwa University Robot for Anti-Landmine Intelligence Design 302
11.2.2 Key Design Features of MURALI Robot 303
11.2.3 Designing the Robot with Solidworks 305
11.2.4 Motherboard to Control the Robot 333
11.2.5 Basics of Sensing and Perception 336
11.2.6 Perception of Environment and Intelligent Path Planning 339
11.3 Combining a Trained Animal with the Robot 346
11.3.1 Basic Background on Animal–Robot Interaction 347
11.3.2 Background on Reward-Based Learning 348
11.3.3 Experience with Training a Mongoose 348
11.3.3.1 Phase 1: Conditioning Smell, Reward, and Sound 348
11.3.3.2 Phase 2: Learning in a Paradigm Where the Degree of Difficulty of Correct Classification Was Progressively Increased 349
11.4 Simulations on Multirobot Approaches to Landmine Detection 353
References 359

12 Robotic Swarms for Mine Detection System of Systems Approach 363
12.1 Introduction 363
12.1.1 Swarm Intelligence 363
12.1.2 Robotic Swarms 364
12.1.3 System of Systems 365
12.2 SoS Approach to Robotic Swarms 366
12.2.1 Interoperability 366
12.2.2 Integration 367

- 12.3 Designing System of Swarm Robots: GroundScouts 368
  - 12.3.1 Hardware Modularity 369
    - 12.3.1.1 Locomotion 370
    - 12.3.1.2 Control 371
    - 12.3.1.3 Sensor 372
    - 12.3.1.4 Communication 373
    - 12.3.1.5 Actuation 373
  - 12.3.2 Software Modularity 374
    - 12.3.2.1 Operating System 375
    - 12.3.2.2 Dynamic Task Uploading 375
  - 12.3.3 Communication Protocol: Adaptive and Robust 376
    - 12.3.3.1 Physical Layer 377
    - 12.3.3.2 MAC Layer 377
    - 12.3.3.3 Implementation Results 381
- 12.4 Mine Detection with Ant Colony-Based Swarm Intelligence 382
  - 12.4.1 Simulation of Mine Detection with ACO 383
    - 12.4.1.1 Ant Colony Optimization 383
    - 12.4.1.2 The Mine Detection Problem 385
    - 12.4.1.3 ACO Model Used for Mine Detection 387
  - 12.4.2 Implementation of ACO-Based Mine Detection Algorithm on GroundScouts 397
    - 12.4.2.1 Implementation of Mines 398
    - 12.4.2.2 Implementation of the Scent 399
    - 12.4.2.3 Main Program 400
    - 12.4.2.4 Graphical User Interface 403
  - 12.4.3 Experimental Results 406
- 12.5 Conclusion 409
- Acknowledgment 410
- References 410

Index 417

Preface

In recent years, there has been a growing recognition that significant changes need to be made in the way different modules interact with each other in industries, especially in the aerospace and defense sector. The aerospace industry, at least in the United States, is undergoing a transformation or evolution. Today, major aerospace and defense manufacturers, including (but not limited to) Boeing, Lockheed-Martin, Northrop-Grumman, Raytheon, BAE Systems, etc., all include some version of "large-scale systems integration" as a key part of their business strategies. In some cases, these companies have even established entire business units dedicated to systems integration activities, implying an emerging engineering field called system of systems (SoS) engineering.

SoS are super systems composed of other elements that themselves are independent complex operational systems interacting among themselves to achieve a common goal. Each element of an SoS achieves well-substantiated goals even if they are detached from the rest of the SoS. SoS exhibit behavior, including emergent behavior, not achievable by the component systems acting independently. SoS are considered metasystems that are diverse in their components' technologies, context, operation, geography, and conceptual framework. For example, a Boeing 747 airplane, as an element of an SoS, is not an SoS, but an airport is an SoS, or a rover on Mars is not an SoS, but a robotic colony (or a robotic swarm) exploring the red planet is an SoS.

In the space research and development domain, National Aeronautics and Space Administration's International Space Station (ISS), which is considered the largest and most complex international scientific project in history and Boeing's largest venture into space to date, is an SoS. When the ISS is completed around 2010, it will be comprised of more than 100 major components carried aloft during 88 space flights to assemble the space station. The ISS is a space example of an SoS, from both technology and planning viewpoints. More than 16 nations (and their agencies) are contributing to the ISS. As an SoS, the ISS has many systems. The various systems developed by Boeing include thermal control; life support; guidance, navigation, and control; data handling; power systems; and communications and tracking. Boeing also integrates vehicle elements provided by the international partners. International partner components include a Canadian-built, 55-ft-long robotic arm and mobile servicing system used for assembly and maintenance tasks on the space station; a pressurized European laboratory called Columbus and logistics transport vehicles; a Japanese laboratory called Kibo, with an attached exposed exterior platform for experiments, as well as logistics transport vehicles; and two Russian research modules, an early living quarters called the Zvezda Service Module with its own life support and habitation systems, logistics transport vehicles, and Soyuz spacecraft for crew return and transfer.

Associated with SoS are numerous problems and open-ended issues needing a substantial number of fundamental advances in theory and verifications. In fact, there is not even a universal definition in the system engineering community. However, there are a number of areas where exciting engineering innovations can take place in an SoS. This book makes an attempt to discuss some fundamentals in the areas of dynamic systems, control, neuroscience, soft computing, signal processing, and systems integration that can potentially make the core of this emerging field of SoS engineering.

MATLAB® is a registered trademark of The MathWorks, Inc. For product information, please contact:

The MathWorks, Inc.

3 Apple Hill Drive

Natick, MA 01760-2098, USA

Tel: 508-647-7000

Fax: 508-647-7001

E-mail: info@mathworks.com

The authors wish to thank Ms. Roberta Bauer of the University of Texas, San Antonio for her assistance in preparing and checking the index.

Thrishantha Nanayakkara
Ferat Sahin
Mo Jamshidi

About the Authors

Thrishantha Nanayakkara was born in 1970 in Galle, Sri Lanka. He obtained his BS degree in electrical engineering from the University of Moratuwa in 1996 and earned his MS in electrical engineering in 1998 and PhD in systems control and robotics from Saga University in 2001. From 2001 to 2003, he was a postdoctoral research fellow at the Department of Biomedical Engineering, School of Medicine, Johns Hopkins University, Baltimore, Maryland. From 2003 to 2007, he was a faculty member at the University of Moratuwa, Sri Lanka, and was the principal investigator of the "Laboratory for Intelligent Field Robots" at the Department of Mechanical Engineering. He received an award as outstanding researcher at the University of Moratuwa in 2006, and has published 4 book chapters, 7 journal articles, and 26 international conference papers. Dr. Nanayakkara was also the founding general chair of the International Conference on Information and Automation, and is an associate editor of the *Journal of Control and Intelligent Systems*. At present, he is a fellow in the Radcliffe Institute, Harvard University, and a research affiliate of the Computer Science and Artificial Intelligence Laboratory of the Massachusetts Institute of Technology, Cambridge, Massachusetts, where his work is partly sponsored by the Harvard Committee on Human Rights. His current research interests are in robotic locomotion, odor-guided behavior of rodents and animal–robot–human combined systems.

Ferat Sahin received his BS degree in electronics and communications engineering from Istanbul Technical University, Turkey, in 1992, and his MS and PhD degrees from the Virginia Polytechnic Institute and State University in 1997 and 2000, respectively. In September 2000, he joined the Rochester Institute of Technology (RIT), where he is an associate professor. He is also the director of the Multi Agent Bio-Robotics Laboratory at RIT. In 2006, he was at the University of Texas San Antonio for his sabbatical as visiting research associate professor. His current research interests are system of systems simulation and modeling, swarm intelligence, robotics, microelectromechanical systems (MEMS) materials modeling, MEMS-based microrobots, microactuators, distributed computing, decision theory, pattern recognition, distributed multiagent systems, and structural Bayesian network learning. In addition to conference and journal publications in these areas, he is also the coauthor of *Experimental and Practical Robotics*. and has been a reviewer of leading journals and conferences for both IEEE and other organizations. Dr. Sahin has been an active IEEE member since 1996, and is a member of the IEEE SMC Society, the Robotics and Automation Society, and the Computational Intelligence Society. Locally, he has served as secretary

(2003) and section vice-chair (2004 and 2005) in the IEEE Rochester Section. He was also the faculty advisor of the IEEE student chapter at RIT from 2001 to 2003. He has been active in the IEEE SMC Society since 2000. He served as the student activities chair for the IEEE SMC Society in 2001, 2002, and 2003, and he has been the secretary of the IEEE SMC Society since 2003. He has received an Outstanding Contribution award for his service as the SMC Society secretary. He was also a member of the SMC Strategic Opportunities and Initiatives Committee and the SMC Technical Committee on Robotics and Intelligent Sensing, and was the publication cochair for the IEEE SMC International Conference on System of Systems Engineering (SOSE 2007). Dr. Sahin currently serves as the deputy editor-in-chief for the *International Journal of Computers and Electrical Engineering* and an associate editor for the *IEEE Systems Journal* and *AutoSoft Journal*, he also serves as the technical cochair of the IEEE SMC International Conference on System of Systems Engineering (SOSE 2008 and SOSE 2009).

Mo M. Jamshidi received his BS degree in electrical engineering from the Oregon State University in 1967, and his MS and PhD degrees in electrical engineering from the University of Illinois at Urbana–Champaign in 1969 and 1971, respectively. He holds three honorary doctorate degrees from Azerbaijan National University (Azerbaijan, 1999); the University of Waterloo, Canada; and the Technical University of Crete, Greece. He is currently the Lutcher Brown endowed chair professor at the University of Texas, San Antonio. He has served in various capacities with the U.S. Air Force Research Laboratory, the U.S. Department of Energy, NASA Headquarters, NASA JPL, Oak Ridge National Laboratory, and the Los Alamos National Laboratory. He has also served in various academic and industrial positions at various national and international organizations including IBM and GM Corporation. In 1999, he was a NATO Distinguished Professor in Portugal, and in 2008, he was a UK Royal Academy of Engineering fellow in the UK. He has published close to 600 technical publications, including 63 books and edited volumes. He is the founding editor, cofounding editor, or editor-in-chief of five journals, including the *International Journal of Computers and Electrical Engineering* and *Intelligent Automation and Soft Computing*, and one magazine (*IEEE Control Systems Magazine*). He is also the editor-in-chief of the new *IEEE Systems Journal* (2006–present), and coeditor-in-chief of the *International Journal on Control and Automation*. He has also been on the executive editorial boards of a number of journals and two encyclopedias. Dr. Jamshidi is a fellow of the Institute of Electrical and Electronics Engineers (IEEE), the American Society of Mechanical Engineers (ASME), the American Association for the Advancement of Science (AAAS), the Third World Academy of Science (TWAS), New York Academy of Science (NYAS), as well as the Academy of Sciences for the Developing World (Trieste, Italy). He is also a member of the Russian Academy of Nonlinear Sciences; associate fellow of the AIAA and the Hungarian Academy of Engineering;

About the Authors

Dr. Jamshidi has been the recipient of the IEEE Centennial Medal and IEEE Control Systems Society (CSS) Distinguished Member Award, and the IEEE CSS Millennium Award, the IEEE's Norbert Weiner Research Achievement Award, and the Outstanding Contribution Award of the IEEE Systems, Man, and Cybernetics (SMC) Society, and is the founding director of the International Consortium on System of Systems (www.icsos.org) and a founding chair of IEEE Conference on System of Systems Engineering (http://www.ieeesose2009.org). He is currently leading a global effort on the theory and applications of system of systems engineering.

1
Introduction

This book discusses the foundation for emerging intelligence in a loosely coupled system of systems (SoS) with a set of objectives. The term "intelligence" is a widely used term without a concrete definition. There are many schools of thought on what intelligence is, with arguments for and against each of them [1]. This book looks at intelligence as an emerging property of systems that interact among each other and with the outside world. Systems can be primitive computational elements such as neurons, mathematical equations, or electrical, electronic, and mechanical components, etc., or they can be complex entities such as a robot, a team of robots, a fleet of vessels, individuals in a group, etc. To understand the term "emergence," let us look at a group of people listening to a speaker. As far as the audience is silent, the speaker goes on talking about the topic. Nothing emerges from the audience except for individual thoughts interacting with the speaker. Imagine a break is given for the audience members to talk to each other on the topic being discussed. Then, people go about talking to each other, exchanging ideas with agreements and disagreements. Over time, the individuals who contribute to the discussion change their original opinions and sometimes come up with new ideas triggered by new ways of thought introduced by others. Therefore, the group as a whole "emerges" concepts that were not there at the beginning. This book tries to argue that this fundamental idea of emergence holds powerful meaning in systems of any scale. The only requirement is to design a process that allows subsystems of a system to communicate among each other in order to emerge intelligent behavior envisaged by the designer.

There are several factors affecting the fruitfulness of the above interaction. If we go back to our example, the final outcomes can vary depending on how the discussion was coordinated. If there was an arbitrator who intervened in the way people met each other, for instance, by introducing one to another with some purpose, the type of ideas that would emerge would be different from those that would emerge if the group was allowed to interact randomly. In addition, the character of individuals also matters in the quality of the outcome. Suppose that there were very stubborn people who would not change and tolerate diverse ideas. They may make the discussion very boring. Therefore, internal characteristics of the participating subsystems, such as plasticity, compliance, adaptiveness, stability, and inertia to change, etc., affect the final behavior. The time span of interaction also decides how transient behaviors tend to settle in a stable behavior. Moreover, how information is interpreted by each individual also affects the final outcome. Therefore, it is difficult to pinpoint a single place where intelligence resides. In essence, it is decided by an array of factors, including how goals are set, the inherent characteristics of subsystems, how information is exchanged, how information is processed, the mechanisms available for adaptation, how adaptation is rewarded or punished, etc.

The book is more of a discussion with the reader than a formal presentation of theories. The chapters are organized in the following manner.

In Chapter 2, we discuss the elements of a classical control system. It is important to discuss the elements of a control system because any system can be viewed as a control system with or without a feedback loop [2,3]. A system changes its state when an external stimulus is applied. There are systems that respond to the stimuli without worrying about how the states are being changed. These are called open loop systems. An example of an open loop system is a water tap. The more you open it, the faster the water flow will be. There are other systems that try to add a corrective feedback action on the system by further processing the states being changed by the external input. These are called feedback control systems. An example is an air conditioner. When you set a desired temperature, the air conditioner tries to keep the room temperature at that level when the outside temperature goes through variations. Knowledge of the essential components needed to build a system that renders a desired primitive behavior is important to design more complex systems with more features such as learning and adaptation. Here, we take examples from robotics to understand how these fundamentals can be used to design stable and optimal controllers for dynamical systems found in the industry [4].

Chapter 3 introduces the concept of SoS and the challenges ahead to extend systems engineering to SoS engineering. The birth of a new engineering field may be on the horizon—System-of-Systems Engineering. An SoS is a collection of individual, possibly heterogeneous, but functional systems integrated together to enhance the overall robustness, lower the cost of operation, and increase the reliability of the overall complex (SoS) system. Having said that, the field has a large vacuum from basic definition, to theory, to management and implementation. Many key issues, such as architecture, modeling, simulation, identification, emergence, standards, net-centricity, control (see Chapter 8), etc., are all begging for attention. In this chapter, we will be going briefly through all these issues and bring the challenges to the attention of interested readers [11–18].

This growing interest in SoS as a new generation of complex systems has opened a great many new challenges for systems engineers. Performance optimization, robustness, and reliability among an emerging group of heterogeneous systems in order to realize a common goal has become the focus of various applications, including those in military, security, aerospace,

Introduction

space, manufacturing, service industry, environmental systems, and disaster management.

Chapter 4 discusses the mathematical tools available to observe the internal states of a system not directly measured by physical sensors, and to optimally estimate the real state of the system when the measurements and states themselves are contaminated with noise. Consider, for instance, a pendulum that swings back and forth under gravity. You have an angle sensor attached to the pivotal point of the pendulum around which it swings. therefore, you get a direct measurement of the angle of the pendulum at any given time. yet, if you see the voltage signal coming from the angle sensor on an oscilloscope, you will notice that the angle signal is contaminated with noise. The level of noise depends on the quality of the sensor on one hand and the firmness of the mechanical assembly of the pendulum on the other. We call The former type of noise the measurement noise, and the latter the process noise. The Next question is how to estimate the state of true angle, and the other higher-order states such as angular velocity of the pendulum using this noisy measurement. To understand how to solve engineering problems like this, we first discuss the basic concepts of observing based on our knowledge about state space modeling and feedback control systems. then, we go on to discuss the mathematical derivation and application of kalman filters and particle filters that are extensively used in various industrial applications [9,10]. Knowledge about these filtering and state observing techniques is important in the systems integration stage in order to insulate one system from noise injected from another system.

Chapter 5 discusses how linguistic information can be modeled to improve the performance of a system or a combination of independent systems. Here, we introduce the concept of "membership" of a measurement in a given set. For instance, if we see a 5.7-ft-tall man, how do we categorize him in the classes of "short" and "tall"? This is the type of classification where we use the notion of memberships in a class. In this particular example, we may assign a membership of 0.7 in the "tall" class and 0.3 in the "short" class. All our subsequent decisions about this person will depend on this interpretation of his height. This conversion of a measurement to membership in linguistic classes is important to exploit linguistic rules being used by people to express their experience on controlling complex systems. For instance, a linguistic rule such as *"if* the road is slippery and the traffic is high, *then* keep your speed low."* This type of inferencing based on linguistic labels is known as fuzzy inferencing [11,12]. This rule may help numerous sensors such as speedometers, cameras, inertial navigation systems, brake sensors, engine power sensors, etc., mounted on intelligent cars to coordinate among each other to do a meaningful job.

Chapter 6 discusses some techniques available to approximate nonlinear static and dynamic systems using a set of primitive functions. This is inspired by the manner in which biological brains approximate the function of complex systems. If we take a simple example, let us consider the nonlinear function given by $y = \exp(x^3 + 2x^2 + x + 5)$. If we are only concerned about doing what this function does in terms of mapping the value of the variable x to the corresponding value of y, do we really have to know the exact equation $y = \exp(x^3 + 2x^2 + x + 5)$? The answer is no. It is enough to know the landscape created by this function in the x-y coordinate system to do this mapping. Similarly, our brain receives sensory information from five sensors, and the processed information is often mapped to actions through muscle commands. This mapping should take place to suit the task being done. For instance, if we are required to learn how to ride a bicycle, the brain has to construct a model of the dynamics of the bicycle in order to perform smooth maneuvering. The brain does this by constructing a neural network that approximates the dynamics of the bicycle. This can be done by combining a set of primitive mathematical landscapes found in special cells knows as neurons. Therefore, in this chapter, we demonstrate how such primitive mathematical functions known as artificial neurons can be used to approximate complex systems. This type of approximation can be useful to complement crude mathematical models of subsystems of a large system or to understand complex dynamics emerging from a system of simpler subsystems of which the dynamics are fully or partially known.

Chapter 7 discusses an SoS [5,6], simulation framework, related case studies, and agent-in-the-loop simulation for SoS test and evaluation. First, it introduces the SoS concepts and recent work on SoS simulation and modeling. Then, it explores an SoS simulation framework based on discrete event specification tools and Extensible Markup Language (XML) [7]. Three case studies (scenarios) are simulated with the SoS simulation framework on robust threat detection and data aggregation using heterogeneous systems of rovers (systems). Finally, a real-time SoS simulation framework is introduced in agent-in-the-loop setting for testing and evaluating systems in an SoS. With the agent-in-the-loop simulation, a robust threat detection scenario is simulated with four virtual (simulated) robots and one real robot. Continuity models for the real and virtual robots are presented in addition to a communication interface between the real robot and the virtual ones [8].

Chapter 8 discusses one of the key theoretical open problems of SoS—control design. Among all open questions in engineering of SoS, control and sensing are among the most important ones. From the control design viewpoint, the difficulty arises that each system's control strategy cannot solely depend on its own onboard sensory information, but also due to communication links among all the neighboring systems or between sensor, controllers, and actuators. The main challenge in the design of a controller for SoS is the difficulty or impossibility of developing a comprehensive SoS model, either analytically or through simulation, by and large; SoS control remains an open problem and is, of course, different for each application domain. Should a mathematical model be available, several control paradigms are available, which will be the focus of this chapter. Moreover, real-time control—which is required in almost all application domains—of interdependent systems poses an especially difficult problem. Nevertheless, several potential control paradigms are briefly considered in this chapter. The control paradigms discussed in Chapter 8 are hierarchical, decentralized, consensus-based, cooperative, and networked. Simulation results and design algorithms are presented in this chapter.

Chapter 9 discusses how adaptive systems can optimize their behaviors using external rewards and punishments. This is an important area of machine learning. The basic idea is to use qualitative feedback as to whether the performance is good or bad to retune various system and control parameters. For instance, let us consider a robotic head that is supposed to look at a human and react with facial expressions like the Kismet robot at Massachusetts Institute of Technology. Here, there are no quantitative measures such as position or velocity error to correct the reaction of the robot. The human can say whether the robot's reaction is close to being natural or not. Perhaps the human can give a score between 0 and 100. Then, the robot can use a strategy to change the behavior to obtain higher scores. This strategy is called a reinforcement-based learning strategy [19]. This chapter discusses the fundamentals of how such a learning system can be designed. Reward-based learning systems are very important to future large-scale systems because the engineered systems undergo many internal and external changes (e.g., friction, electrical impedance, etc.). It is very difficult to design controllers to be optimal and stable for a wide class of environments. If subsystems of a large system can stay tuned to their respective goals, the control of the large system can become simpler.

Chapter 10 discusses an industrial application of an SoS design. The problem to be solved was given by a garment manufacturing company in Sri Lanka. The factory employs more than 400 manual workers to operate modern sewing machines. In a given session, the whole group sews one particular piece of garment such as a sleeve or a collar of a shirt. The optimality of the machine operating behavior of a worker is measured using a set of criteria such as the requirement to minimize the idling time while sewing a given piece, minimizing jerk (time derivative of acceleration) of the machine, and minimizing the time taken to finish a piece. In the absence of precise measurements, the supervisors evaluate the effectiveness of workers by observing the quality of the finish of the pieces sewn by a worker. This process is quite inefficient in a factory with a large number of workers manufacturing a wide variety of garments. Furthermore, the workers themselves do not have an objective method to evaluate their own performance. The proposed solution was to provide a system where each worker can instantly see his/her machine's spin velocity profile compared against that of the best worker at any given time. This allowed each worker to match their performance against an elite template and derive their own interpretations of the possible causes that make the difference. This evaluation helps them to experiment with their own motor skills and come up with innovative ideas to improve their own performance [20]. The best worker was selected by a manager who watched the machine spin profiles of each worker from a remote computer. It was observed that workers dynamically improved their skills to beat elite performers, and sometimes, new elite performers emerged with minimum involvement from the management. In this chapter, we discuss the fundamentals of motor learning concepts and evaluate several hypotheses that can explain the emerging behavior of a system of live systems.

Chapter 11 discusses an application where a field robot was designed and tested to support a trained animal, a mongoose, to detect landmines [21]. A trained human operator could influence the animal's odor-guided behavior to locate landmines by controlling the field robot from a remote location. Therefore, this system is a coexistence of an animal, a human, and a field robot to achieve a common goal of detecting landmines. The chapter presents real-world experimental results. Furthermore, we will go through the steps of designing the robot using Solidworks. It is very important for a modern engineer to be able to use a computer-aided design software to sketch, simulate, and visualize designs before fabricating real hardware. A good feature of Solidworks is that we have the opportunity to visualize the drawings in a three-dimensional world, add mathematical relationships among various elements of the design, edit dimensions at any stage of the development, and finally simulate to see how different components interact among each other. In the beginning, we will try to go through detailed steps. However, later on, we will skip those principles learned earlier. We will not discuss the drawing of the complete robot since the purpose is to give training on basic design using Solidworks.

Chapter 12 discusses robotic swarms for mine detection in the context of SoS [13,14]. The chapter discusses swarm intelligence, robot swarm, and SoS characteristics. Then, it draws parallels between robotic swarms and SoS by evaluating common characteristics such as interoperability, integration, and adaptive communications. With these common characteristics in mind, the chapter evaluates a swarm intelligence technique, ant colony optimization (ACO), and applies it to a mine detection problem. The ACO-based mine detection algorithm is then developed and simulated in MATLAB® with good results. Finally, the ACO-based mine detection algorithm is implemented and tested on micromodular swarm robots, called GroundScouts, in real life. The experiments are carried out in a basketball court with infrared emitting mines. The chapter also discusses detailed hardware and software design approaches for GroundScouts in order to make them suitable for swarm intelligence applications.

References

[1] Steels, L., and Brooks, R. (Eds.), *An Artificial life Route to Artificial Intelligence—Building Embodied Situated Agents*, Lawrence Erlbraum Associates, New Haven, CT, 1995.
[2] Ogata, K., *Modern Control Engineering*, 4th edition, Pearson Education Inc., New Delhi, 2002.
[3] Nise, N. S., *Control Systems Engineering*, 3rd edition, John Wiley & Sons Inc., New York, NY, 2000.
[4] Craig, J. J., *Introduction to Robotics—Mechanics and Control*, 2nd edition, Pearson Education Asia, Singapore, 1989.
[5] Jamshidi, M. (Ed.), *System of Systems—Innovations for the 21st Century*, Wiley & Sons, New York, NY, 2009.
[6] Jamshidi, M. (Ed.), *System of Systems Engineering*, CRC Press, Boca Raton, FL, 2008.
[7] Parisi, C., Sahin, F., and Jamshidi, M., "A discrete event XML based system of systems simulation for robust threat detection and integration," in *IEEE International Conference on System of Systems*, Monterey, CA, June 2008.
[8] Hu, X., and Zeigler B., "Model continuity in the design of dynamic distributed real-time systems," *IEEE Transactions on Systems, Man and Cybernetics, Part A* 35(6), 867–878, 2005.
[9] Chui, C. K., and Chen, G., *Kalman Filering with Real-Time Applications*, 2nd edition, Springer Verlag, Berlin, 1991.
[10] Ristic, B., Arulampalam, S., and Gordon N., *Beyond the Kalman Filter—Particle Filters for Tracking Applications*, Artech House, Boston, MA, 2004.
[11] Zadeh, L. A., "Fuzzy sets," *Information and Control*, 8, 338–353, 1965.
[12] Baldwin, J. F., "Fuzzy logic and fuzzy reasoning," in *Fuzzy Reasoning and Its Applications*, E. H. Mamdani and B. R. Gaines, Eds., Academic Press, New York, NY, 1981.
[13] Crossley, W. A., "System of systems: An introduction of Purdue University Schools of Engineering's signature area," Engineering Systems Symposium, March 29–31 2004, Tang Center, Wong Auditorium, MIT, 2004.
[14] Abel, A., and Sukkarieh S., "The coordination of multiple autonomous systems using information theoretic political science voting models," *Proceedings of IEEE International Conference on System of Systems Engineering*, Los Angeles, April 2006.
[15] Azarnoosh, H., Horan, B., Sridhar, P., Madni, A. M., and Jamshidi, M., "Towards optimization of a real-world robotic-sensor system of systems," in *Proceedings of World Automation Congress (WAC) 2006*, Budapest, Hungary, July 24–26 2006.
[16] DiMario, M. J., "System of systems interoperability types and characteristics in joint command and control," *Proceedings of IEEE International Conference on System of Systems Engineering*, Los Angeles, April 2006.
[17] Lopez, D., "Lessons learned from the front lines of the aerospace," *Proceedings of IEEE International Conference on System of Systems Engineering*, Los Angeles, April 2006.

[18] Wojcik, L. A., and Hoffman, K. C., "Systems of systems engineering in the enterprise context: a unifying framework for dynamics," *Proceedings of IEEE International Conference on System of Systems Engineering*, Los Angeles, April 2006.
[19] Sutton, R. S., and Barto, A. G., *Reinforcement Learning—An Introduction*, MIT Press, Cambridge, MA, 2000.
[20] Shadmehr, R., and Wise, S. P., *The Computational Neurobiology of Reaching and Pointing—A Foundation for Motor Learning*, MIT Press, Cambridge, MA, 2005.
[21] Habib, M. K. (Ed.), *Humanitarian Demining—Innovative Solutions and the Challenges of Technology*, Online book: http://www.intechweb.org/books.php?sid=11&content=subject.

2

Elements of a Classical Control System

2.1 Introduction

We are living in an era where man has been able to design and implement precise control systems not only in factory applications, such as robotic welding, assembling, and painting, etc., but also in military and space applications such as antimissile guard systems, coordination and stabilizing the modules in the International Space Station, and robots that explore other planets in the solar system. The history of modern control systems dates back to the seventeenth century. One of the best-known engineering achievements is the centrifugal governor designed by James Watt in 1769 based on a suggestion made by his industrial partner, Mathew Boulton. Before J. C. Maxwell gave the first rigorous mathematical foundations of automatic control in 1868 [1], the mechanical feedback systems proposed by James Watt contributed to the industrial revolution. We have to keep in mind that these developments took place when there were no digital computers. However, the early control engineers had carefully accounted for the essential features of a dynamical system before devising a control system to keep the dynamic system on track. With the advent of digital computers, the theoretical basis and the range of applications of control systems has witnessed dramatic improvements in the past few decades [2,3]. Figure 2.1 shows the concept of the centrifugal governor. Let us discuss the elements of a classical controller using Figure 2.1.

2.1.1 Plant

The plant is a set of machine parts that performs the intended task. In this case, the turbine and the generator together deliver electricity. The turbine spins when the nozzle injects water or steam toward the turbine blades. We derive some useful output such as electricity by connecting a generator to the turbine shaft. Therefore, the turbine and the generator will make up most of the plant needed to generate electricity. Yet, for the electricity to be useful, we have to maintain the frequency of the voltage waveform at a fixed value (e.g., 50 or 60 Hz). This is achieved by controlling the system variables.

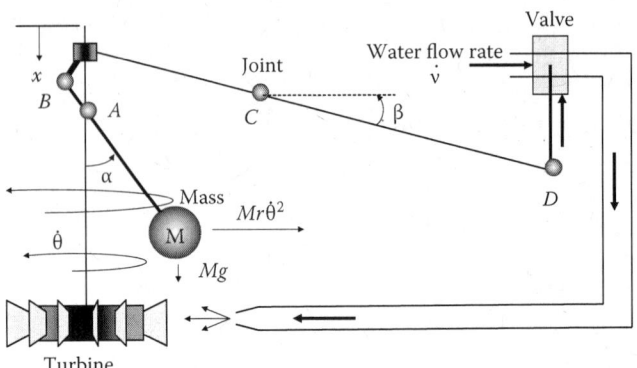

FIGURE 2.1
Elements of the centrifugal governor in a water turbine control system.

2.1.2 Variables

There are two sets of variables—the controlled variables and the manipulated variables. The purpose of a control system is to keep the controlled variables close to the desired values. In the example above, the controlled variable is the angular speed of the turbine, $\dot{\theta}$. The manipulated variable is the one we change to keep the controlled variable close to a desired value. Therefore, the manipulated variable and the controlled variable should have some relationship. In this case, the manipulated variable is the water or steam flow rate $\dot{v}$ through the valve. In order to relate the flow rate through the valve $\dot{v}$ to the angular speed of the turbine $\dot{\theta}$, the system has designed a mass M hung on the turbine shaft so that the mass rotates at the same angular speed as the turbine. When a mass rotates, it produces a centrifugal force $Mr\dot{\theta}^2$, where M and r are the mass, and the radius of the locus of the mass. Therefore, when $\dot{\theta}$ increases, the elevated centrifugal force increases α. This makes the lever mechanism reduce the opened area of the valve. This leads to a reduction in the water/steam flow rate that directly results in a reduction in the energy given to the turbine. Therefore, the turbine slows down. On the other hand, the valve opens more when $\dot{\theta}$ drops and gives more energy to increase $\dot{\theta}$. We can change the relationship between the energy injection through the valve and the angular speed of the turbine $\dot{\theta}$ by changing the kinematics of the lever system.

2.1.3 System

A system is a collection of elements working together to achieve a set of objectives. In the above example, the mass, lever arrangement, and the valve work together to regulate the water/steam flow rate to control the angular speed of the turbine. Moreover, the turbine and the generator make up a system to convert potential energy to electric energy. Therefore, we can identify two systems in an automatically controlled system. One is the system being controlled and the other is the controller. We can also consider the two systems together to form one system of systems.

2.1.4 Disturbances

In any natural environment, it is very hard to isolate a system from other systems. Therefore, variables in the control system will be affected by various signals. A disturbance is such a signal coming from either outside or inside the system that influences the variables of the control system. Sometimes, the disturbances can be detrimental to the performance of the control system. An example of a destructive disturbance is a pressure change or the turbulence in the water flow that makes the energy injected into the turbine irregular. There are very special cases where the artificial disturbances are added to improve the system. An example is landing an unmanned aerial vehicle (UAV). When a UAV is being landed, the remote operators tend to misestimate the ground clearance when it is very low. Sometimes, this leads to plane crashes. Adding a jitter proportional to the inverse of the ground clearance to the joystick helps the operator to change his/her hand stiffness and improve the landing performance.

The example in Figure 2.1 does not use a dynamic model of the plant (turbine and the generator) to control the angular speed of the turbine. Instead, it has introduced another dynamic system that reacts to the controlled variable of the system. However, there are applications where the controlled variable changes its value very fast or, in another case, coupling another physical system to the main plant may be practically impossible. For example, in the case of a robot manipulator tracking a contour in a welding task, the joints move only a little. It does not produce enough energy to drive another dynamic system like the rotating mass in the above example. Therefore, we have to measure weak variables and amplify them to see the states of the dynamic system. In other cases, the controller and the dynamic system are located far apart, so that the measurement of the controlled variable arrives at the controller after some time. Therefore, in such cases, the model of the dynamic system would help the controller predict the response of the dynamic system. Therefore, we discuss some useful concepts in model-based control in the following sections.

2.2 How the Model of a Dynamic System Can Help to Control It

2.2.1 Newtonian Dynamics

The dynamic model of a mechanical system refers to a function that maps the positions, velocities, angles, and angular velocities to forces and torques. For instance, the dynamic model of a mass m is given by Newton's equation:

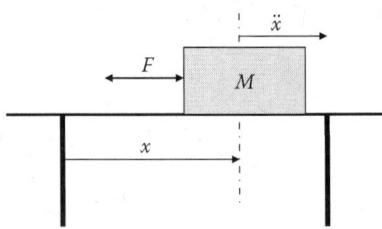

FIGURE 2.2
A force applied to a mass causes the mass to accelerate.

$F = \frac{mv_t - mv_{t-1}}{T}$, where F is the external force measured in Newtons applied on the mass, v_t is the velocity measured in meters per second at time t, v_{t-1} is the velocity at time $t - 1$, and T is the difference between the two time steps in seconds. The term mv_t refers to the momentum of the mass at time t. Therefore, Newton's equation states that the force measured in Newtons applied on a mass measured in kilograms is equal to its rate of change of momentum.

2.2.2 Control Experiments on a Point Mass

Let us focus on a very simple control problem. We want to control the speed of a mass on the horizontal plane as shown in Figure 2.2. Now, we know how the force applied on the mass is related to its velocity. Before testing how this knowledge will help us to solve the above speed control problem, let us review a common method being adopted in the industry. The current practice in most industrial controllers is to use a simple proportional and derivative controller (PD controller), as shown in Figure 2.3. The control force is calculated by $F = P * E_v + D * \frac{dE_v}{dt}$, where $E_v = v^* - v$ is the velocity error (v^* is the desired velocity and v is the actual velocity), P is the proportional gain, and the D is the derivative gain. This way of controlling requires you to tune the P and D parameters to deliver the best control performance. Please refer to the MATLAB® program at Chap2/PD_mass_vel_control.m, where we have tested how different combinations of P and D parameters perform. Results are shown in Figure 2.4.

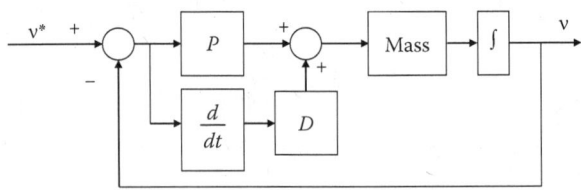

FIGURE 2.3
The PD controller for velocity control of a mass.

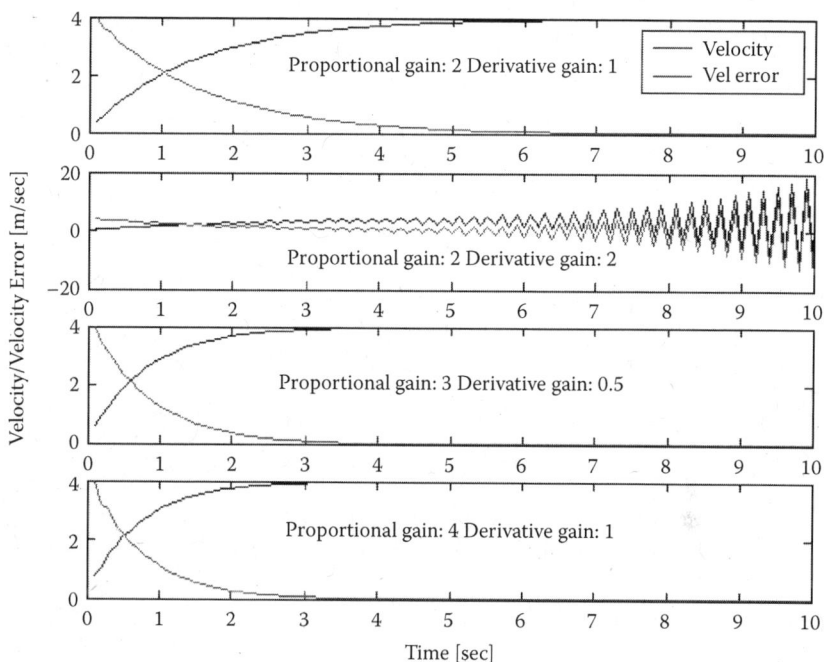

FIGURE 2.4
How a slight change in the feedback gain parameters affects the performance of a controller in the absence of a model.

Careful observation of Figure 2.4 shows that a parameter change from $P = 2$, $D = 1$ to $P = 2$, $D = 2$ can drive the system from a stable system to an unstable system. Yet, a parameter set such as $P = 4$, $D = 1$ does not do that. Therefore, it is very confusing to understand the effect of P and D parameters on the behavior of the system. In fact, tuning the P and D parameters that best suit a system that changes over time due to wear and tear, friction, temperature, humidity, etc., takes a large amount of time.

Now, let us look at how the forces were applied in each case. In the first, third, and fourth cases where the system is stable, we can note that differences in the P and D parameters cause a difference in the manner in which forces are applied on the mass to bring it to the target velocity. It seems that forces do not go beyond 20 N in a stable case, where the mass reaches the target velocity and forces converge to zero. Therefore, the practical implementation of the controller will not expect anything beyond 100 N. In the second case, where the system goes unstable, it is noticeable that forces grow in the range of 1000 N. Although this is the theoretical range, the controller will electrically and mechanically fail well before it can generate such extreme forces. If it is a motor that should produce the force, the large currents drawn through the windings will burn the motor. Even if the motor is not burned in

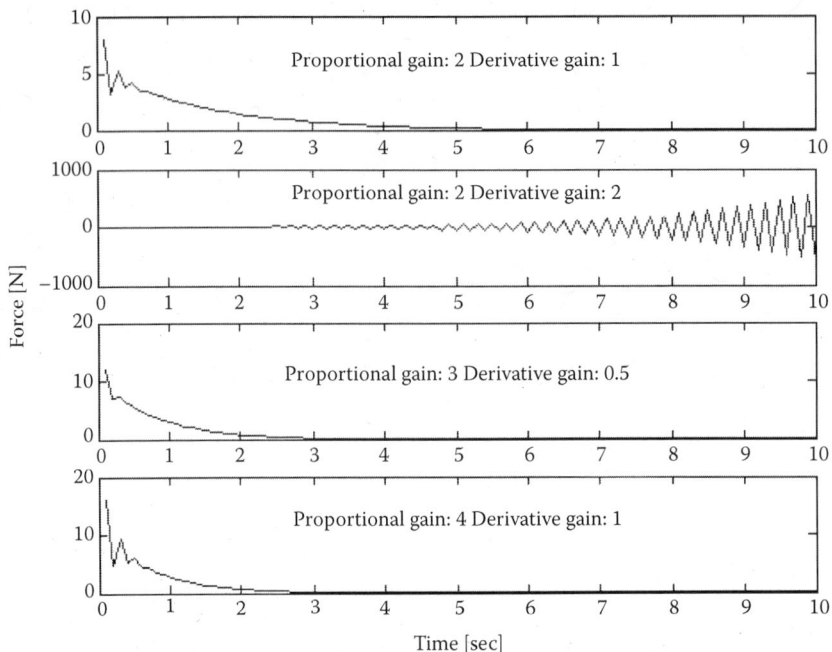

FIGURE 2.5
How the force behaves during the control time span.

an extreme case, the mechanical components will fail. Therefore, we should do whatever is possible to avoid an unstable situation (Figure 2.5).

Now, let us consider a simple model-based controller as shown in Figure 2.6. We use the same feedback control architecture found in Figure 2.3. The only difference is that we use the rules of physics this time. We know from Newton's laws that the acceleration $\ddot{x}$ of a mass M and the force F applied on it are related through the equation given by $F = M\ddot{x}$. In this case, we calculate a desired acceleration at each step by assuming that we want to reach the desired velocity in 2 more seconds. Then, the force is calculated by multiplying this desired acceleration by the estimated mass ($F = M\ddot{x}$). Figures 2.7

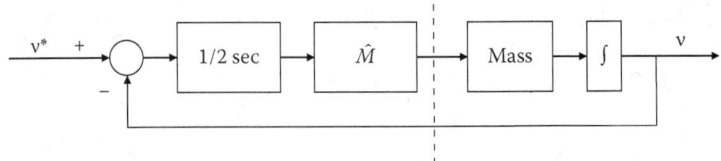

FIGURE 2.6
Model-based controller for velocity control of a mass.

Elements of a Classical Control System

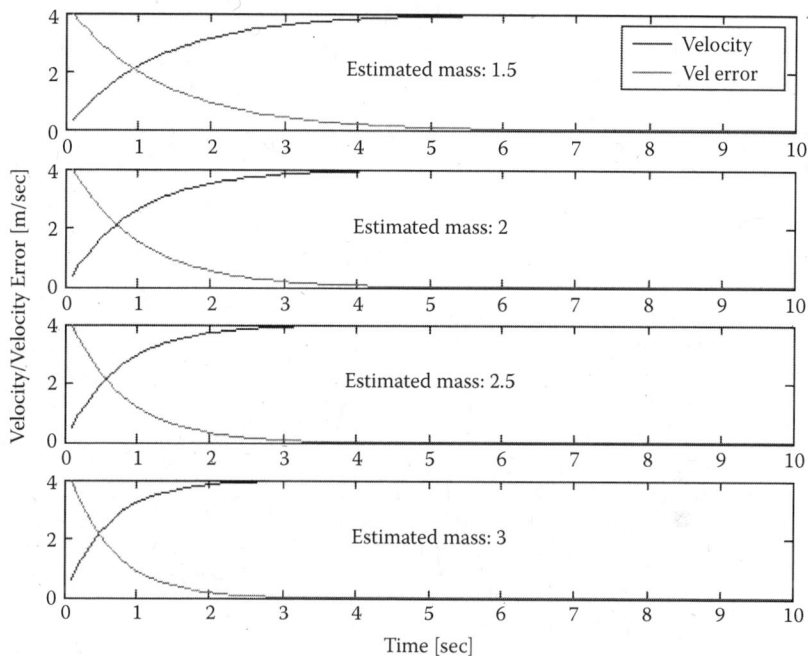

FIGURE 2.7
Velocity profiles for different model estimates.

and 2.8 were obtained by running the program Chap2/Model_mass_vel_control.m. You may notice in Figure 2.7 that this controller is stable for different estimated values of the mass ($\hat{M} = 1.5, 2, 2.5,$ and 3 kg), when the actual mass is 2 kg. Forces shown in Figure 2.8 are not only within the acceptable range, they are also smoother than those of the PD controller.

We have to keep in mind that this is a linear model. In a nonlinear model, modeling errors would give rise to different results. Yet, it is clear that some model of the system, although it may not be perfect, will help to control the system to make it perform satisfactorily. In practice, we can compensate for the unmodeled dynamics by using a PD controller in parallel.

2.2.3 Projects

Change the Chap2/PD_mass_vel_control.m program to include an integral control part so that the control command will be given by $F = P* E_v + D* \frac{dE_v}{dt} + I* \int E_v dt$, where the velocity error is given by $E_v = v^* - v$. Plot how the velocity of the mass and the applied force vary for four sets of $P, I,$ and D parameters. Compare the effect of the integral parameter by repeating the above experiment with the same set of P and D parameters with the integral parameter set to zero.

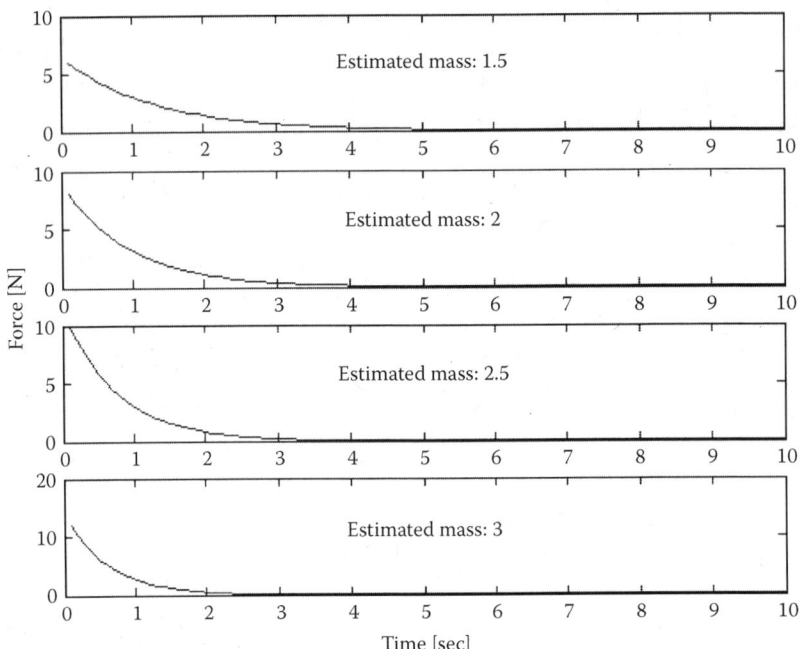

FIGURE 2.8
Forces applied on the mass for different estimated models.

Change the Chap2/Model_mass_vel_control.m program to include a PD controller to work in parallel with the model-based controller so that the force applied on the mass will be the summation of the individual forces calculated by the PD controller and the model-based controller. Observe the behavior of the combined controller in terms of forces applied and the stability. For further reading, readers are referred to the work of Ogata [2] and Nise [3].

2.3 Control of Robot Manipulators

2.3.1 Derivation of Dynamics of a Robot Manipulator

In the previous chapter, we saw that Newton's equation $F = M\ddot{x}$ states that the force (N) applied on a mass (kg) is equal to the rate of change of momentum of the mass. The same can be said of rotating systems. The torque τ (N m) applied on a system with moment of inertia I_w (kg m^2), is related to the angular acceleration $\ddot{\theta}$ (rad/s^2), as given by the equation $\tau = I_w \ddot{\theta}$. Therefore, the rate of change of angular momentum is equal to the torque applied.

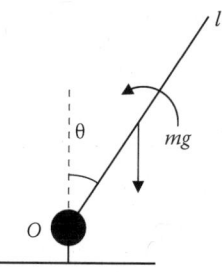

FIGURE 2.9
Single link manipulator.

The generalized forces or torques applied on a rigid body, or the generalized momentums in a generalized coordinate system, are given by the Lagrange equation:

$$B_i = \frac{d}{dt}\left(\frac{dL}{d\dot{\theta}_i}\right) - \frac{dL}{d\theta_i} \tag{2.1}$$

where B_i is the generalized force at the ith coordinate frame where θ_i is measured, L is the Lagrangian (the kinetic potential) given by the difference between the kinetic energy and the potential energy, and $p_i = (\frac{dL}{d\dot{\theta}_i})$ is the generalized momentum.

Now, let us look at a single link manipulator of length l as shown in Figure 2.9. Assume that the rod's moment of inertia around joint O is I. Then, the kinetic energy of the rotating rod is $\frac{1}{2}I\dot{\theta}^2$. The potential energy of the rod is $\frac{1}{2}mgl\cos\theta$. Therefore, the kinetic potential is given by:

$$L = \frac{1}{2}I\dot{\theta}^2 - \frac{1}{2}mgl\cos\theta \tag{2.2}$$

Now, let us use the Lagrange equation to derive the dynamics of the rod. $\frac{dL}{d\dot{\theta}_i} = I\dot{\theta}$, $\frac{d}{dt}(\frac{dL}{d\dot{\theta}_i}) = I\ddot{\theta}$, and $\frac{dL}{d\theta_i} = -\frac{1}{2}mgl\sin\theta$. Therefore, according to the Lagrange equation, the torque applied on the rod at joint O is given by $\tau = I\ddot{\theta} + \frac{1}{2}mgl\sin\theta$. Therefore, this is the simplest form of the generic dynamic equation of a rigid body manipulator given by

$$\tau = M(\theta)\ddot{\theta} + V(\theta,\dot{\theta}) + G(\theta) \tag{2.3}$$

In this case, there are no Coriolis and centrifugal parts because there are no rotating parts with a translation.

Now, let us consider a two-link robot manipulator as shown in Figure 2.10. We have to derive the kinetic energy and potential energy of the manipulator

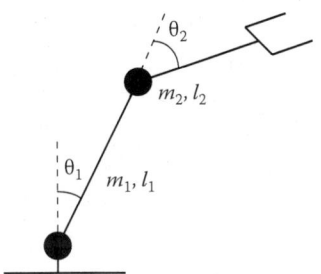

FIGURE 2.10
A two-link manipulator.

when the two links rotate at $\dot{\theta}_1$ and $\dot{\theta}_2$, respectively. The kinetic energy has two parts: one is due to the translation of the center of mass and the other component is due to the rotation around an axis. To calculate the linear velocities, let us first calculate the coordinates of the center of gravity of each link given by (x_1, y_1) and (x_2, y_2), respectively.

$$x_1 = \frac{1}{2}l_1 \sin\theta_1, \; y_1 = \frac{1}{2}l_1 \cos\theta_1, \; x_2 = l_1 \sin\theta_1 + \frac{1}{2}l_2 \sin(\theta_1 + \theta_2)$$

$$y_2 = l_1 \cos\theta_1 + \frac{1}{2}l_2 \cos(\theta_1 + \theta_2) \tag{2.4}$$

Therefore, the linear velocities of the two links will be given by

$$\dot{x}_1 = \frac{1}{2}l_1\dot{\theta}_1 \cos\theta_1, \; \dot{y}_1 = -\frac{1}{2}l_1\dot{\theta}_1 \sin\theta_1, \; \dot{x}_2 = l_1\dot{\theta}_1 \cos\theta_1 + \frac{1}{2}l_2(\dot{\theta}_1 + \dot{\theta}_2)\cos(\theta_1 + \theta_2)$$

$$\dot{y}_2 = -l_1\dot{\theta}_1 \sin\theta_1 - \frac{1}{2}l_2(\dot{\theta}_1 + \dot{\theta}_2)\sin(\theta_1 + \theta_2) \tag{2.5}$$

The angular velocities of the two centers of gravity are given by

$$\omega_1 = \dot{\theta}_1, \; \omega_2 = (\dot{\theta}_1 + \dot{\theta}_2) \tag{2.6}$$

Therefore, the total kinetic energy of the manipulator is given by

$$KE = \frac{1}{2}m_1\left(\dot{x}_1^2 + \dot{y}_1^2\right) + \frac{1}{2}m_2\left(\dot{x}_2^2 + \dot{y}_2^2\right) + \frac{1}{2}I_1\dot{\theta}_1^2 + \frac{1}{2}I_2(\dot{\theta}_1 + \dot{\theta}_2)^2 \tag{2.7}$$

The potential energy of the system is given by

$$PE = \frac{1}{2}m_1 g l_1 \cos\theta_1 + m_2 g \left(l_1 \cos\theta_1 + \frac{1}{2}l_2 \cos(\theta_1 + \theta_2) \right) \tag{2.8}$$

Elements of a Classical Control System

Therefore, the Lagrange variable or the "kinetic potential" is given by

$$L = KE - PE \tag{2.9}$$

$$\frac{dL}{d\dot{\theta}_1} = \frac{1}{2}m_1 l_1(\dot{x}_1 + \dot{y}_1)(\cos\theta_1 - \sin\theta_1) + m_2(\dot{x}_2 + \dot{y}_2)(l_1(\cos\theta_1 - \sin\theta_1)$$

$$+ \frac{1}{2}l_2(\cos(\theta_1 + \theta_2) - \sin(\theta_1 + \theta_2))) + I_1\dot{\theta}_1 + I_2(\dot{\theta}_1 + \dot{\theta}_2)$$

$$- \frac{dPE}{d\dot{\theta}_1} \tag{2.10}$$

$$\frac{d}{dt}\left(\frac{dL}{d\dot{\theta}_1}\right) = \frac{1}{2}m_1 l_1[(\dot{x}_1 + \dot{y}_1)(\cos\theta_1 - \sin\theta_1)\dot{\theta}_1 + (\cos\theta_1 - \sin\theta_1)(\ddot{x}_1 + \ddot{y}_1)]$$

$$+ m_2(\ddot{x}_2 + \ddot{y}_2)(l_1(\cos\theta_1 - \sin\theta_1) + \frac{1}{2}l_2(\cos(\theta_1 + \theta_2) - \sin(\theta_1 + \theta_2)))$$

$$+ m_2(\dot{x}_2 + \dot{y}_2)(l_1(\cos\theta_1 - \sin\theta_1)\dot{\theta}_1 + \frac{1}{2}l_2(\cos(\theta_1 + \theta_2)$$

$$- \sin(\theta_1 + \theta_2))(\dot{\theta}_1 + \dot{\theta}_2)) + I_1\ddot{\theta}_1 + I_2(\ddot{\theta}_1 + \ddot{\theta}_2) \tag{2.11}$$

$$\frac{dL}{d\dot{\theta}_2} = m_2(\dot{x}_2 + \dot{y}_2)(l_1(\cos\theta_1 - \sin\theta_1)$$

$$+ \frac{1}{2}l_2(\cos(\theta_1 + \theta_2) - \sin(\theta_1 + \theta_2))) + I_2(\dot{\theta}_1 + \dot{\theta}_2)$$

$$- \frac{dPE}{d\dot{\theta}_2} \tag{2.12}$$

$$\frac{d}{dt}\left(\frac{dL}{d\dot{\theta}_2}\right) = m_2(\ddot{x}_2 + \ddot{y}_2)(l_1(\cos\theta_1 - \sin\theta_1) + \frac{1}{2}l_2(\cos(\theta_1 + \theta_2) - \sin(\theta_1 + \theta_2)))$$

$$+ I_2(\ddot{\theta}_1 + \ddot{\theta}_2) + m_2(\dot{x}_2 + \dot{y}_2)(l_1(\cos\theta_1 - \sin\theta_1)\dot{\theta}_1$$

$$+ \frac{1}{2}l_2(\cos(\theta_1 + \theta_2) - \sin(\theta_1 + \theta_2))(\dot{\theta}_1 + \dot{\theta}_2)) \tag{2.13}$$

$$\frac{dL}{d\theta_1} = -\frac{1}{2}m_1 l_1 \dot{\theta}_1(\dot{x}_1 + \dot{y}_1)(\sin\theta_1 + \cos\theta_1)$$

$$+ m_2(\dot{x}_2 + \dot{y}_2)(-l_1\dot{\theta}_1(\sin\theta_1 + \cos\theta_1) + \frac{1}{2}l_2(\dot{\theta}_1 + \dot{\theta}_2)(\sin(\theta_1 + \theta_2) + \cos(\theta_1 + \theta_2)))$$

$$+ \frac{1}{2}m_1 g l_1 \sin\theta_1 + m_2 g \left(l_1 \sin\theta_1 + \frac{1}{2}l_2 \sin(\theta_1 + \theta_2)\right) \tag{2.14}$$

$$\frac{dL}{d\theta_2} = m_2(\dot{x}_2+\dot{y}_2)\left(\frac{1}{2}(\dot{\theta}_1+\dot{\theta}_2)(l_1\sin(\theta_1+\theta_2)-l_2\cos(\theta_1+\theta_2))\right)+\frac{1}{2}m_2gl_2\sin(\theta_1+\theta_2))$$

(2.15)

We know that the torques applied on the two joints are given by $\tau_i = \frac{d}{dt}\left(\frac{dL}{d\dot{\theta}_i}\right) - \frac{dL}{d\theta_i}$. Substituting the above results in the Lagrange equation and simplifying will yield

$$\tau = M(\theta)\ddot{\theta} + V(\theta,\dot{\theta}) + G(\theta)$$

where

$$M(\theta) = \begin{bmatrix} I_1 + I_2 + \frac{1}{4}M_1l_1^2 + m_2\left(\frac{1}{4}l_2^2 + l_1^2 + l_1l_2\cos\theta_2\right) & I_2 + \frac{1}{2}m_2l_2\left(\frac{1}{2}l_2 + l_1\cos\theta_2\right) \\ I_2 + \frac{1}{2}m_2l_2\left(\frac{1}{2}l_2 + l_1\cos\theta_2\right) & I_2 + \frac{1}{4}m_2l_2^2 \end{bmatrix}$$

(2.16)

is the positive definite inertia matrix,

$$V(\theta,\dot{\theta}) = \begin{bmatrix} -\frac{1}{2}m_2l_1l_2\sin\theta_2(2\dot{\theta}_1\dot{\theta}_2 + \dot{\theta}_2^2) \\ \left(\frac{1}{2}m_2l_1l_2\sin\theta_2\right)\dot{\theta}_1^2 \end{bmatrix}$$

(2.17)

is the Coriolis and centrifugal matrix, and

$$G(\theta) = \begin{bmatrix} \frac{1}{2}m_2l_2g\sin(\theta_1+\theta_2) + \left(\frac{1}{2}m_1l_1 + m_2l_1\right)g\sin\theta_1 \\ \frac{1}{2}m_2l_2g\sin(\theta_1+\theta_2) \end{bmatrix}$$

(2.18)

is the gravity matrix.

I am sure now that you know how to derive the dynamics of a rigid body system. All that it demands is patience. However, it should be clear from this exercise that the complexity of the differential equations grows exponentially with the degrees of freedom of the manipulator [4]. We will discuss other simpler methods to estimate the dynamic models of physical systems in Chapter 6. Now, we will do a few projects based on the dynamics derived above.

Elements of a Classical Control System

TABLE 2.1
Parameters of a Two-Link Robot Manipulator

m_1	m_2	l_1	l_2
0.4	0.1	0.3	0.2

Assume the center of gravity of each link is located at the center of each link.

2.3.2 Worked Out Project 1: Manipulator Dynamics

The parameters of a robot manipulator are given in Table 2.1.

- Calculate the elements of the inertia, Coriolis and centrifugal, and gravity matrices.
- Test whether the inertia matrix is symmetric and positive definite.
- Construct a model of the above manipulator by introducing random a 10% offset to the parameters given in the above Table 2.1.
- Use the estimated model constructed in Table 2.1 to control the above robot manipulator along the desired trajectory given by

$$\theta_{1d} = \frac{\pi}{6} + \frac{\pi}{4}\sin\left(2\pi ft - \frac{\pi}{6}\right), \quad \theta_{2d} = -\frac{\pi}{6} + \frac{\pi}{6}\sin\left(2\pi ft + \frac{\pi}{6}\right)$$
$$f = 0.2 Hz, \, t = [0, 2] \sec$$

- Implement the controller for two cases: (a) The model used for the controller is 100% accurate. (b) The estimated link parameters in the model used for the controller have 10% error. Assume a sampling interval of 2 ms.

Answer:

1.1 The inertia matrix for a two-link manipulator with the center of gravity located at the center of each link is given by Equation (2.16). Substituting the above parameters in Equation (2.16) yields

$$M_{11} = \frac{1}{3}(0.4)(0.3)^2 + \frac{1}{3}(0.1)(0.2)^2 + \frac{1}{4}(0.4)(0.3)^2$$
$$+ (0.1)\left(\frac{1}{4}(0.2)^2 + (0.3)^2 + (0.3)(0.2)\cos\theta_2\right)$$

$$M_{12} = M_{21} = \frac{1}{3}(0.1)(0.2)^2 + \frac{1}{2}(0.1)(0.2)\left(\frac{1}{2}(0.1) + (0.3)\cos\theta_2\right)$$

$$M_{22} = \frac{1}{3}(0.1)(0.2)^2 + \frac{1}{4}(0.1)(0.2)^2$$

$$M_{11} = 0.0363 + 0.006 * \cos\theta_2$$
$$M_{12} = M_{21} = 0.00135 + 0.0003 * \cos\theta_2$$
$$M_{22} = 0.0023 \tag{2.19}$$

The Coriolis and centrifugal force matrices are given by Equation (2.17)

$$V(\theta, \dot\theta) = \begin{bmatrix} -\frac{1}{2}(0.1)(0.2)(0.3)\sin\theta_2(2\dot\theta_1\dot\theta_2 + \dot\theta_2^2) \\ \left(\frac{1}{2}(0.1)(0.2)(9.81)\sin\theta_2\right)\dot\theta_1^2 \end{bmatrix}$$

$$= \begin{bmatrix} -0.003\sin\theta_2\left(2\dot\theta_1\dot\theta_2 + \dot\theta_2^2\right) \\ (0.0981\sin\theta_2)\dot\theta_1^2 \end{bmatrix} \tag{2.20}$$

$$G(\theta) = \begin{bmatrix} 0.0981\sin(\theta_1 + \theta_2) + 0.8829*\sin\theta_1 \\ 0.0981*\sin(\theta_1 + \theta_2) \end{bmatrix} \tag{2.21}$$

We can note from Equation (2.19) that $M_{11}, M_{22} > 0, \forall \theta_2 \in [-\pi, \pi]$. Figure 2.11 shows how M_{11} is behaving. Since $M_{12} = M_{21}$, the inertia matrix is positive definite.

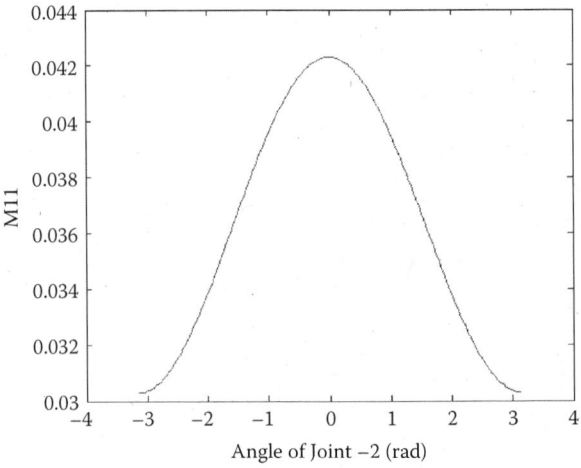

FIGURE 2.11
The behavior of M_{11} for varying θ_2.

TABLE 2.2
Estimated Parameters of the Manipulator

m_1	m_2	l_1	l_2
0.44	0.09	0.27	0.18

1.3 We introduce an uncertainty of 10% to each parameter in Table 2.1 to simulate the error of estimation of parameters. The estimated parameters with 10% error are given in Table 2.2.

2.3.3 Worked Out Project 2: Manipulator Dynamics

$$M_{11} = \frac{1}{3}(0.44)(0.27)^2 + \frac{1}{3}(0.09)(0.18)^2 + \frac{1}{4}(0.44)(0.27)^2$$

$$+ (0.09)\left(\frac{1}{4}(0.18)^2 + (0.27)^2 + (0.27)(0.18)\cos\theta_2\right)$$

$$M_{12} = M_{21} = \frac{1}{3}(0.09)(0.18)^2 + \frac{1}{2}(0.09)(0.18)\left(\frac{1}{2}(0.09) + (0.27)\cos\theta_2\right)$$

$$M_{22} = \frac{1}{3}(0.09)(0.18)^2 + \frac{1}{4}(0.09)(0.18)^2$$

$$M_{11} = 0.027 + 0.0044\cos\theta_2$$
$$M_{12} = M_{21} = 0.0011 + 0.0022\cos\theta_2$$
$$M_{22} = 0.0017$$

The estimated Coriolis and centrifugal force matrices are:

$$V(\theta, \dot\theta) = \begin{bmatrix} -\frac{1}{2}(0.09)(0.18)(0.27)\sin\theta_2 \left(2\dot\theta_1\dot\theta_2 + \dot\theta_2^2\right) \\ \left(\frac{1}{2}(0.09)(0.18)(9.81)\sin\theta_2\right)\dot\theta_1^2 \end{bmatrix}$$

$$= \begin{bmatrix} -0.0022\sin\theta_2\left(2\dot\theta_1\dot\theta_2 + \dot\theta_2^2\right) \\ (0.0795\sin\theta_2)\dot\theta_1^2 \end{bmatrix}$$

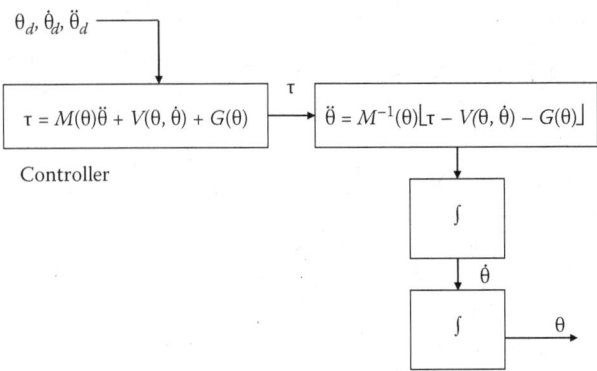

FIGURE 2.12
Model-based controller with no uncertainty in link parameters.

The estimated gravity matrix:

$$G(\theta) = \begin{bmatrix} \frac{1}{2}(0.09)(0.18)(9.81)\sin(\theta_1 + \theta_2) + \left(\frac{1}{2}(0.44)(0.27) + (0.09)(0.27)\right) g \sin\theta_1 \\ \frac{1}{2}(0.09)(0.18)(9.81)\sin(\theta_1 + \theta_2) \end{bmatrix}$$

$$= \begin{bmatrix} 0.0795\sin(\theta_1 + \theta_2) + 0.2978\sin\theta_1 \\ 0.0795\sin(\theta_1 + \theta_2) \end{bmatrix}$$

2.3.4 Effect of Parameter Uncertainty on Model-Based Control Performance

It can be seen from these results that a 10% uncertainty in the estimation of link parameters can lead to unproportional changes in the elements in the dynamic equation of the manipulator. However, this can be a practical scenario. Now, we are going to develop a model-based controller for the above manipulator.

To develop a model-based controller, we first study how we can design a simple model-based controller with the dynamic equations derived above. Figure 2.12 shows how the model computes the torque commands for the two joints of the robot manipulator (please refer to Chap2\model_control_no_uncertainty.m). The resulting angular acceleration $\ddot{\theta}$ is integrated to estimate the angular velocity vector $\dot{\theta}$ and the angle vector θ. The integration is done via a fourth-order Runge-Kutta integration algorithm after rearranging the dynamic equation to form an ordinary differential equation given by

$$\begin{pmatrix} \dot{\theta}_1 \\ \dot{\theta}_2 \\ \ddot{\theta}_1 \\ \ddot{\theta}_2 \end{pmatrix} = \begin{bmatrix} 0 & 0 & 1 & 0 \\ 0 & 0 & 0 & 1 \\ 0 & 0 & 0 & 0 \\ 0 & 0 & 0 & 0 \end{bmatrix} \begin{pmatrix} \theta_1 \\ \theta_2 \\ \dot{\theta}_1 \\ \dot{\theta}_2 \end{pmatrix} + \begin{bmatrix} 0 \\ 0 \\ M^{-1}(\theta)[\tau - V(\theta, \dot{\theta}) - G(\theta)] \end{bmatrix}$$

$$\Rightarrow \dot{x} = f(x)$$

$$x = \begin{bmatrix} \theta \\ \dot{\theta} \end{bmatrix}$$

Then the Runge-Kutta integration algorithm can be used to integrate the angular acceleration $\ddot{\theta}$ and angular velocity $\dot{\theta}$ as given by

$$k_1 = f([\theta \quad \dot{\theta}]^T)$$

$$k_2 = f\left([\theta \quad \dot{\theta}]^T + \frac{h}{2} k_1\right)$$

$$k_3 = f\left([\theta \quad \dot{\theta}]^T + \frac{h}{2} k_2\right)$$

$$k_4 = f([\theta \quad \dot{\theta}]^T + h k_3)$$

$$\times \begin{bmatrix} \theta_1(t+1) \\ \theta_2(t+1) \\ \dot{\theta}_1(t+1) \\ \dot{\theta}_2(t+1) \end{bmatrix} = \begin{bmatrix} \theta_1(t) \\ \theta_2(t) \\ \dot{\theta}_1(t) \\ \dot{\theta}_2(t) \end{bmatrix} + \frac{h}{6}(k_1 + 2k_2 + 2k_3 + k_4)$$

The results given in Figures 2.13 and 2.14 suggest that the performance of the controller with perfect estimation of the dynamics of the robot manipulator gives excellent results with joint angle error less than 0.002 (rad). One would question why there is an error even of this small magnitude when the model is perfect. The reason is that we had a finite sampling interval of 2 (ms). This results in some error in the integration of angular acceleration and angular velocities to estimate the joint angles. This error could have been improved by reducing the sampling interval (please try this). You may obviously ask the question of why we cannot reduce the sampling interval indefinitely. This can be done in a simulation, but not in a real-world experiment because there is a finite number of instructions that the processor has to execute within two sampling intervals. Depending on the complexity of the control algorithm and the optimality of the software code, the number of

26   *Intelligent Control Systems*

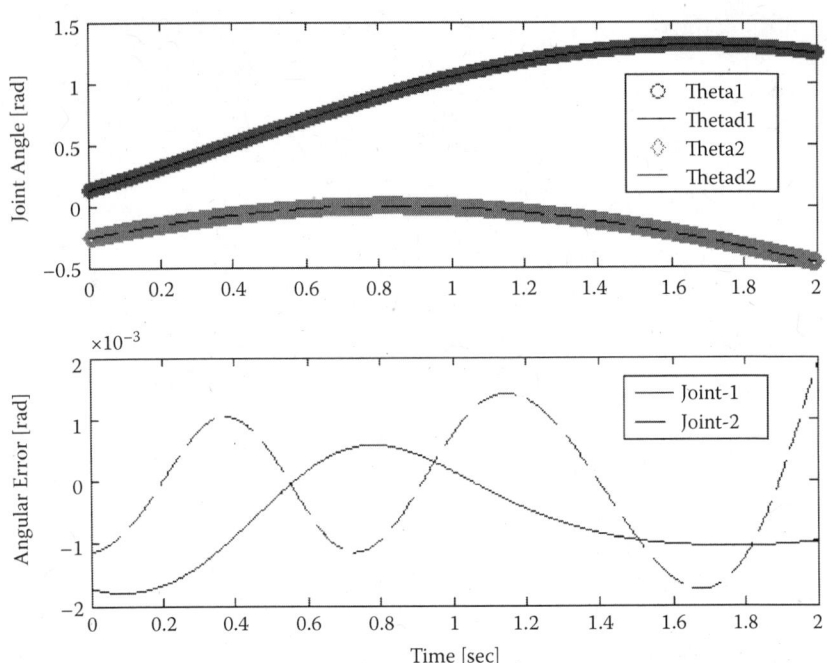

FIGURE 2.13
Joint angle trajectories and angular errors for the accurate model.

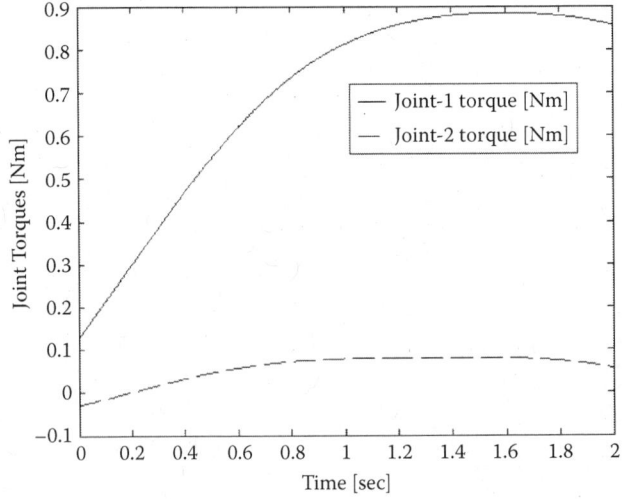

FIGURE 2.14
Joint torques produced by the accurate model.

Elements of a Classical Control System

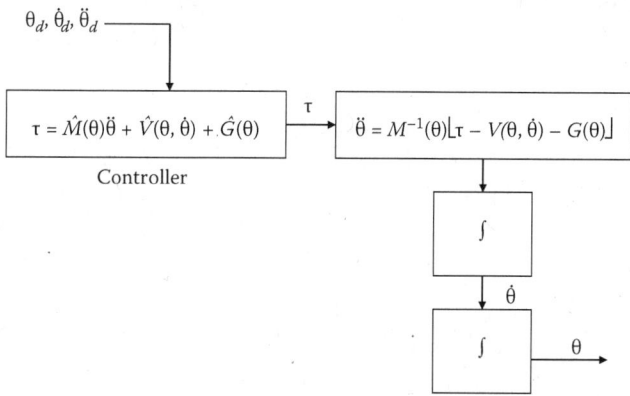

FIGURE 2.15
Model-based controller with 10% error in link parameter estimation.

instructions to be executed between two sampling points will vary. Therefore, the minimum sampling interval depends on the time taken by the processor to complete executing an instruction, the complexity of the control algorithm, and the optimality of the code.

Figure 2.15 shows how the model with 10% error in link parameter estimation is used to control the robot manipulator. Figure 2.16 shows the effect of the error on the estimation of model parameters. We can see that the tracking performance has deteriorated even though it stays stable. In this case, there is no feedback mechanism to correct the error. We can notice this in Figure 2.17, where the torque calculated by the model remains smooth even though the error has some fluctuations (please use the MATLAB® code Chap2\model_control_with_uncertainty.m to study the effect of the model estimation error on different trajectories).

2.3.5 Worked Out Project 3: Hybrid Controllers

Develop a PD controller to work in parallel with the model-based controller with 10% error in the estimated link parameters discussed in project 1.

Case 1: $K_p = \begin{bmatrix} 2 & 0 \\ 0 & 2 \end{bmatrix}, K_v = \begin{bmatrix} 0.5 & 0 \\ 0 & 0.5 \end{bmatrix}$

Case 2: $K_p = \begin{bmatrix} 4 & 0 \\ 0 & 2 \end{bmatrix}, K_v = \begin{bmatrix} 0.5 & 0 \\ 0 & 0.5 \end{bmatrix}$

Case 3: $K_p = \begin{bmatrix} 4 & 0 \\ 0 & 2 \end{bmatrix}, K_v = \begin{bmatrix} 1 & 0 \\ 0 & 1 \end{bmatrix}$

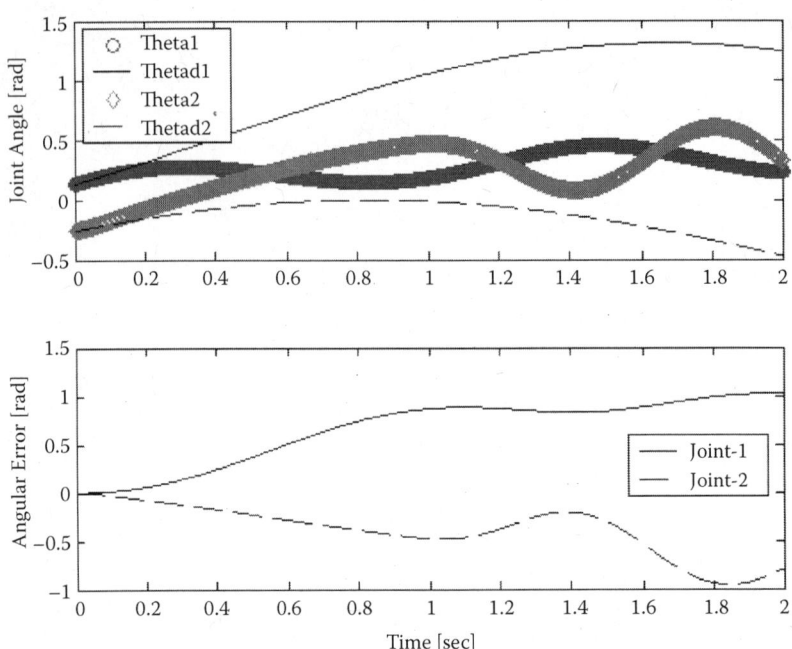

FIGURE 2.16
Joint angle trajectories and angular errors for the model with 10% error in link parameter estimation.

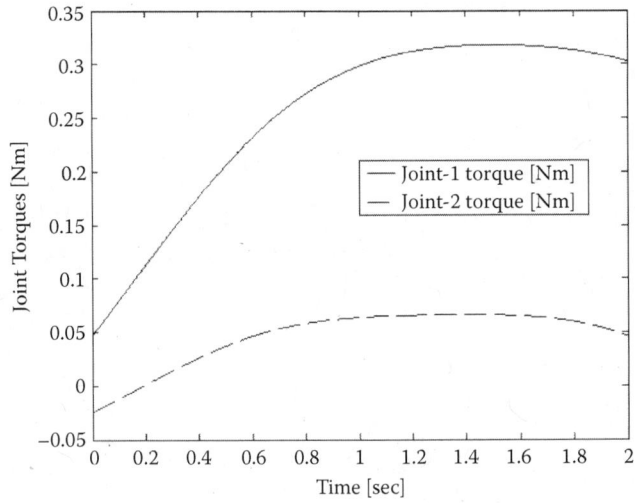

FIGURE 2.17
Joint torques generated by the model with 10% error in link parameter estimation.

Elements of a Classical Control System

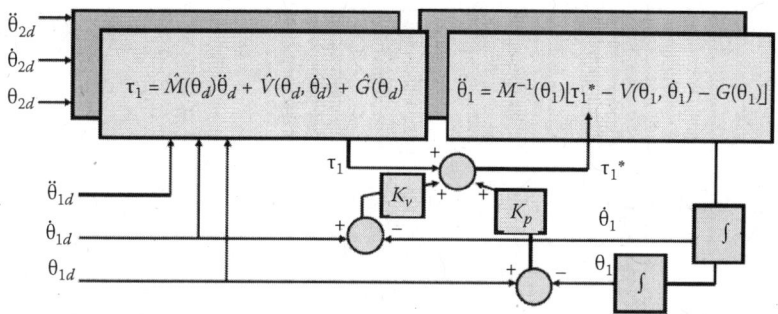

FIGURE 2.18
Model-based controller supported by a proportional-derivative controller.

Answer:

We construct a controller as shown in Figure 2.18. Results for case 1 are shown in Figures 2.19 and 2.20. Compare the results with those in Figures 2.15 and 2.16, where the estimated model controlled the manipulator without any feedback correction. It can be seen that the PD controller has helped the

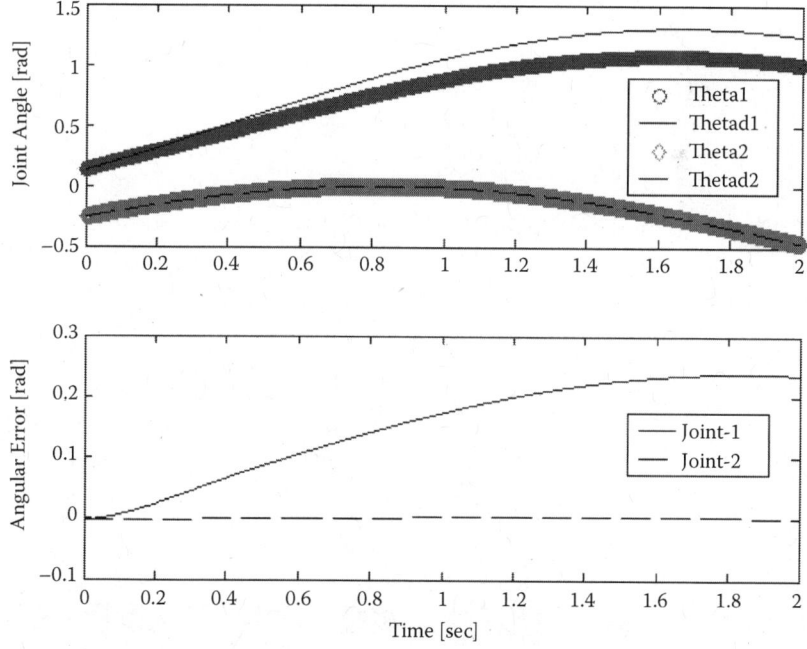

FIGURE 2.19
Tracking performance for case 1.

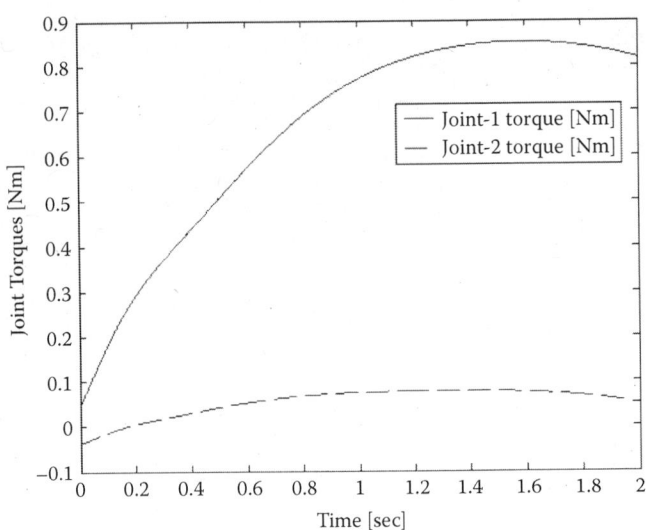

FIGURE 2.20
Joint torques for case 1.

model-based controller to improve the tracking performance significantly. Therefore, we can conclude that the PD controller can complement for the uncompensated dynamics due to the error in the link parameter estimates.

Case 1: $K_p = \begin{bmatrix} 2 & 0 \\ 0 & 2 \end{bmatrix}, K_v = \begin{bmatrix} 0.5 & 0 \\ 0 & 0.5 \end{bmatrix}$

Please refer to the Chap2\Model_PD_control_with_uncertainty.m program.

Case 2: $K_p = \begin{bmatrix} 4 & 0 \\ 0 & 2 \end{bmatrix}, K_v = \begin{bmatrix} 0.5 & 0 \\ 0 & 0.5 \end{bmatrix}$

Observe the difference between case 1 and case 2. The only difference in the controller is in the P gain for joint 1. We can see from Figures 2.18 and 2.20 that the tracking error of joint 1 has dropped almost 50% from case 1 to case 2. It is also evident from Figures 2.19 and 2.21 that the peak torque has not significantly changed across the two cases.

Case 3: $K_p = \begin{bmatrix} 4 & 0 \\ 0 & 2 \end{bmatrix}, K_v = \begin{bmatrix} 1 & 0 \\ 0 & 1 \end{bmatrix}$

The difference in the controller from case 2 to case 3 is that the derivative gains in both joints have doubled. From Figures 2.20 and 2.21, we cannot

Elements of a Classical Control System

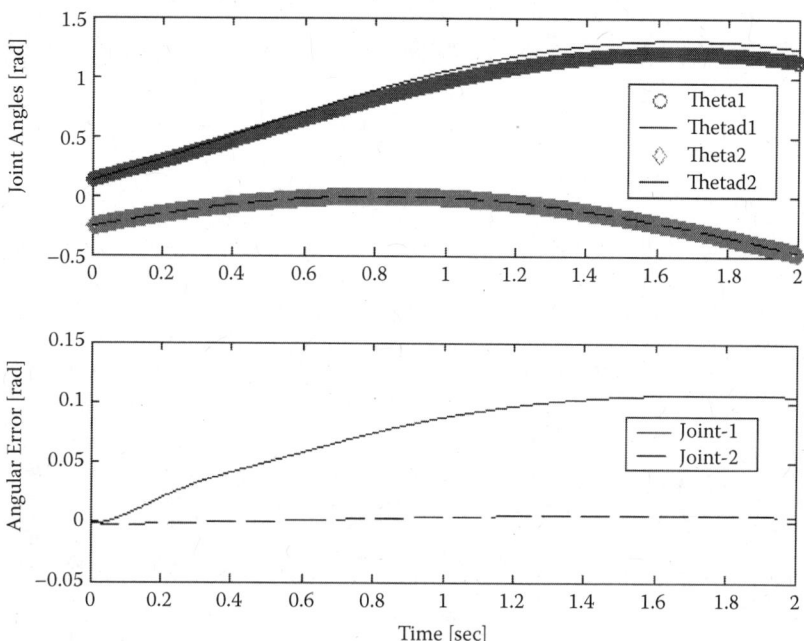

FIGURE 2.21
Tracking performance for case 2.

observe a significant change in the tracking performance. This comes from the fact that the rate of change of error in this case is very low. Please use the MATLAB® code given in Chap2\Model_PD_control_with_uncertainty.m to test the performance for another trajectory (Figures 2.22, 2.23, and 2.24).

2.4 Stability

2.4.1 Equilibrium Points and Their Stability

Stability refers to the behavior of a perturbed state in the neighborhood of an equilibrium point. An equilibrium point is a state where the system rests in the absence of a perturbation. A stable equilibrium is a state where the system returns to the original equilibrium state when it is disturbed slightly. Please have a look at Figure 2.25, where three identical masses are kept at three different points in a vertical landscape. In this sense, point A is an *unstable equilibrium point* because a disturbance of the resting mass will cause its position to diverge from point A. Point C is a *stable equilibrium point* because

32   *Intelligent Control Systems*

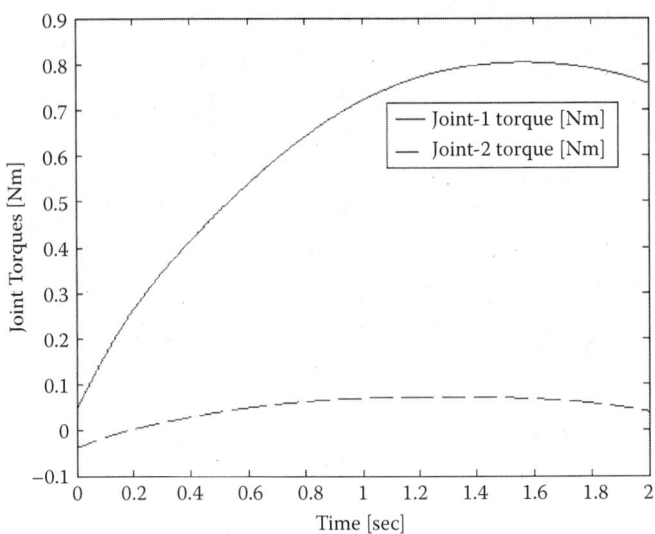

FIGURE 2.22
Joint torques calculated by the controller in case 2.

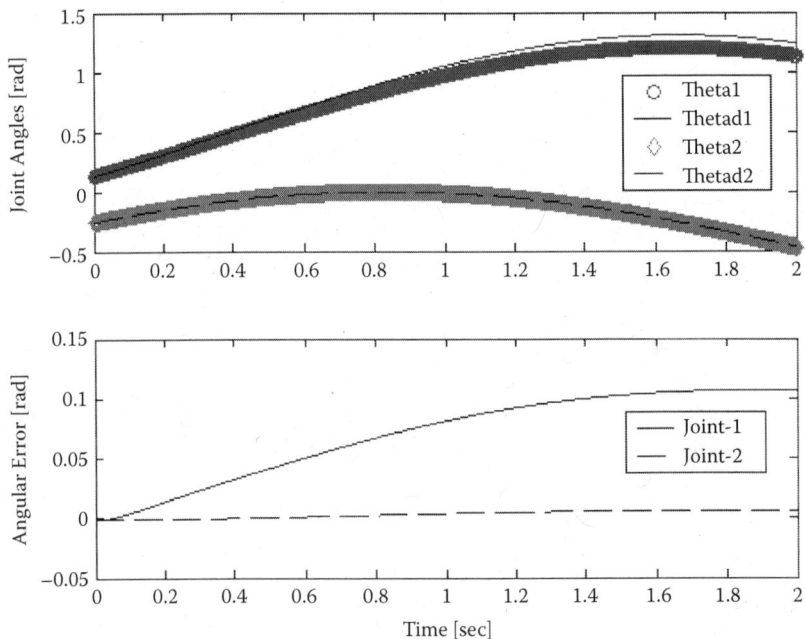

FIGURE 2.23
Tracking performance for case 3.

Elements of a Classical Control System 33

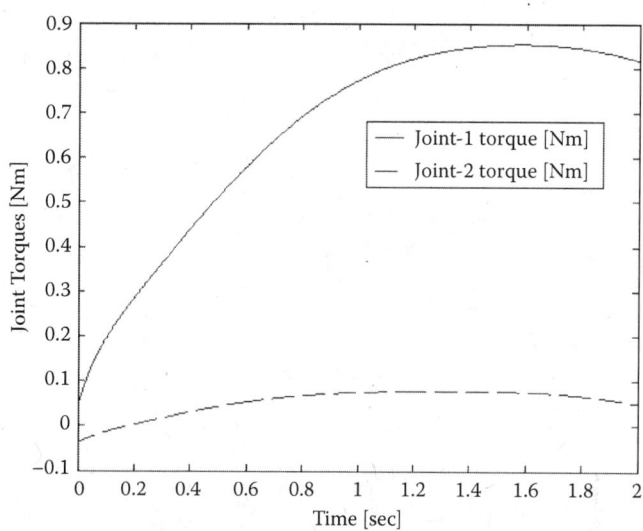

FIGURE 2.24
Joint torques calculated by the controller in case 3.

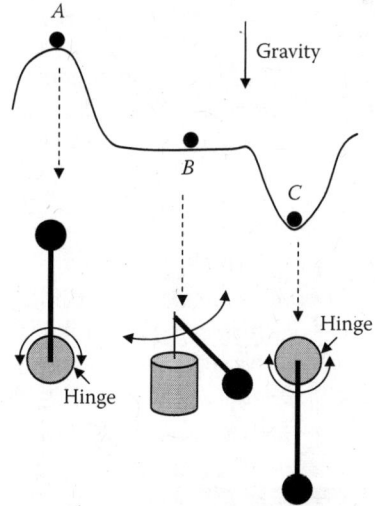

FIGURE 2.25
Stability of fixed points.

the position of the mass will converge back to C if it is perturbed at C. Point B is a *neutral equilibrium point*, where the state does not return or flies out to be unstable but remains conservative. The equivalent state of a pendulum is shown under each point. A stable equilibrium point is always an attracter. Point C is one. All points around C that will cause the mass to converge to C are called the basin of attraction of point C.

Therefore, in mathematical terms, we can say that a dynamic system given by $\dot{x} = f(x,t)$, $x \in \Re^n$ is in an equilibrium state if $f(x, t) = 0$ for some $x = x^*$. In mathematical terms, an equilibrium point x^* is stable if, for $\varepsilon > 0$, there exists a $\delta = \delta(\varepsilon, t_0) > 0$ such that $\|x(t) - x^*\| < \delta$ means $\|x(t) - x^*\| < \varepsilon$ at any given time t. This is also known as the Lyapunov stability requirement. You can imagine this condition as a point originating within an arbitrarily small bounded neighborhood of x^* at time t_0 staying within another bounded neighborhood around x^* at any given time $t > t_0$.

Uniform stability is achieved if, for $\varepsilon > 0$, there exists a $\delta = \delta(\varepsilon) > 0$ independent of t_0 such that $\|x(t) - x^*\| < \delta$ means $\|x(t) - x^*\| < \varepsilon$ at any given time t. Please note that this is stricter than the previous condition because it requires the system to be stable all the time.

Asymptotic stability is achieved if the equilibrium point is stable and there is a positive constant $\mu = \mu(t_0)$ such that $x(t) \to x^*$ as $t \to \infty$ for all $\|x^*\| < \mu(t_0)$. The equilibrium point x^* is exponentially stable if there exist scalars $\alpha, \beta, \varepsilon > 0$ such that $\|x(t)\| \leq \beta e^{-\alpha(t-t_0)} \|x(t_0)\|$, $\forall \|x(t_0)\| \leq \varepsilon, t \geq t_0$. Instability occurs if the equilibrium point is not stable.

2.4.2 Lyapunov Stability

The direct method to see if a dynamic system given by $\dot{x} = f(x,t)$, $x \in \Re^n$ is stable is to integrate it for $t \to \infty$. Yet, it is practically infeasible. Lyapunov's direct method allows us to check the stability of a system by looking at the properties of an energy function without having to integrate the original dynamic system over the entire time span.

DEFINITION: **Locally positive definite functions**
First, we have to choose a locally positive definite function to represent the energy of the system. A continuous function V is locally positive definite if, for some continuously increasing function, $a(\|x\|)$, $V(0,t) = 0$, and $V(x,t) \geq a(\|x\|) \ \forall \|x\| \leq \varepsilon, \forall t \geq 0$ for some $\varepsilon > 0$.

DEFINITION: **Positive definite functions**
A continuous function V is positive definite if, for some continuously increasing function, $a(\|x\|)$, $V(0,t) = 0$, and $V(x,t) \geq a(\|x\|) \ \forall \|x\| \leq \varepsilon, \forall t \geq 0$ for some $\varepsilon > 0$, and $a(\|x\|) \to \infty$ when $\|x\| \to \infty$.

Elements of a Classical Control System

DEFINITION: **Decrescent functions**
A continuous function V is decrescent if, for some $\varepsilon > 0$ and some continuous strictly increasing function β, $V(x,t) \leq \beta(\|x\|) \forall \|x\| \leq \varepsilon, \forall t \geq 0$.

Once we have chosen a suitable positive definite function of the state variables of the system, we can judge how the energy dissipates or grows by taking the derivative of the function V along the trajectories of the system given by

$$\dot{V} = \frac{\partial V}{\partial t} + \frac{\partial V}{\partial x}\frac{dx}{dt}$$

$$= \frac{\partial V}{\partial t} + \frac{\partial V}{\partial x} f$$

Next, the Lyapunov stability criterion states that, if $V(x, t)$ is locally positive definite and $\dot{V}(x,t) \leq 0 \forall \|x\| \leq \varepsilon, \forall t \geq 0$, then the origin of the system is locally stable in the sense of Lyapunov.

If $V(x, t)$ is locally positive definite and decrescent, and $\dot{V}(x,t) \leq 0$ $\forall \|x\| \leq \varepsilon, \forall t \geq 0$, then the origin of the system is uniformly locally stable in the sense of Lyapunov. If $V(x, t)$ is locally positive definite and decrescent, and $-\dot{V}(x,t)$ is locally positive definite $\forall \|x\| \leq \varepsilon, \forall t \geq 0$, then the origin of the system is uniformly locally asymptotically stable in the sense of Lyapunov. If $V(x, t)$ is locally positive definite and decrescent, and $-\dot{V}(x,t)$ is positive definite, then the origin of the system is globally uniformly asymptotically stable in the sense of Lyapunov.

2.4.3 Worked Out Example

Let us consider the dynamics of a robot manipulator given by $\tau = M(\theta)\ddot{\theta} + V(\theta,\dot{\theta}) + G(\theta)$. Assume that the robot is controlled by a PD controller given by $\tau = K_p e + K_d \dot{e} + G(\theta)$, $e = \theta_d - \theta$, where θ_d is the desired state of the joint angle vector θ and K_p, K_d are diagonal positive definite gain matrices.

Therefore, we can rewrite the dynamics of the close loop system as

$$M(\theta)\ddot{\theta} + V(\theta,\dot{\theta}) + K_p e + K_d \dot{e} = 0$$

Then, we design a positive definite Lyapunov function given by

$$V = \frac{1}{2}\dot{\theta}^T M(\theta)\dot{\theta} + e^T K_p e$$

By taking the derivative along the state trajectories, we obtain

$$\dot{V} = \dot{\theta}^T M(\theta)\ddot{\theta} + \dot{\theta}^T \dot{M}(\theta)\dot{\theta} - e^T K_p \dot{\theta}$$

$$= \dot{\theta}^T [K_p e + K_d \dot{e} - V(\theta,\dot{\theta})] + \dot{\theta}^T \dot{M}(\theta)\dot{\theta} - e^T K_p \dot{\theta}$$

Since $\dot{\theta}^T \dot{M}(\theta)\dot{\theta} = V(\theta,\dot{\theta})$ and $\dot{\theta}^T K_p e = e^T K_p \dot{\theta}$, $\dot{V} = -\dot{\theta}^T K_d \dot{\theta}$, which is negative definite because K_d is a positive definite matrix. Therefore, we can conclude that a controller given by $\tau = K_p e + K_d \dot{e} + G(\theta)$ is stable as far as K_p, K_d are positive definite matrices. For further reading on dynamics, kinematics, and stability of PD-controlled robot manipulators, please refer to the work of Craig [4].

References

[1] J. C. Maxwell, "On governors," *Proceedings of the Royal Society of London* 16, 270–283, 1868.

[2] Ogata, K., *Modern Control Engineering*, 4th edition, Pearson Education Inc., 2002.

[3] Nise, N. S., *Control Systems Engineering*, 3rd edition, John Wiley & Sons Inc., New York, 2000.

[4] J. J. Craig, *Introduction to Robotics—Mechanics and Control*, 2nd edition, Pearson Education Asia, Singapore, 1989.

3

Introduction to System of Systems

3.1 Introduction

This chapter introduces the concept of a system of systems (SoS) and the challenges ahead to extend systems engineering (SE) to SoS engineering. The birth of a new engineering field may be on the horizon—system of systems engineering (SoSE). An SoS is a collection of individual, possibly heterogeneous, but functional systems integrated together to enhance the overall robustness, lower the cost of operation, and increase the reliability of the overall complex (SoS) system. Having said that, the field has a large vacuum from basic definition, to theory, to management and implementation. Many key issues, such as architecture, modeling, simulation, identification, emergence, standards, net-centricity, control (see Chapter 8), etc., are all begging for attention. In this review chapter, we will be going through all these issues briefly and bring out the challenges to the attention of interested readers.

This growing interest in SoS as a new generation of complex systems has opened a great many new challenges for systems engineers. Performance optimization, robustness, and reliability among an emerging group of heterogeneous systems in order to realize a common goal has become the focus of various applications, including military, security, aerospace, space, manufacturing, service industry, environmental systems, and disaster management, to name a few [1–3]. There is an increasing interest in achieving synergy between these independent systems to achieve the desired overall system performance, as shown by Azarnoosh et al. [4]. In the literature, researchers have addressed the issue of coordination and interoperability in an SoS [5,6]. SoS technology is believed to more effectively implement and analyze large, complex, independent, and *heterogeneous* systems working (or made to work) cooperatively [5]. The main thrust behind the desire to view the systems as an SoS is to obtain higher capabilities and performance than would be possible with a traditional system view. The SoS concept presents a high-level viewpoint and explains the interactions between each of the independent systems. However, the SoS concept is still in its developing stages [7–8].

The next section will present some definitions out of many possible definitions of SoS. However, a practical definition may be that an SoS is a "super system" consisting of other elements that themselves are independent complex operational systems and interact among themselves to achieve a common goal. Each element of an SoS achieves well-substantiated goals even if they are detached from the rest of the SoS. For example, a Boeing 747 airplane, as an element of an SoS, is not SoS, but an airport is an SoS, or a rover on Mars is not an SoS, but a robotic colony (or a robotic swarm) exploring the red planet, or any other place, is an SoS. As will be illustrated shortly, associated with SoS, there are numerous problems and open-ended issues that need a great number of fundamental advances in theory and verifications. It is hoped that this chapter will be a first effort toward bridging the gaps between an *idea* and a *practice*.

3.2 Definitions of SoS

Based on the literature survey on SoS, there are numerous definitions whose detailed discussion is beyond the space allotted to this topic [7–10]. In this work, we enumerate only six of many potential definitions:

Definition 1: SoS exist when there is a presence of a majority of the following five characteristics: operational and managerial independence, geographic distribution, emergent behavior, and evolutionary development [7].

Definition 2: SoS are large-scale concurrent and distributed systems consisting of complex systems [7,8].

Definition 3: Enterprise SoSE is focused on coupling traditional SE activities with enterprise activities of strategic planning and investment analysis [7].

Definition 4: SoS integration is a method to pursue development, integration, interoperability, and optimization of systems to enhance performance in future battlefield scenarios [9].

Definition 5: SoSE involves the integration of systems into SoS that ultimately contribute to the evolution of the social infrastructure [14].

Definition 6: In relation to joint war-fighting, SoS is concerned with interoperability and synergism of command, control, computers, communications, and information (C4I) and intelligence, surveillance, and reconnaissance (ISR) systems [10].

Detailed literature survey and discussions on these definitions are given by Jamshidi [7,11]. Various definitions of SoS have their own merits, depending on their application. The favorite definition of this author and the volume's editor is "Systems of systems are large-scale integrated systems which are heterogeneous and independently operable on their own, but are networked together for a common goal." The goal, as mentioned before, may be cost, performance, robustness, etc.

3.3 Challenging Problems in SoS

In the realm of open problems in SoS, just about anywhere one touches, there is an unsolved problem and immense attention is needed by many engineers and scientists. No engineering field is more urgently needed in tackling SoS problems than SE. On top of the list of engineering issues in SoS is the "engineering of SoS," leading to a new field of SoSE [13]. How does one extend SE concepts such as analysis, control, estimation, design, modeling, controllability, observability, stability, filtering, simulation, etc., to be applied to SoS? Among numerous open questions is how can one model and simulate such systems by Mittal et al. [14]? In almost all cases, a chapter in this volume will accommodate the topic raised.

3.3.1 Theoretical Problems

In this section, a number of urgent problems facing SoS and SoSE are discussed. The major issue here is that a merger between SoS and engineering needs to be made. In other words, SE needs to undergo a number of innovative changes to accommodate and encompass SoS.

3.3.1.1 Open Systems Approach to SoSE

Azani [15] discusses an open-systems approach to SoSE. The author notes that SoS exists within a continuum that contains *ad hoc*, short-lived, and relatively speaking simple SoS on one end, and long-lasting, continually evolving, and complex SoS on the other end of the continuum. Military operations and less sophisticated biotic systems (e.g., bacteria and ant colonies) are examples of *ad hoc*, simple, and short-lived SoS, whereas galactic and more sophisticated biotic systems (e.g., ecosystems, human colonies) are examples of SoS at the opposite end of the SoS continuum. The engineering approaches used by galactic SoS are, at best, unknown and perhaps forever inconceivable. However, biotic SoS seem to follow, relatively speaking, less complicated engineering and development strategies, allowing them to continually learn and adapt, grow and evolve, resolve emerging conflicts, and have more predictable behavior. Based on what the author already knows about biotic SoS, it is apparent that these systems use robust reconfigurable architectures enabling them to effectively capitalize on open systems development principles and strategies such as modular design, standardized interfaces, emergence, natural selection, conservation, synergism, symbiosis, homeostasis, and self-organization. Azani [15] provides further elaboration on open systems development strategies and principles used by biotic SoS, discusses their implications for engineering of man-made SoS, and introduces an integrated SoS development methodology for engineering and development of adaptable, sustainable, and interoperable SoS based on open systems principles and strategies.

3.3.1.2 Engineering of SoS

Emerging needs for a comprehensive look at the applications of classical SE issue in SoSE will be discussed in this volume. The thrust of the discussion will concern the reality that the technological, human, and organizational issues are each far different when considering an SoS or federation of systems and that these needs are very significant when considering SoSE and management.

As we have noted, today, there is much interest in the engineering of systems composed of other component systems, and where each of the component systems serves organizational and human purposes. These systems have several principal characteristics that make the system family designation appropriate: operational independence of the individual systems; managerial independence of the systems; often large geographic and temporal distribution of the individual systems; and emergent behavior, in which the system family performs functions and carries out purposes that do not reside uniquely in any of the constituent systems but which evolve over time in an adaptive manner and where these behaviors arise as a consequence of the formation of the entire system family and are not the behavior of any constituent system. The principal purposes supporting engineering of these individual systems and the composite system family are fulfilled by these emergent behaviors. Thus, an SoS is never fully formed or complete. Development of these systems is evolutionary and adaptive over time, and structures, functions, and purposes are added, removed, and modified as experience of the community with the individual systems and the composite system grows and evolves. The SE and management of these system families pose special challenges. This is especially the case with respect to the federated systems management principles that must be used to deal successfully with the multiple contractors and interests involved in these efforts. Please refer to the reports of Sage and Biemer [16] and De Larentis et al. [17] for the creation of an SoS Consortium [International Consortium on System of Systems (ICSoS)] of concerned individuals and organizations by the author of this chapter.

3.3.1.3 Standards of SoS

SoS literature, definitions, and perspectives are marked with great variability in the engineering community. Viewed as an extension of SE to a means of describing and managing social networks and organizations, variations in perspectives lead to difficulty in advancing and understanding the discipline. Standards have been used to facilitate a common understanding and approach to align disparities of perspectives to drive a uniform agreement to definitions and approaches. Having the ICSoS [17] present to the Institute of Electrical and Electronics Engineers and the International Consortium on System of Systems Engineering for support of technical committees to derive standards for SoS will help unify and advance the discipline for engineering, health care, banking, space exploration, and all other disciplines requiring interoperability among disparate systems [17].

3.3.1.4 SoS Architecting

Dagli and Kilicay-Ergin [18] provide a framework for SoS Architectures. As the world is moving toward a networked society, the authors assert, the business and government applications require integrated systems that exhibit intelligent behavior. The dynamically changing environmental and operational conditions necessitate a need for system architectures that will be effective for the duration of the mission but evolve to new system architectures as the mission changes. This new challenging demand has led to a new operational style: Instead of designing or subcontracting systems from scratch, business or government gets the best systems that the industry develops and focuses on becoming the lead system integrator to provide an SoS. SoS is a set of interdependent systems that are related or connected to provide a common mission. In the SoS environment, architectural constraints imposed by existing systems have a major effect on the system capabilities, requirements, and behavior. This fact is important, because it complicates the system architecting activities. Hence, architecture becomes a dominating but confusing concept in capability development. There is a need to push system architecting research to meet the challenges imposed by new demands of the SoS environment. This chapter focuses on SoS architecting in terms of creating meta-architectures from collections of different systems. Several examples are provided to clarify the SoS architecting concept. Since the technology base, organizational needs, and human needs are changing, SoS architecting becomes an evolutionary process. Components and functions are added, removed, and modified as owners of the SoS experience and use the system. Finally, the authors discuss the possible use of artificial life tools for the design and architecting of SoS. Artificial life tools such as swarm intelligence, evolutionary computation, and multiagent systems have been successfully used for the analysis of complex adaptive systems. The potential use of these tools for SoS analysis and architecting are discussed, by the authors, using several domain application specific examples. Figure 3.1 shows meta-architecture generation for financial markets [18].

3.3.1.5 SoS Simulation

Sahin et al. [19] have presented an SoS architecture based on Extensible Markup Language (XML) to wrap data coming from different systems in a common fashion. XML can be used to describe each component of the SoS and their data in a unifying manner. If XML-based data architecture is used in an SoS, the SoS components are the requirement to understand/parse XML files received from the components of the SoS. In XML, data

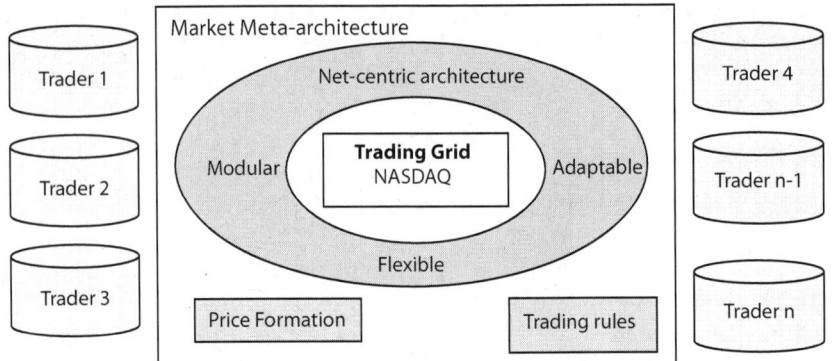

FIGURE 3.1
Meta-architecture generation for financial markets [22].

can be represented in addition to the properties of the data such as source name, data type, importance of the data, and so on. Thus, it does not only represent data, but it also gives useful information that can be used in the SoS to take better action and to understand the situation better. XML language has a hierarchical structure, where an environment can be described with a standard and without a huge overhead. Each entity can be defined by the user in XML in terms of its visualization and functionality. As a case study in this effort, see the report of Mittal et al. [14], where discrete event systems (DEVS) are presented as a platform for modeling and simulation of SoS. Also presented there is the architecture for a master-scout rover combination representing an SoS where a sensor detects a fire in a field. The fire is detected by the master rover, which commands the scout rover to verify the existence of the fire. It is important to note that such an architecture and simulation does not need any mathematical model for members of the systems. Figure 3.2 shows a DEVS-XML simulation framework for a system of robots seeking to warn of a disaster waiting to occur [19].

3.3.1.6 SoS Integration

Integration is probably the key viability of any SoS. Integration of SoS implies that each system can communicate and interact (control) with the SoS regardless of their hardware, software characteristics, or nature. This means that they need to have the ability to communicate with the SoS or a part of the SoS without compatibility issues, such as operating systems, communication hardware, and so on. For this purpose, an SoS needs a common language that the systems of SoS can speak. Without a common language, the systems of any SoS cannot be fully functional and the SoS cannot be adaptive in the sense that new components cannot be integrated into it without major effort. Integration also implies the control aspects of the SoS because systems

Introduction to System of Systems

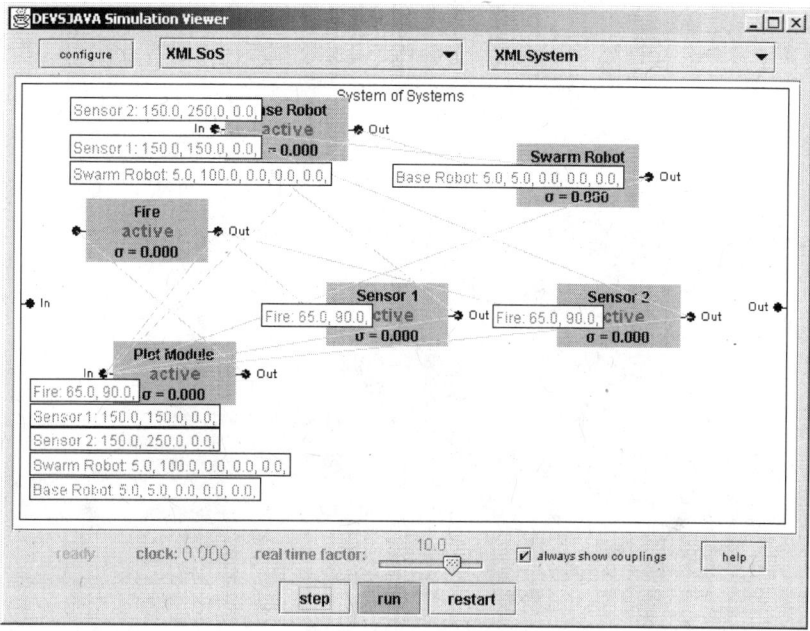

FIGURE 3.2
A DEV-XML simulation framework for a system of robots finding start of a fire [23].

need to understand each other in order to take commands or signals from other SoS systems. Refer to Cloutier et al. [21] for more on the network-centric architecture of SoS. Figure 3.3 shows a net-centric community-of-interest SoS ecology [21].

3.3.1.7 Emergence in SoS

Emergent behavior of an SoS resembles the slowing down of the traffic going through a tunnel, even in the absence of any lights, obstacles, or accidents. A tunnel, automobiles, and the highway, as systems of an SoS, have an emergent behavior or property in slowing down [25]. Fisher [22] has noted that whether an SoS can achieve its goals depends on its emergent behaviors. The author *explores* "interdependencies among systems, emergence, and interoperation" and develops maxim-like findings such as the following: (1) Because they cannot control one another, autonomous entities can achieve goals that are not local to themselves only by increasing their influence through cooperative interactions with others. (2) Emergent composition is often poorly understood and sometimes misunderstood because it has few analogies in traditional SE. (3) Even in the absence of accidents, tight coupling can ensure that an SoS is unable to satisfy its objectives. (4) If it is to remain scalable and affordable

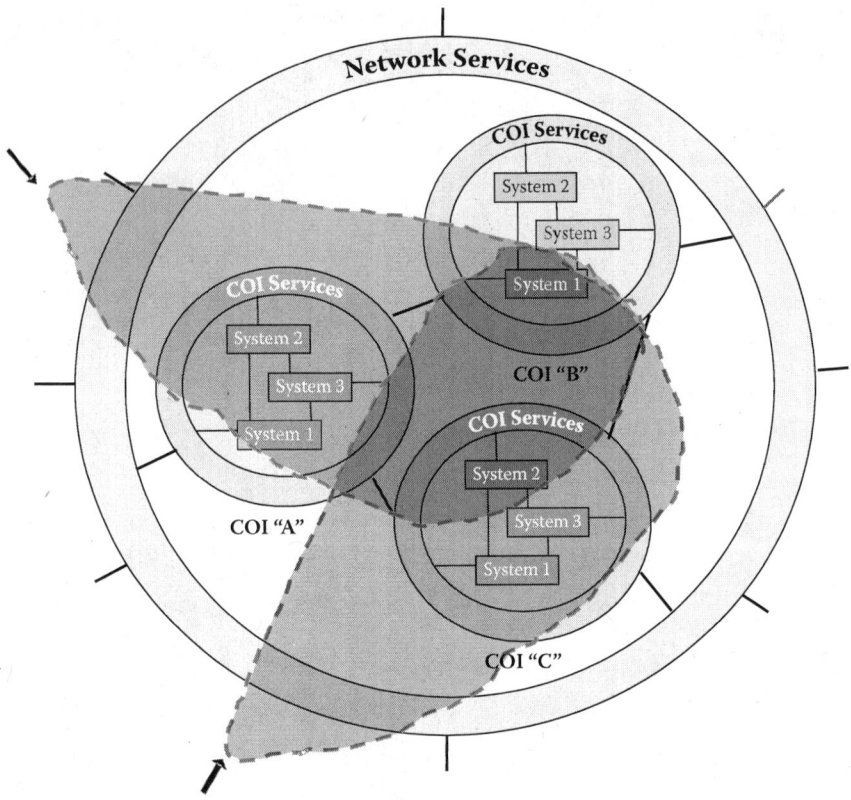

FIGURE 3.3
A netcentric community-of-interest SoS ecology. (Image courtesy of R. Cloutier, M.J. Dimano and H.W. Polzer [21] used with permission.)

regardless of how large it may become, a system's cost per constituent must grow less linearly with its size. (5) Delay is a critical aspect of SoS. Keating [23] provides a detailed perspective into the emergence property of SoS.

3.3.1.8 SoS Management: Governance of Paradox

Sauser and Boardman [25] present an SoS approach to the management problem. They note that the study of SoS has moved many to support their understanding of these systems through the groundbreaking science of networks. The understanding of networks and how to manage them may give one the fingerprint that is independent of the specific systems that exemplify this complexity. The authors point out that it does not matter whether they are studying the synchronized flashing of fireflies, space stations, the structure of the human brain, the Internet, the flocking of birds, a future combat system, or the behavior of red harvester ants. The same emergent principles apply:

large is really small, weak is really strong, significance is really obscure, little means a lot, simple is really complex, and complexity hides simplicity. The conceptual foundation of complexity is a paradox, which leads us to a paradigm shift in the SE body of knowledge.

Paradox exists for a reason, and there are reasons for systems engineers to appreciate paradox even though they may be unable to resolve them because they would a problem specification into a system solution. Hitherto, paradoxes have confronted current logic only to yield at a later date to more refined thinking. The existence of a paradox is always the inspirational source for seeking new wisdom, attempting new thought patterns, and ultimately building systems for the "flat world." It is our ability to govern, not control, these paradoxes that will bring new knowledge to our understanding on how to manage the emerging complex systems called SoS.

Sauser and Boardman [25] establish a foundation in what has been learned about how one practices project management, establish several key concepts and challenges making the management of SoS different from our fundamental practices, present an intellectual model for how they classify and manage an SoS, appraise this model with recognized SoS, and conclude with grand challenges for how they may move their understanding of SoS management beyond the foundation.

In the previous section, a brief introduction was presented for six theoretical issues of SoS, that is, integration, engineering, standards, open and other architectures, modeling, infrastructure, and simulation. These topics are discussed in great detail by a number of experts in the in two recent books by Jamshidi [12,24].

3.3.2 Implementation Problems for SoS

Aside from many theoretical and essential difficulties with SoS, there are many implementation challenges facing SoS. In this work, some of these implementation problems are briefly discussed and references are made to some with their full coverage.

3.3.2.1 SE for the Department of Defense SoS

Dahmann [26] has addressed the national defense aspects of SoS. Military operations are the synchronized efforts of people and systems toward a common objective. In this way, from an operational perspective, defense is essentially a "systems of systems" enterprise. However, although today almost every military system is operated as part of an SoS, most of these systems were designed and developed without the benefit of SE at the SoS level factoring the role the system will play in the broader SoS context. With changes in operations and technology, the need for systems that work effectively together is increasingly visible. Dahmann [26] outline the changing situation in the defense department and the challenges it poses for SE.

(a)

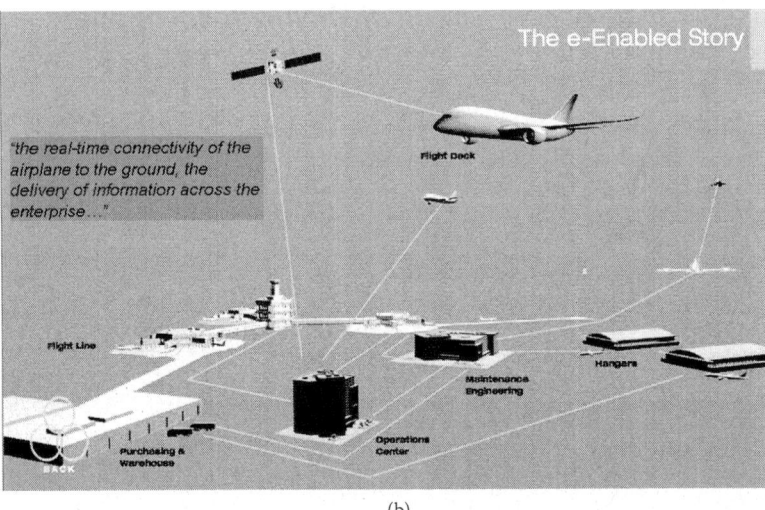

"the real-time connectivity of the airplane to the ground, the delivery of information across the enterprise..."

(b)

FIGURE 3.4
E-enabling of Boeing 787 Dream liner via SoS. (a) A photo of the new SoS E-enabled Boeing 787 (courtesy of Boeing Company; see also Wilber [27]). (b) Connectivity of the B-787 dream liner to the ground delivery of information across the enterprise [27].

3.3.2.2 E-enabling and SoS Aircraft Design via SoSE

A case of aeronautical application of SoS worth noting is that of E-enabling in aircraft design as a system of an SoS at Boeing Commercial Aircraft Division [27]. The project focused on developing a strategy and technical architecture to facilitate making the airplane (Boeing 787, see Figure 3.4) network-aware

Introduction to System of Systems

and capable of leveraging computing and network advances in industry. The project grew to include many ground-based architectural components at the airlines and at the Boeing factory, as well as other key locations such as the airports, suppliers, and terrestrial Internet service suppliers.

Wilber [27] points out that the E-enabled project took on the task of defining an SoSE solution to the problem of interoperation and communication with the existing, numerous, and diverse elements making up the airlines' operational systems (flight operations and maintenance operations). The objective has been to find ways of leveraging network-centric operations to reduce production, operations, and maintenance costs for both Boeing and the airline customers.

> One of the key products of this effort is the "e-Enabled Architecture." The e-Enabling Architecture is defined at multiple levels of abstraction. There is a single top-level or "Reference Architecture" that is necessarily abstract and multiple "Implementation Architectures." The implementation architectures map directly to airplane and airline implementations and provide a family of physical solutions that all exhibit common attributes and are designed to work together and allow reuse of systems components. The implementation architectures allow for effective forward and retrofit installations addressing a wide range of market needs for narrow and wide-body aircraft.
>
> The 787 "Open Data Network" is a key element of one implementation of this architecture. It enabled on-board and off-board elements to be networked in a fashion that is efficient, flexible, and secure. The fullest implementations are best depicted in Boeing's GoldCare Architecture and design.

Wilber [28] presents architecture at the reference level and how it has been mapped into the 787 airplane implementation. The *GoldCare* environment is described and is used as an example of the full potential of the current E-enabling (see Figure 3.4).

3.3.2.3 An SoS Perspective on Infrastructures

Thissen and Herder [51] touch on a very important application in the service industry (also see Tien [31]). Infrastructure systems (or infrasystems) providing services such as energy, transport, communications, and clean and safe water are vital to the functioning of modern society. Key societal challenges with respect to our present and future infrastructure systems relate to, among other things, safety and reliability, affordability, and transitions to sustainability. Infrasystem complexity precludes simple answers to these challenges. Although each of the infrasystems can be seen as a complex SoS in itself, increasing interdependency among these systems (both technologically and institutionally) adds a layer of complexity.

One approach to increased understanding of complex infrasystems that has received little attention in the engineering community thus far is to focus on the commonalities of the different sectors, and to develop generic theories and approaches such that lessons from one sector could easily be applied to other sectors. The SoS paradigm offers interesting perspectives in this respect. The authors present, as an initial step in this direction, a fairly simple three-level model distinguishing the physical/technological systems, the organization and management systems, and the systems and organizations providing infrastructure-related products and services. The authors use the model as a conceptual structure to identify a number of key commonalities and differences between the transport, energy, drinking water, and information and communications technology sectors. Using two energy-related examples, the authors further illustrate some of the SoS-related complexities of analysis and design at a more operational level. The authors finally discuss a number of key research and engineering challenges related to infrastructure systems, with a focus on the potential contributions of SoS perspectives.

3.3.2.4 Sensor Networks

The main purpose of sensor networks is to utilize the distributed sensing capability provided by tiny, low-powered, and low-cost devices. Multiple sensing devices can be used cooperatively and collaboratively to capture events or monitor space more effectively than a single sensing device [31,32]. The realm of applications for sensor networks is quite diverse, which includes military, aerospace, industrial, commercial, environmental, and health monitoring, to name a few. Applications include traffic monitoring of vehicles, cross-border infiltration detection and assessment, military reconnaissance and surveillance, target tracking, habitat monitoring and structure monitoring, etc.

The communication capability of these small devices, which often have heterogeneous attributes, makes them good candidates for SoS. Numerous issues exist with sensor networks, such as data integrity, data fusion and compression, power consumption, multidecision making, and fault tolerance; all these issues make this SoS very challenging, just like other SoS. Thus, it is necessary to devise a fault-tolerant mechanism with a low computation overhead to validate the integrity of the data obtained from the sensors ("systems"). Moreover, a robust diagnostics and decision-making process should aid in the monitoring and control of critical parameters to efficiently manage the operational behavior of a deployed sensor network. Specifically, Azarnoosh et al. [4] have focused on innovative approaches to deal with multivariable, multispace problem domains, as well as other issues, in wireless sensor networks within the framework of an SoS. Figure 3.5 shows that the components in the SoS, which are themselves systems, are sufficiently *complex*, as shown in Exhibit 1.

3.3.2.5 An SoS View of Services

Tien [31] covers a very important application of SoS in today's global village—the *service industry*. The services sector employs a large and growing proportion of workers in industrialized nations, and it is increasingly dependent on information technology. Although the interdependences,

Introduction to System of Systems

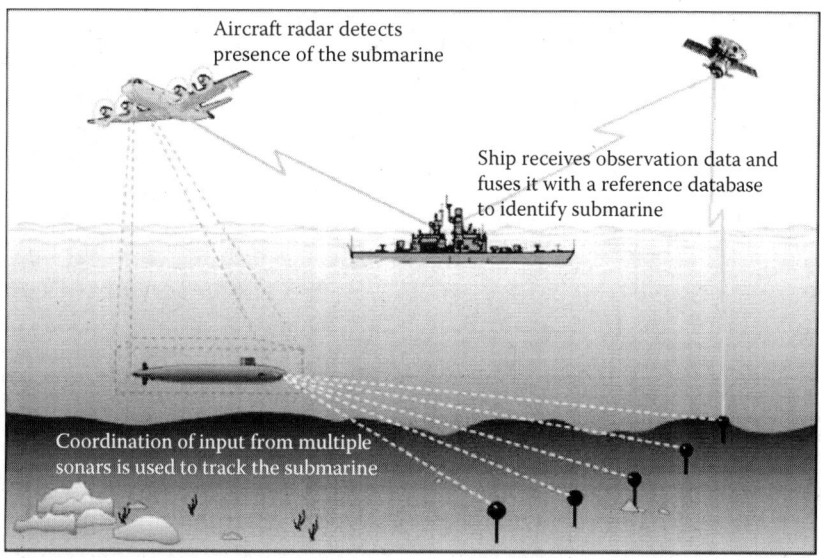

FIGURE 3.5
A classical SoS application (image courtesy of Sridhar et al. [29]).

similarities, and complementarities of manufacturing and services are significant, there are considerable differences between goods and services, including the shift in focus from mass production to mass customization (whereby a service is produced and delivered in response to a customer's stated or imputed needs). In general, a service system can be considered to be a combination or recombination of three essential components—people (characterized by behaviors, attitudes, values, etc.), processes (characterized by collaboration, customization, etc.), and products (characterized by software, hardware, infrastructures, etc.). Furthermore, inasmuch as a service system is an integrated system, it is, in essence, an SoS, the objectives of which are to enhance its efficiency (leading to greater interdependency), effectiveness (leading to greater usefulness), and adaptiveness (leading to greater responsiveness). The integrative methods include a component's design, interface, and interdependency; a decision's strategic, tactical, and operational orientation; and an organization's data, modeling, and cybernetic consideration. A number of insights are also provided, including an alternative SoS view of services; the increasing complexity of systems (especially service systems), with all the attendant life-cycle design, human interface, and system integration issues; the increasing need for real-time, adaptive decision making within such SoS; and the fact that modern systems are also becoming increasingly more human-centered, if not human-focused—thus, products and services are becoming more complex and more personalized or customized.

3.3.2.6 SoSE in Space Exploration

Jolly and Muirhead [32] cover SoSE topics that are largely unique for space exploration with the intent of providing the reader with a discussion of the key issues, the major challenges of the twenty-first century in moving from SE to SoSE, potential applications in the future, and the current state-of-the-art. Specific emphasis is placed on how software and electronics are revolutionizing the way space missions are being designed, including both the capabilities and vulnerabilities introduced. The role of margins, risk management, and interface control are all critically important in current space mission design and execution—but in SoSE applications, they become paramount. Similarly, SoSE space missions will have extremely large, complex, and intertwined command and control and data distribution ground networks, most of which will involve extensive parallel processing to produce tera- to petabytes of products per day and distribute them worldwide. Figure 3.6 indicates the National Aeronautic and Space Administration's (NASA) space constellation project as an SoS.

3.3.2.7 Communication and Navigation in Space SoS

Bahsin and Hayden [33] have taken on the challenges in communication and navigation for space SoS. They indicate that communication and navigation

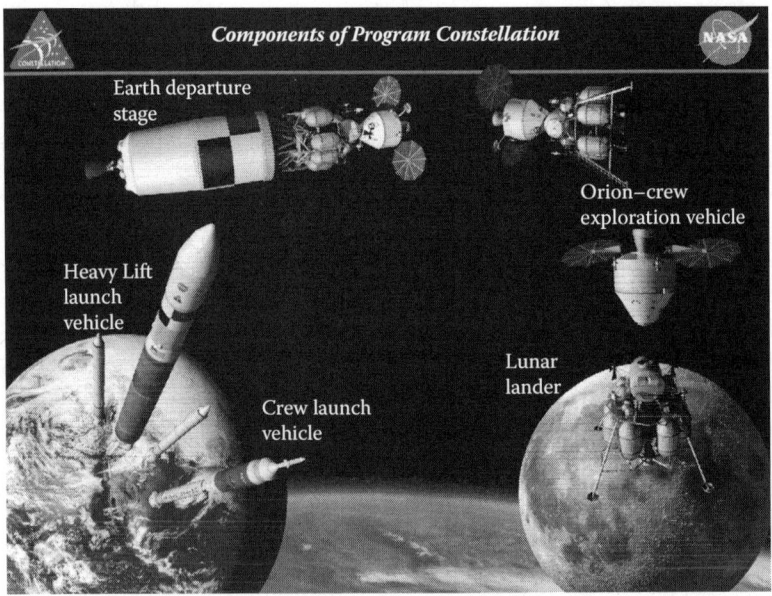

FIGURE 3.6
NASA's Space Constellation Project as an SoS (photo courtesy of NASA [32]).

networks provide critical services in the operation, system management, information transfer, and situation awareness to the space SoS. In addition, space SoS are requiring system interoperability, enhanced reliability, common interfaces, dynamic operations, and autonomy in system management. New approaches to communications and navigation networks are required to enable the interoperability needed to satisfy the complex goals and dynamic operations and activities of the space SoS. Historically, space systems had direct links to Earth ground communication systems, or they required a space communication satellite infrastructure to achieve higher coverage around the Earth. It is becoming increasingly apparent that many SoS may include communication networks that are also SoS. These communication and navigation networks must be as nearly ubiquitous as possible and accessible on the demand of the user, much like the cell phone link is available at any time to an Earth user in range of a cell tower. The new demands on communication and navigation networks will be met by space Internet technologies. It is important to bring Internet technologies, Internet protocols, routers, servers, software, and interfaces to space networks to enable as much autonomous operation of those networks as possible. These technologies provide extensive savings in reduced cost of operations. The more these networks can be made to run themselves, the less humans will to schedule and control them. Internet technologies also bring with them a very large repertoire of hardware and software solutions to communication and networking problems that would be very expensive to replicate under a different paradigm. Higher bandwidths are needed to support the expected voice, video, and data transfer traffic for the coordination of activities at each stage of an exploration mission.

Existing communications, navigation, and networking have grown in an independent fashion, with experts in each field solving the problem just for that field. Radio engineers designed the payloads for today's "bent pipe" communication satellites. The global positioning satellite (GPS) system design for providing precise Earth location determination is an extrapolation of the long-range navigation technique of the 1950s, where precise time is correlated to a precise position on the Earth. Other space navigation techniques use artifacts in the radio frequency (RF) communication path (Doppler shift of the RF and transponder-reflected ranging signals in the RF) and time transfer techniques to determine the location and velocity of a spacecraft within the solar system. Networking in space today is point-to-point among ground terminals and spacecraft, requiring most communication paths to/from space to be scheduled such that communication is available only on an operational plan and is not easily adapted to handle multidirectional communications under dynamic conditions.

Bahsin and Hayden [33] begin with a brief history of the communications, navigation, and networks of the 1960s and 1970s in use by the first SoS, the NASA Apollo missions; it is followed by short discussions of the communication and navigation networks and architectures that the Department of

Defense and NASA used from the 1980s onward. Next is a synopsis of the emerging space SoS that will require complex communication and navigation networks to meet their needs. Architecture approaches and processes being developed for communication and navigation networks in the emerging space system and systems are also described. Several examples are given of the products generated in using the architecture development process for space exploration systems. The architecture addresses the capabilities to enable voice, video, and data interoperability needed among the explorers during exploration, while in habitat, and with Earth operations. Advanced technologies are then described that will allow space SoS to operate autonomously or semiautonomously. Korba and Hiskins [34] end with a summary of the challenges and issues raised in implementing these new concepts.

3.3.2.8 Electric Power Systems Grids as SoS

Korba and Hiskins [34] provide an overview of the SoS that are fundamental to the operation and control of electrical power systems. Perspectives are drawn from industry and academia, and reflect theoretical and practical challenges that are facing power systems in an era of energy markets and increasing utilization of renewable energy resources (see also [35]). Power systems cover extensive geographical regions and are composed of many diverse components. Accordingly, power systems are large-scale, complex, dynamical systems that must operate reliably to supply electrical energy to customers. Stable operation is achieved through extensive monitoring systems, and a hierarchy of controls, that together seek to ensure that total generation matches consumption, and that voltages remain at acceptable levels. Safety margins play an important role in ensuring reliability, but they tend to incur economic penalties. Significant effort is therefore being devoted to the development of demanding control and supervision strategies enabling the reduction of these safety margins, with consequent improvements in transfer limits and profitability. Recent academic and industrial research activities in this field are addressed by Korba and Hiskins [34]. Figure 3.7 shows a wide-area monitoring of electrical power systems using synchronized phasor measurement units [35].

3.3.2.9 SoS Approach for Renewable Energy

Duffy et al. [35] have applied the SoS approach to the sustainable supply of energy. They note that over one-half of the petroleum consumed in the United States is imported, and that percentage is expected to rise to 60% by 2025. America's transportation SoS relies almost exclusively on refined petroleum products, accounting for over two-thirds of the oil used. Each day, more than 8 million barrels of oil are required to fuel more than 225 million vehicles that constitute the U.S. light-duty transportation fleet. The gap between U.S. oil production and transportation oil needs is projected to

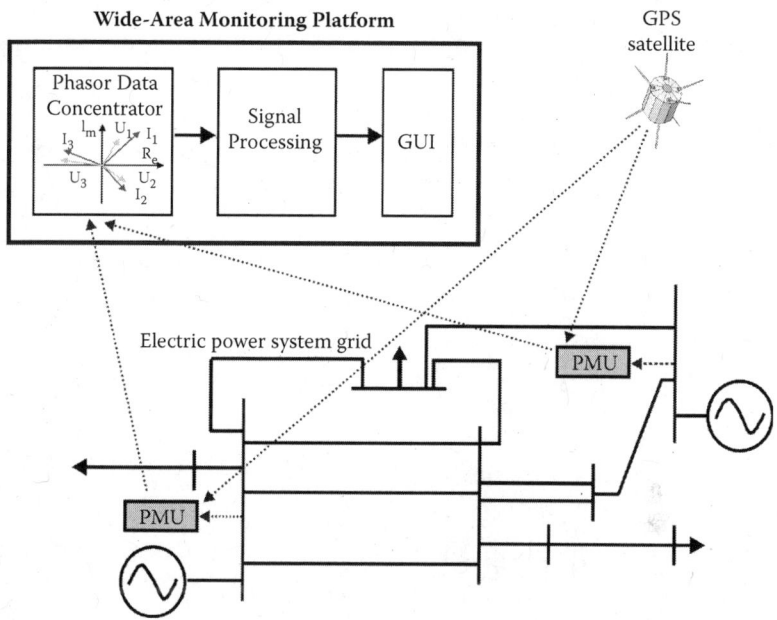

FIGURE 3.7
A wide-area monitoring of electrical power systems using synchronized phasor measurement.

grow, and the increase in the number of light-duty vehicles will account for most of that growth. On a global scale, petroleum supplies will be in increasingly higher demand as highly populated developing countries expand their economies and become more energy-intensive. Clean forms of energy are needed to support sustainable global economic growth while mitigating impacts on air quality and the potential effects of greenhouse gas emissions. The growing dependence of the United States on foreign sources of energy threatens its national security. The authors assert that, as a nation, we must work to reduce our dependence on foreign sources of energy in a manner that is affordable and preserves environmental quality. Figure 3.8 shows the existing petroleum-based transportation SoS [36].

3.3.2.10 Sustainable Environmental Management from an SoSE Perspective

Hipel et al. [36] provide a rich range of decision tools from the field of SE that are described for addressing complex environmental SoS problems in order to obtain sustainable, fair, and responsible solutions to satisfy the value systems of stakeholders as much as possible, including the natural environment and future generations who are not even present at the bargaining table. To better understand the environmental problem being investigated and, thereby, eventually reach more informed decisions, the insightful paradigm

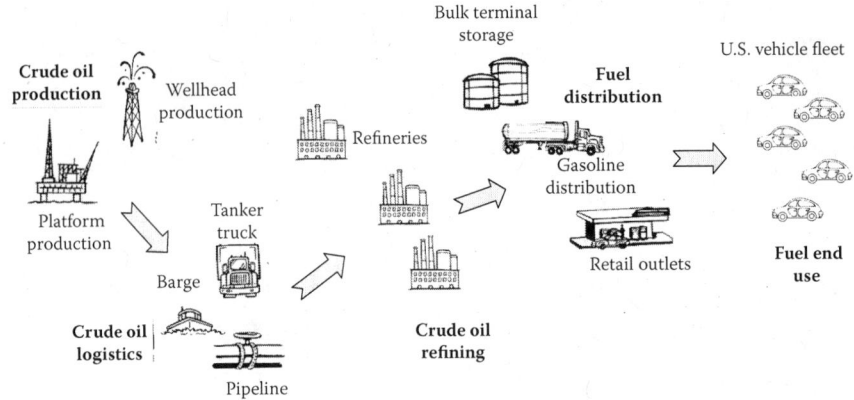

FIGURE 3.8
The existing petroleum-based transportation SoS [35].

of an SoS can be readily utilized. For example, when developing solutions to global warming problems, one can envision how societal systems, such as agricultural and industrial systems, interact with the atmospheric SoS, especially at the tropospheric level. The great importance of developing a comprehensive toolbox of decision methodologies and techniques is emphasized by pointing out many current pressing environmental issues, such as global warming and its potential adverse affects and the widespread pollution of our land, water, and air SoS. To tackle these large-scale complex SoS problems, SE decision techniques that can take into account multiple stakeholders having multiple objectives are explained according to their design and capabilities. To illustrate how systems decision tools can be used in practice to assist in reaching better decisions for benefiting society, different decision tools are applied to three real-world SoS environmental problems. Specifically, the graph model for conflict resolution is applied to the international dispute over the utilization of water in the Aral Sea Basin; a large-scale optimization model founded upon concepts from cooperative game theory, economics, and hydrology is used for systematically investigating the fair allocation of scarce water resources among multiple users in the South Saskatchewan River Basin in western Canada; and multiple criteria decision analysis methods are used to evaluate and compare solutions to handling fluctuating water levels in the five Great Lakes located along the border of Canada and the United States [37]. Figure 3.9 shows the interactions among energy and atmospheric SoS [37].

3.3.2.11 Robotic Swarms as an SoS

As another application of SoS, a robotic swarm, is considered by Sahin [37]. Here, a robotic swarm based on ant colony optimization and artificial immune

Introduction to System of Systems

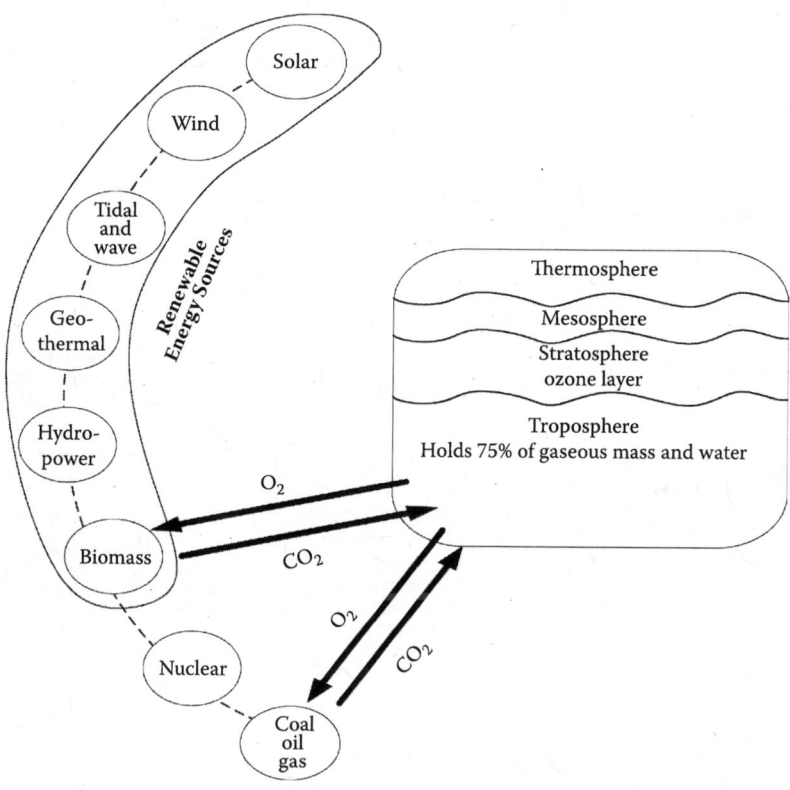

FIGURE 3.9
Interactions among energy and atmospheric SoS [36].

systems is considered. In the ant colony optimization, the author has developed a multiagent system model based on the food gathering behaviors of the ants. Similarly, a multiagent system model is developed based on the human immune system. These multiagent system models are then tested on the mine detection problem. A modular microrobot is designed to emulate the mine detection problem in a basketball court. The software and hardware components of the modular robot are designed to be modular so that robots can be assembled using hot swappable components. An adaptive TDMA (time division multiple access) communication protocol is developed in order to control connectivity among the swarm robots without user intervention. Figure 3.10 shows a robotic swarm isolating a mine as an SoS at Rochester Institute of Technology.

3.3.2.12 Transportation Systems

The National Transportation System (NTS) can be viewed as a collection of layered networks composed of heterogeneous systems, for which the Air

FIGURE 3.10
A robotic swarm isolating a mine as an SoS.

Transportation System (ATS) and its National Airspace System are one part. At present, research on each sector of the NTS is generally conducted independently, with infrequent and/or incomplete consideration of scope dimensions (e.g., multimodal impacts and policy, societal, and business enterprise influences) and network interactions (e.g., layered dynamics within a scope category). This isolated treatment does not capture the higher-level interactions seen at the NTS or ATS architecture level; thus, modifying the transportation system based on limited observations and analyses may not necessarily have the intended effect or impact. A systematic method for modeling these interactions with an SoS approach is essential to the formation of a more complete model and understanding of the ATS, which would ultimately lead to better outcomes from high-consequence decisions in technological, socioeconomic, operational, and political policy-making contexts [40]. This is especially vital as decision-makers in both the public and private sector, for example, at the interagency Joint Planning and Development Office, which is charged with transformation of air transportation, are facing problems of increasing complexity and uncertainty in attempting to encourage the evolution of superior transportation architectures [41]. De Laurentis [41] has addressed this application. Figure 3.11 shows an entity-centric abstraction model in transportation SoS with two pairs of entity descriptors emerging from the abstraction process: explicit-implicit and endogenous-exogenous [41].

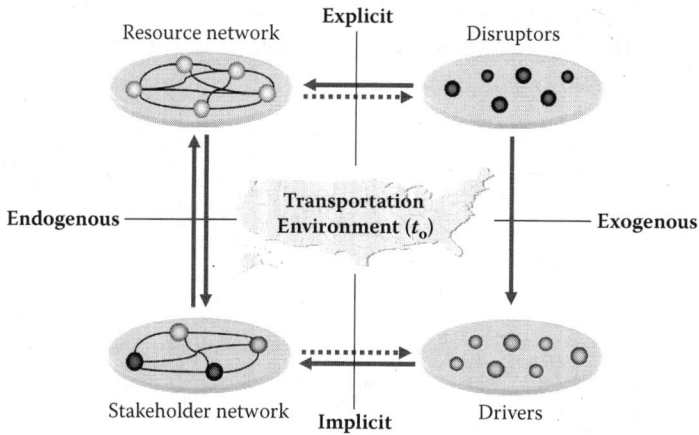

FIGURE 3.11
Entity-centric abstraction model in transportation SoS with two pairs of entity descriptors emerge: explicit-implicit and endogenous-exogenous [41].

3.3.2.13 Health Care Systems

Under a 2004 Presidential Order, the U.S. Secretary of Health has initiated the development of a National Healthcare Information Network (NHIN), with the goal of creating a nationwide information system that can build and maintain electronic health records (EHRs) for all citizens by 2014. The NHIN system architecture currently under development will provide a near-real-time heterogeneous integration of disaggregated hospital, departmental, and physician patient care data, and will assemble and present a complete current EHR to any physician or hospital a patient consults [43]. The NHIN will rely on a network of independent regional health care information organizations (RHIOs) that are being developed and deployed to transform and communicate data from the hundreds of thousands of legacy medical information systems presently used in hospital departments, physician offices, and telemedicine sites into NHIN-specified metaformats that can be securely relayed and reliably interpreted anywhere in the country. The NHIN "network of networks" will clearly be a very complex SoS, and the performance of the NHIN and RHIOs will directly affect the safety, efficacy, and efficiency of health care in the United States. Simulation, modeling, and other appropriate SoSE tools are under development to help ensure reliable, cost-effective planning, configuration, deployment, and management of the heterogeneous, life-critical NHIN and RHIO systems and subsystems [43]. ICSoS represents an invaluable opportunity to access and leverage SoSE expertise already under development in other industry and academic sectors. ICSoS also represents an opportunity to discuss the positive and negative emergent behaviors that can significantly affect personal and public health status and the costs of health care in the United States [44]. The reader may also see the report of Wickramasinghe et al. [42] on how the health care system can be treated as an SoS.

3.3.2.14 Global Earth Observation SoS

Global Earth Observation SoS (GEOSS) is a global project enjoining more than 60 nations whose purpose is to address the need for timely, quality, long-term, global information as a basis for sound decision making [45–47]. Its objectives are (1) improved coordination of strategies and systems for Earth observations to achieve a comprehensive, coordinated, and sustained Earth observation system or systems; (2) a coordinated effort to involve and assist developing countries in improving and sustaining their contributions to observing systems, their effective utilization of observations, and the related technologies; and (3) the exchange of observations recorded from *in situ*, air full, and open manner with minimum time delay and cost. In GEOSS, the "SoSE Process provides a complete, detailed, and systematic development approach for engineering systems of systems. Boeing's new architecture-centric, model-based systems engineering process emphasizes concurrent development of the system architecture model and system specifications. The process is applicable to all phases of a system's lifecycle. The SoSE Process is a unified approach for system architecture development that integrates the views of each of a program's participating engineering disciplines into a single system architecture model supporting civil and military domain applications" [47]. ICSoS will be another platform for all concerned around the globe to bring the progress and principles of GEOSS to formal discussions and examination on an annual basis.

Shibasaki and Pearlman [47] have presented a detailed description of the GEOSS system, its background, and objectives and challenges.

> The authors note that the first step is to understand the Earth system—its weather, climate, oceans, atmosphere, water, land, geodynamics, natural resources, ecosystems, and natural and human-induced hazards—is crucial to enhancing human health, safety, and welfare, alleviating human suffering including poverty, protecting the global environment, reducing disaster losses, and achieving sustainable development. Observations of the Earth system and the information derived from these observations provide critical inputs for advancing this understanding.
>
> The GEO (Group on Earth Observations), a voluntary partnership of governments and international organizations, was established at the Third Earth Observation Summit in February 2005 to coordinate efforts to build a Global Earth Observation System of Systems (GEOSS). As of November 2007, GEO's Members include 72 Governments and the European Commission. In addition, 46 intergovernmental, international, and regional organizations with a mandate in Earth observation or related issues have been recognized as Participating Organizations.

Introduction to System of Systems

The 10-Year Implementation Plan Reference Document of GEOSS states the importance of the Earth observation and the challenges to enhance human and societal welfare. This Implementation Plan, for the period 2005 to 2015, provides a basis for GEO to construct GEOSS. The Plan defines a vision statement for GEOSS, its purpose and scope, and the expected benefits. Prior to its formal establishment, the *Ad Hoc* GEO (established at the First Earth Observation Summit in July 2003) met as a planning body to develop the GEOSS 10-Year Implementation Plan.

The purpose of GEOSS is to achieve comprehensive, coordinated, and sustained observations of the Earth system to meet the need for timely, quality long-term global information as a basis for sound decision making, initially in nine societal benefit are:

(1) Reducing loss of life and property from natural and human-induced disasters
(2) Understanding environmental factors affecting human health and well-being
(3) Improving management of energy resources
(4) Understanding, assessing, predicting, mitigating, and adapting to climate variability and change
(5) Improving water resource management through better understanding of the water cycle
(6) Improving weather information, forecasting, and warning
(7) Improving the management and protection of terrestrial, coastal, and marine ecosystems
(8) Supporting sustainable agriculture and combating desertification
(9) Understanding, monitoring, and conserving biodiversity

Figure 3.12 shows the SoS approach in the GEOSS Project.

3.3.2.15 Deepwater Coastguard Program

One of the earliest realizations of an SoS in the United States is the so-called Deepwater Coastguard Program shown in Figure 3.13. As seen here, the program takes advantage of all the necessary assets at their disposal, for example, helicopters, aircrafts, cutters, satellite (GPS), ground stations, people, computers, etc.—all systems of the SoS integrated together to react unforeseen circumstances to secure the coastal borders of the southeastern United States, for example, the Florida Coast. The Deepwater program is making progress in the development and delivery of mission-effective command, control, communications, computers, intelligence, surveillance, and reconnaissance (C4ISR) equipment [29]. The SoS approach, the report goes on, has "improved the operational capabilities of legacy cutters and aircraft, and will provide even more functionality when the next generation of surface and air platforms arrives in service." The key feature of the system is its ability to interoperate among all Coast Guard mission assets

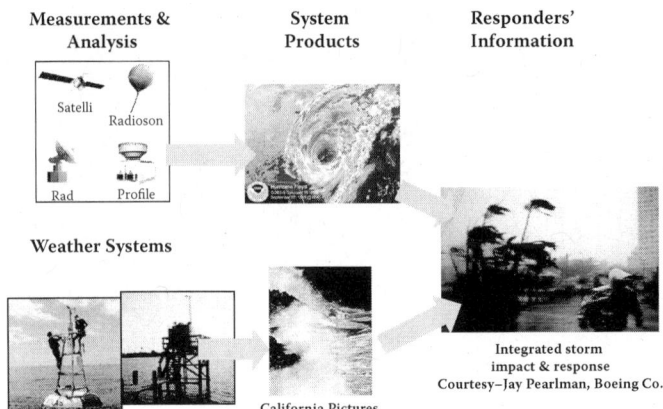

FIGURE 3.12
GEOSS Project Systems in SoS framework. (Courtesy of J. Pearlman [44]. Used with permission).

and capabilities with those of appropriate authorities at both local and federal levels.

3.3.2.16 *Future Combat Missions*

Another national security or defense application of SoS is the future combat mission (FCM). Figure 3.14 shows one of numerous possible configurations

FIGURE 3.13
A security example of an SoS—Deepwater coastguard configuration in the United States.

FIGURE 3.14
A defense example of an SoS (courtesy of Don Walker, Aerospace Corporation).

of an FCM. The FCM system is "envisioned to be an ensemble of manned and potentially unmanned combat systems, designed to ensure that the Future Force is strategically responsive and dominant at every point on the spectrum of operations from nonlethal to full-scale conflict. FCM will provide a rapidly deployable capability for mounted tactical operations by conducting direct combat, delivering both line-of-sight and beyond-line-of-sight precision munitions, providing variable lethal effect (nonlethal to lethal), performing reconnaissance, and transporting troops. Significant capability enhancements will be achieved by developing multifunctional, multimission, and modular features for system and component commonality that will allow for multiple state-of-the-art technology options for mission tailoring and performance enhancements. The FCM force will incorporate and exploit information dominance to develop a common, relevant operating picture and achieve battle space situational understanding" [48,49]. See also Dahmann [26] for insights on this and other defense applications.

3.3.2.17 National Security

Perhaps one of the most talked-about application areas of SoSE is national security. After many years of discussion of the goals, merits, and attributes of SoS, very few tangible results or solutions have appeared in this or other areas of this technology. It is commonly believed that "Systems engineering tools, methods, and processes are becoming inadequate to perform the tasks needed to realize the SoS envisioned for future human endeavors. This is especially becoming evident in evolving national security capabilities realizations for large-scale, complex space and terrestrial military endeavors. Therefore the development of SoSE tools, methods and processes is imperative to enable the realization of future national security capabilities" [50]. In most SoSE applications, heterogeneous systems (or communities) are brought together to cooperate for a common good and to enhance robustness and performance. "These communities range in focus from architectures, to lasers, to complex systems, and will eventually cover each area involved in aerospace related national security endeavors. These communities are not developed in isolation in that cross-community interactions on terminology, methods, and processes are done" [50]. The key is to have these communities work together to guarantee the common goal of making our world a safer place for all. See Dahmann [26] for insights on this and other security applications.

3.4 Conclusions

This chapter was written to serve as an introduction to SoSE. The subject matter of this chapter is an unsettled topic in engineering in general and in SE in particular. An attempt has been made to cover as many open questions in both theory and applications of SoS and SoSE. It is our intention that this chapter would be a small beginning of much debate and challenges among and for the readers of this new area of engineering.

References

[1] Crossley, W. A., "System of systems: An introduction of Purdue University Schools of Engineering's signature area," in *Engineering Systems Symposium*, Tang Center, Wong Auditorium, MIT, March 29–31 2004.

[2] Lopez, D., "Lessons learned from the front lines of the aerospace," in *Proceedings of the IEEE International Conference on System of Systems Engineering*, Los Angeles, April 2006.

[3] Wojcik, L., A., and Hoffman, K. C., "Systems of systems engineering in the enterprise context: A unifying framework for dynamics," in *Proceedings of the IEEE International Conference on System of Systems Engineering*, Los Angeles, April 2006.

[4] Azarnoosh, H., Horan, B., Sridhar, P., Madni, A. M., and Jamshidi, M., "Towards optimization of a real-world robotic-sensor system of systems," in *Proceedings of World Automation Congress (WAC) 2006*, Budapest, July 24–26 2006.

[5] Abel, A., and Sukkarieh, S., "The coordination of multiple autonomous systems using information theoretic political science voting models," in *Proceedings of the IEEE International Conference on System of Systems Engineering*, Los Angeles, April 2006.

[6] DiMario, M. J., "System of systems interoperability types and characteristics in joint command and control," in *Proceedings of the IEEE International Conference on System of Systems Engineering*, Los Angeles, April 2006.

[7] Jamshidi, M., "Theme of the IEEE SMC 2005, Waikoloa, Hawaii, USA," http://ieeesmc2005.unm.edu/.

[8] Carlock, P. G., and Fenton, R. E., "System of systems (SoS) enterprise systems for information-intensive organizations," *Systems Engineering* 4(4), 242–261, 2001.

[9] Pei, R. S., "Systems of systems integration (SoSI)—A smart way of acquiring Army C4I2WS systems," in *Proceedings of the Summer Computer Simulation Conference*, pp. 134–139, 2000.

[10] Manthorpe, W. H., "The emerging joint system of systems: A systems engineering challenge and opportunity for APL," *John Hopkins APL Technical Digest*, 17(3), 305–310, 1996.

[11] Jamshidi, M., *System of Systems Engineering—Principles and Applications*, Taylor Francis CRC Publishers, Boca Raton, FL, 2008.

[12] Jamshidi, M. (Ed.), *System of Systems Engineering—Innovations for the 21st Century*, Wiley & Sons, Inc., New York, 2008.

[13] Wells, G. D., and Sage, A. P., "Engineering of a system of systems," *System of Systems Engineering—Innovations for the 21st Century, John Wiley Series on Systems Engineering* (M. Jamshidi, Ed.). Wiley, New York, 2008.

[14] Mittal, S. B. P., Zeigler, J. L. R., and Sahin, F., "Modeling and simulation for systems of systems engineering," *Systems Engineering—Innovations for the 21st Century, John Wiley Series on Systems Engineering* (M. Jamshidi, Ed.). Wiley, New York, 2008.

[15] Azani, C., "An open systems approach to system of systems engineering" *System of Systems Engineering—Innovations for the 21st Century, John Wiley Series on Systems Engineering* (M. Jamshidi, Ed.). Wiley, New York, 2008.

[16] Sage, A. P., and Biemer, S. M., "Processes for system family architecting, design, and integration," *IEEE Systems Journal* ISJ1-1, 5–16, 2007.

[17] De Larentis, D., Dickerson, C., Di Mario, M., Gartz, P., Jamshidi, M., Nahavandi, S., Sage, A. P., Sloane, E., and Walker, D., "A case for an international consortium on system of systems engineering," *IEEE Systems Journal* 1(1), 68–73, 2007.

[18] Dagli, C. H., and Kilicay-Ergin, N., "System of systems architecting," *System of Systems Engineering—Innovations for the 21st Century, John Wiley Series on Systems Engineering*, Chapter 4 (M. Jamshidi, Ed.). Wiley, New York, 2008.

[19] Sahin, F., Jamshidi, M., and Sridhar, P., "A discrete event XML based simulation framework for system of systems architectures," in *Proceedings the IEEE International Conference on System of Systems*, April 2007.

[20] Morley, J., "Five maxims about emergent behavior in systems of systems," http://www.sei.cmu.edu/news-at-sei/features/2006/06/feature-2-2006-06.htm, 2006.

[21] Cloutier, R. DiMario, M. J., and Polzer, H. W., "Net-centricity and system of systems," *System of Systems Engineering—Innovations for the 21st Century, John Wiley Series on Systems Engineering* (M. Jamshidi, Ed.). Wiley, New York, 2008.

[22] Fisher, D., *An Emergent Perspective on Interoperation in Systems of Systems*, (CMU/SEI-2006-TR- 003), Software Engineering Institute, Carnegie Mellon University, Pittsburgh, 2006.

[23] Keating, C. B., "Emergence in system of systems" *System of Systems Engineering—Innovations for the 21st Century, John Wiley Series on Systems Engineering* (M. Jamshidi, Ed.). Wiley, New York, 2008.

[24] Jamshidi, M. (Ed.), *System of Systems Engineering—Principles and Applications*, CRC Publishers, Boca Raton, FL, 2008.

[25] Sauser, B., and Boardman, J., "System of systems management," *System of Systems Engineering—Innovations for the 21st Century, John Wiley Series on Systems Engineering*, Chapter 8 (M. Jamshidi, Ed.). Wiley, New York, 2008.

[26] Dahmann, J., "Systems engineering for department of defense systems of systems," *System of Systems Engineering—Innovations for the 21st Century, John Wiley Series on Systems Engineering*, Chapter 9 (M. Jamshidi, Ed.). Wiley, New York, 2008.

[27] Wilber, F. R., "A system of systems approach to e-Enabling the commercial airline applications from an airframer's perspective," Keynote presentation, *2007IEEE SoSE Conference*, San Antonio, 18 April 2007.

[28] Wilber, F. R., "Boeing's SOSE approach to E-enabling commercial airlines," *System of Systems Engineering—Innovations for the 21st Century, John Wiley Series on Systems Engineering* (M. Jamshidi, Ed.). Wiley, New York, 2008.

[29] Sridhar, P., Madni, A. M., and Jamshidi, M., "Multi-criteria decision making and behavior assignment in sensor networks," *IEEE Instrumentation and Measurement Magazine* 11(1), 24–29, 2008.

[30] Sridhar, P., Madni, A. M., and Jamshidi, M., "Hierarchical aggregation and intelligent monitoring and control in fault-tolerant wireless sensor networks," *IEEE Systems Journal* 1(1), 38–54, 2007.

[31] Tien, J. M., "A system of systems view of services," *Innovations for the 21st Century, John Wiley Series on Systems Engineering*, Chapter 13 (M. Jamshidi, Ed.). Wiley, New York, 2008.

[32] Jolly, S. D., and Muirhead, B., "System of systems engineering in space exploration," *System of Systems Engineering—Innovations for the 21st Century, John Wiley Series on Systems Engineering*, Chapter 14 (M. Jamshidi, Ed.). Wiley, New York, 2008.

[33] Bahsin, K. B., and Hayden, J. L., "Communication and navigation networks in space system of systems," *System of Systems Engineering—Innovations for the 21st Century, John Wiley Series on Systems Engineering*, Chapter 15 (M. Jamshidi, Ed.). Wiley, New York, 2008.

[34] Korba, P., and Hiskins, I. A., "Operation and control of electrical power systems," *System of Systems Engineering—Innovations for the 21st Century, John Wiley Series on Systems Engineering*, Chapter 16 (M. Jamshidi, Ed.). Wiley, New York, 2008.

[35] Duffy, M., Garrett, B., Riley, C., and Sandor, D., "Future transportation fuel system of systems," *System of Systems Engineering—Innovations for the 21st Century, John Wiley Series on Systems Engineering*, Chapter 17 (M. Jamshidi, Ed.). Wiley, New York, 2008.

[36] Hipel, K., Obeidi, A., Fang, L., and Kilgour, D. M., "Sustainable environmental management from a system of systems engineering perspective," *System of Systems Engineering—Innovations for the 21st Century, John Wiley Series on Systems Engineering*, Chapter 11 (M. Jamshidi, Ed.). Wiley, New York, 2008.

[37] Sahin, F., "Robotic swarm as a system of systems," *System of Systems Engineering—Innovations for the 21st Century, John Wiley Series on Systems Engineering*, Chapter 19 (M. Jamshidi, Ed.). Wiley, New York, 2008.
[38] Wang, L., Fang, L., and Hipel, K. W., "On achieving fairness in the allocation of scarce resources: Measurable principles and multiple objective optimization approaches," *IEEE Systems Journal* 1(1), 17–28, 2007.
[39] De Laurentis, D.A., "Understanding transportation as a system-of-systems design, problem," in *AIAA Aerospace Sciences Meeting and Exhibit*, 10–13 January 2005. AIAA-2005-123, 2005.
[40] De Laurentis, D.A., and Callaway, R.K., "A system-of-systems perspective for future public policy," *Review of Policy Research* 21(6), 2006.
[41] De Laurentis, D., "Understanding transportation as a system-of-systems problem," *System of Systems Engineering—Innovations for the 21st Century, John Wiley Series on Systems Engineering*, Chapter 20 (M. Jamshidi, Ed.). Wiley, New York, 2008.
[42] Wickramasinghe, N., Chalasani, S., Boppana, R. V., and Madni, A. M., "Healthcare system of systems," *System of Systems Engineering—Innovations for the 21st Century, John Wiley Series on Systems Engineering* (M. Jamshidi, Ed.). Wiley, New York, 2008.
[43] Sloane, E., "Understanding the emerging national healthcare IT infrastructure," *24x7 Magazine*, December 2006.
[44] Sloane, E., Way, T., Gehlot, V., and Beck, R., "Conceptual SoS model and simulation systems for a next generation national healthcare information network (NHIN-2)," in *Proceedings of the 1st Annual IEEE Systems Conference*, Honolulu, HI, April 9–12 2007.
[45] Butterfield, M. L., Pearlman, J., and Vickroy, S. C., "System-of-systems engineering in a global environment," in *Proceedings of International Conference on Trends in Product Life Cycle, Modeling, Simulation and Synthesis PLMSS*, 2006.
[46] Pearlman, J., "GEOSS—Global Earth observation system of systems," Keynote presentation, 2006 IEEE SoSE Conference, Los Angeles, CA, April 24 2006.
[47] Shibasaki, R., and Pearlman, J., "Global Earth observation system of systems," *System of Systems Engineering—Innovations for the 21st Century, John Wiley Series on Systems Engineering* (M. Jamshidi, Ed.). Wiley, New York.
[48] Global Security Organization, http://www.globalsecurity.org/military/systems/ground/fcs-back.htm, 2007.
[49] Abbott, R., "Open at the top; open at the bottom; and continually (but slowly) evolving," in *Proceedings of the IEEE International Conference on System of Systems Engineering*, Los Angeles, April 2006.
[50] Walker, D., "Realizing a corporate SOSE environment," Keynote presentation, 2007 IEEE SoSE Conference, San Antonio, April 18 2007.

4

Observer Design and Kalman Filtering

4.1 State Space Methods for Model-Based Control

Figure 4.1 shows the inverted pendulum setup. The control problem is to start the cart from anywhere with the rod in a vertical position and bring the cart to the origin with the rod in an upright position. In other words, make $x = 0, \dot{x} = 0, \theta = 0, \dot{\theta} = 0$.

4.1.1 Derivation of Dynamics for the Cart-Pole Balancing Problem

To formulate the Lagrange dynamic equations for the cart-pole balancing problem, we first calculate the kinetic and potential energy of the system.

$$T_{cart} = \frac{1}{2} M \dot{x}^2$$

$$T_{rod} = \frac{1}{2} m[(\dot{x} + l\dot{\theta} \cos \theta)^2 + (l\dot{\theta} \sin \theta)^2] + \frac{1}{2} I \dot{\theta}^2$$

$$P_{cart} = 0$$

$$P_{rod} = mgl \cos \theta \qquad (4.1)$$

where T_{cart} and T_{rod} are the kinetic energy of the cart and the rod, respectively; P_{cart} and P_{rod} are the potential energy of the cart and the rod, respectively; M and m are the mass of the cart and the rod, respectively; and $x, l, \theta,$ and I are the cart position, the half-length of the rod, the angle of the rod, and the moment of inertia of the rod around its center of gravity, respectively.

Therefore, the kinetic potential or the Langrange variable is given by

$$L = T_{cart} + T_{rod} - P_{cart} - P_{rod} \qquad (4.2)$$

$$L = \frac{1}{2} M \dot{x}^2 + \frac{1}{2} m[(\dot{x} + l\dot{\theta} \cos \theta)^2 + (l\dot{\theta} \sin \theta)^2] + \frac{1}{2} I \dot{\theta}^2 - mgl \cos \theta$$

$$= \frac{1}{2} (M+m) \dot{x}^2 + \frac{1}{2} m(2l\dot{x}\dot{\theta} \cos \theta) + \frac{1}{2} ml^2 \dot{\theta}^2 \cos^2 \theta + \frac{1}{2} I \dot{\theta}^2 - mgl \cos \theta \qquad (4.3)$$

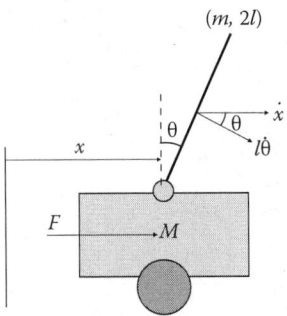

FIGURE 4.1
Inverted pendulum arrangement. The cart is free to move in one direction. The pendulum is mounted on the cart with a free joint.

Taking the joint angle as one variable, we can write one relationship given by

$$\frac{d}{dt}\left(\frac{\partial L}{\partial \dot\theta}\right) = ml\ddot x \cos\theta - ml\dot x\dot\theta \sin\theta - ml^2\dot\theta^2 \cos\theta \sin\theta + (I + ml^2 \cos^2\theta)\ddot\theta$$

$$\frac{\partial L}{\partial \theta} = -ml\dot x\dot\theta \sin\theta - ml^2\dot\theta^2 \cos\theta \sin\theta - mgl \sin\theta$$

$$\frac{d}{dt}\left(\frac{\partial L}{\partial \dot\theta}\right) - \frac{\partial L}{\partial \theta} = 0$$

$$\Rightarrow ml\ddot x \cos\theta + (I + ml^2 \cos^2\theta)\ddot\theta + mgl \sin\theta = 0 \tag{4.4}$$

Taking the position of the cart as the other variable, we can write the second relationship given by

$$\frac{d}{dt}\left(\frac{\partial L}{\partial \dot x}\right) = (M+m)\ddot x + ml[\ddot\theta \cos\theta - \dot\theta^2 \sin\theta]$$

$$\frac{\partial L}{\partial x} = 0$$

$$\frac{d}{dt}\left(\frac{\partial L}{\partial \dot x}\right) - \frac{\partial L}{\partial x} = F$$

$$\Rightarrow (M+m)\ddot x + ml[\ddot\theta \cos\theta - \dot\theta^2 \sin\theta] = F - b\dot x \tag{4.5}$$

Combining Equations (4.4) and (4.5), we can derive the dynamics of the system given by

$$\begin{pmatrix} 0 \\ F \end{pmatrix} = \begin{pmatrix} I + ml^2 \cos^2 \theta & ml \cos \theta \\ ml \cos \theta & M + m \end{pmatrix} \begin{pmatrix} \ddot{\theta} \\ \ddot{x} \end{pmatrix} + \begin{pmatrix} 0 \\ -ml\dot{\theta}^2 \sin \theta + b\dot{x} \end{pmatrix} + \begin{pmatrix} mgl \sin \theta \\ 0 \end{pmatrix} \quad (4.6)$$

Please compare the format of this equation with the dynamic equation we derived for the two-link manipulator. You will see that the format is the same. In the case of the manipulator, we worked with this nonlinear equation because the robot is supposed to move in the whole workspace. Yet, in this cart-rod balancing exercise, we start the cart with the rod in a vertical position and the cart is moved with the rod swinging around the vertical position. We can take advantage of this context to simplify the equation further.

In this scenario, we can assume the following:

$$\sin \theta \approx \theta, \cos \theta \approx 0, \dot{\theta}^2 = 0 \quad (4.7)$$

Therefore, Equation (3.6) can be written as

$$\begin{pmatrix} 0 \\ F \end{pmatrix} = \begin{pmatrix} I + ml^2 & ml \\ ml & M + m \end{pmatrix} \begin{pmatrix} \ddot{\theta} \\ \ddot{x} \end{pmatrix} + \begin{pmatrix} 0 \\ b\dot{x} \end{pmatrix} + \begin{pmatrix} -mgl\theta \\ 0 \end{pmatrix}$$

Now, we want to reformat this equation to obtain the format known as the state space equation given by $\dot{X} = AX + Bu$, where X is the state vector describing the situation of the dynamic system at a given time, u is the external force applied on the system, and A and B are two matrices that determine how the state transition occurs from one state to another given an external force.

Therefore, we can rearrange the equation as follows:

$$\begin{pmatrix} \ddot{\theta} \\ \ddot{x} \end{pmatrix} = \begin{pmatrix} I + ml^2 & ml \\ ml & M + m \end{pmatrix}^{-1} \begin{pmatrix} mgl\theta \\ -b\dot{x} \end{pmatrix} + \begin{pmatrix} I + ml^2 & ml \\ ml & M + m \end{pmatrix}^{-1} \begin{pmatrix} 0 \\ F \end{pmatrix}$$

$$\begin{pmatrix} \ddot{\theta} \\ \ddot{x} \end{pmatrix} = \frac{1}{\det \begin{pmatrix} I + ml^2 & ml \\ ml & M + m \end{pmatrix}} \begin{pmatrix} M + m & -ml \\ -ml & I + ml^2 \end{pmatrix} \left[\begin{pmatrix} mgl\theta \\ -b\dot{x} \end{pmatrix} + \begin{pmatrix} 0 \\ F \end{pmatrix} \right]$$

$$= \begin{pmatrix} \dfrac{(M+m)}{I(M+m)+mMl^2} & -\dfrac{ml}{I(M+m)+mMl^2} \\ -\dfrac{ml}{I(M+m)+mMl^2} & \dfrac{I+ml^2}{I(M+m)+mMl^2} \end{pmatrix} \left[\begin{pmatrix} mgl\theta \\ -b\dot{x} \end{pmatrix} + \begin{pmatrix} 0 \\ F \end{pmatrix} \right] \quad (4.8)$$

We can rewrite the equation given in Equation (4.8) to form the state space model given by

$$\begin{pmatrix}\ddot{\theta}\\ \dot{\theta}\\ \ddot{x}\\ \dot{x}\end{pmatrix} = \begin{pmatrix} 0 & \dfrac{(M+m)mgl}{I(M+m)+mMl^2} & \dfrac{mlb}{I(M+m)+mMl^2} & 0 \\ 1 & 0 & 0 & 0 \\ 0 & \dfrac{-m^2gl^2}{I(M+m)+mMl^2} & \dfrac{-(I+ml^2)b}{I(M+m)+mMl^2} & 0 \\ 0 & 0 & 1 & 0 \end{pmatrix}\begin{pmatrix}\dot{\theta}\\ \theta\\ \dot{x}\\ x\end{pmatrix}$$

$$+ \begin{pmatrix}-\dfrac{ml}{I(M+m)+mMl^2}\\ 0\\ \dfrac{I+ml^2}{I(M+m)+mMl^2}\\ 0\end{pmatrix} F \qquad (4.9)$$

Once we have this format, there are numerous control theories we can apply to drive the states of the system to a desired set of states. Before writing our own code to implement the controller, let us look at what is meant by state space control.

4.1.2 Introduction to State Space Representation of Dynamics

Let us first review some definitions:

Definition 1 State: The state of a dynamic system is the smallest set of variables (called state variables) such that the knowledge of these variables at $t = t_0$, together with the knowledge of the input for $t \geq t_0$, completely determines the behavior of the system for any time $t \geq t_0$. The concept of a state is applicable to mechanical systems and other systems, such as biological, economical, and social systems.

Definition 2 State variables: The state variables of a dynamic system are the variables making up the smallest set of variables that determine the state of the dynamic system.

Definition 3 State space: The n-dimensional space whose coordinate axes consist of the x_1 axis, x_2 axis, ... , x_n axis, where $x_1, x_2, \ldots, x_n$ are the state variables, is called a state space. Any state can be represented by a point in this state space.

Figure 4.2 demonstrates the physical meaning of this definition. In the state space, each axis represents a state variable. Therefore, in this space, a

Observer Design and Kalman Filtering

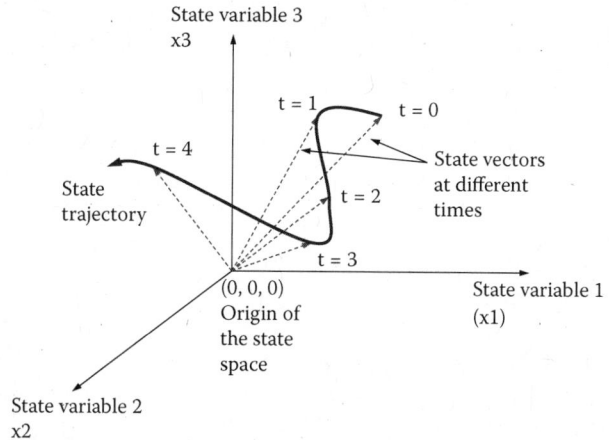

FIGURE 4.2
Concept of a state space.

point is a state of the dynamical system at a given time. Therefore, if we know the state of the system X at time $t = t_0$, together with the control input u for $t \geq t_0$, we can predict the behavior of the system if we know the equation $\dot{X} = AX + Bu$. The objective of the control exercise is to drive the state of the dynamical system to converge toward a desired state at some arbitrary time $t = t_N$ as shown by Figure 4.3.

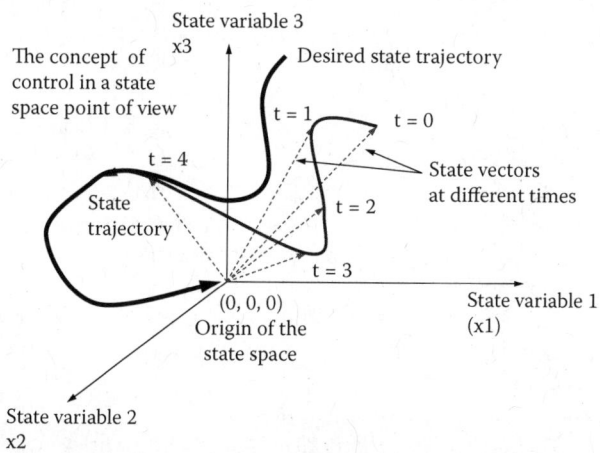

FIGURE 4.3
Objective of the control exercise is to drive the state of the system to converge to a desired state.

Now let us write a MATLAB® code to control our cart-rod system. Please refer to my code at Chap4/pendulum.m. I have used a controller known as the linear regulator. In a linear regulator, the controller should drive the system states from an initial state to a target state. This differs from a tracking controller where you have to worry not only about the target state, but also about the intermediate states between the initial state and the target state.

4.1.3 Regulator Control of State Vectors

A regulator controller computes the control input given by $u = -KX$, where K is the feedback gain vector and X is the state at a given time. We can prove that this controller is stable and drives the states from any initial state to any given target state, providing that some conditions are satisfied.

Proof

We start from the state space equation given by

$$\dot{X} = AX + Bu \qquad (4.10)$$

Substituting $u = -KX$ in Equation (4.10), we obtain

$$\dot{X} = AX - BKX$$

$$= (A - BK)X \qquad (4.11)$$

The solutions of Equation (4.11) for each state variable are given by

$$\begin{pmatrix} x_1(t) \\ x_2(t) \\ \vdots \\ x_n(t) \end{pmatrix} = \begin{bmatrix} e^{\lambda_1 t} & 0 & 0 & 0 \\ 0 & e^{\lambda_2 t} & 0 & 0 \\ 0 & 0 & \ddots & 0 \\ 0 & 0 & 0 & e^{\lambda_n t} \end{bmatrix} \begin{pmatrix} x_1(0) \\ x_2(0) \\ \vdots \\ x_n(0) \end{pmatrix} \qquad (4.12)$$

where λ_i, $i = 1, 2, \ldots, n$ are eigenvalues of $(A - BK)$. Therefore, if all eigenvalues are negative, the states $x_1(t), x_2, \ldots, x_n(t)$ should converge to the origin of the state space when time grows. Therefore, by defining the state as the error between the target state and the current state of the system, we can guarantee that the error vector will converge to the origin. The schematic diagram of the regulator controller is shown in Figure 4.4.

In our case, we define our target state as the origin. The cart starts 1 m away from the origin. It is expected to move into position zero, where the cart should come to a standstill (velocity reaches zero) and the rod is kept at the vertical position with any movement.

We use the MATLAB® library function $[K] = LQR(A, B, Q, R, N)$ to compute the optimum gain vector K^*. It computes the optimum gain K^* so that the total cost of controlling over the time control span $J = \int_{t=t_0}^{t_n} (X^T Q X + u^T R u + 2 X^T N u) dt$ is minimized.

Observer Design and Kalman Filtering

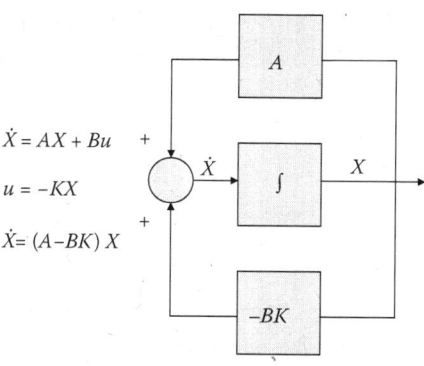

FIGURE 4.4
Schematic diagram of the regulator.

In our case, the weight matrices were:

$$Q = \begin{bmatrix} 5 & 0 & 0 & 0 \\ 0 & 5 & 0 & 0 \\ 0 & 0 & 5 & 0 \\ 0 & 0 & 0 & 5 \end{bmatrix}, R = 5, N = 0$$

The system parameters were: $M = 5$, $m = 0.2$, $I = 0.006$, $g = 9.8$, $l = 0.3$, $b = 0.1$. The sampling interval was 2 ms. The resulting optimal feedback gain vector was $K^* = [-3.9302\ -20.3672\ -2.0408\ -1.0000]^T$. Figure 4.5 shows how the states converged to the origin over time. The control command converged to zero once all the states were converged.

4.2 Observing and Filtering Based on Dynamic Models

Note that the controllers discussed above depend on the full knowledge of the states at any given time. Yet, we do not get sensors to measure all the states in a practical industrial application. Even if sensors are available, we try to avoid expensive sensors. Now, let us see how we can omit sensing some of the states of the system, but mathematically observe them given the model of the system. This is a good way to bring down the cost of building a controller and to be immune to the noise generated by most of the sensors.

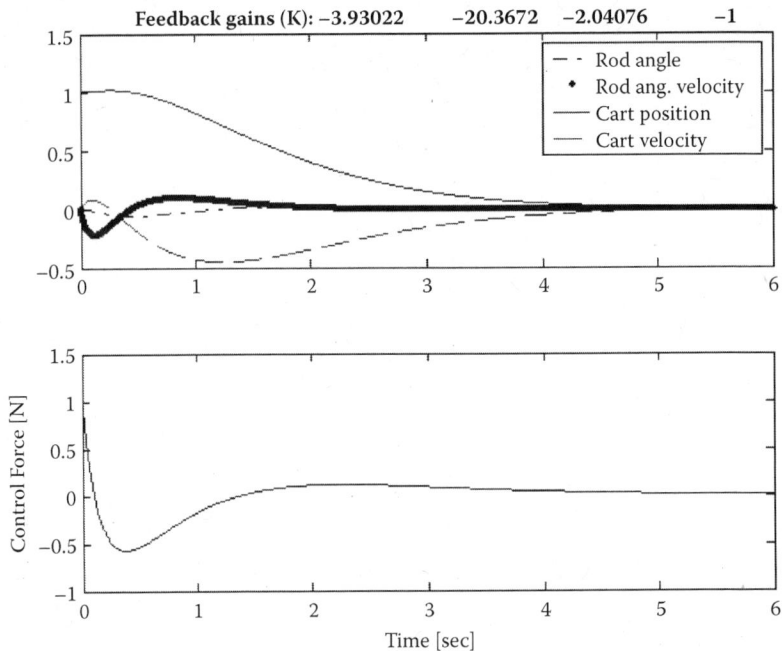

FIGURE 4.5
Control performance of the linear regulator for the inverted pendulum example.

4.2.1 Why Observers?

Let us do some thinking before going into details. Our requirement is to estimate the states that are not sensed using those that are sensed. In the inverted pendulum problem, assume that the cart position and the angle of the rod are the only two states that are sensed. Cart velocity and angular velocity of the rod have to be observed. Can we use any of the knowledge we gained from the above calculation in Equations (4.10) to (4.12) on the principle of convergence of states to a desired state to estimate the other two states here? What type of assumptions do we have to make and what type of new parameters do we have to introduce? Please make your own observations before reading any further.

4.2.2 Elements of an Observer

Assuming you have done some research on the above problem, let us discuss how we can solve the problem. Let us rewrite the situation.

Equation (4.9) characterizing the physical system does not change. It is only that the sensed states are reduced. Therefore, we introduce a new vector y of

Observer Design and Kalman Filtering

sensed variables given by:

$$\begin{pmatrix} \ddot{\theta} \\ \dot{\theta} \\ \ddot{x} \\ \dot{x} \end{pmatrix} = \begin{pmatrix} 0 & \dfrac{(M+m)mgl}{I(M+m)+mMl^2} & \dfrac{mlb}{I(M+m)+mMl^2} & 0 \\ 1 & 0 & 0 & 0 \\ 0 & \dfrac{-m^2gl^2}{I(M+m)+mMl^2} & \dfrac{-(I+ml^2)b}{I(M+m)+mMl^2} & 0 \\ 0 & 0 & 1 & 0 \end{pmatrix} \begin{pmatrix} \dot{\theta} \\ \theta \\ \dot{x} \\ x \end{pmatrix}$$

$$+ \begin{pmatrix} -\dfrac{ml}{I(M+m)+mMl^2} \\ 0 \\ \dfrac{I+ml^2}{I(M+m)+mMl^2} \\ 0 \end{pmatrix} F$$

$$y = C \begin{pmatrix} \dot{\theta} \\ \theta \\ \dot{x} \\ x \end{pmatrix}, C = \begin{bmatrix} 0 & 1 & 0 & 0 \\ 0 & 0 & 0 & 1 \end{bmatrix} \quad (4.13)$$

Now, we can use this model in parallel with the real physical system to derive an error vector. If we find a state space equation for that error vector and a new feedback parameter vector so that the error of estimation converges to zero much faster than the convergence rate of the regulator controller, we achieve our purpose.

We set the sensed states as references for the output of the model to calculate a measuring error given by

$$e = y - \hat{y}$$
$$y = CX$$
$$\hat{y} = C\hat{X}$$
$$\dot{\hat{X}} = A\hat{X} + Bu \quad (4.14)$$

where y is the vector of sensed variables using physical sensors and $\hat{y}$ is the estimate of the sensed variables using our model given in Equation (4.13). If the model is accurate, the measurement error e should be zero.

Now, we can write a new state space equation of estimated states given by

$$\dot{\hat{X}} = A\hat{X} + Bu + L(y - \hat{y}) \quad (4.15)$$

If we compare this equation with estimated states with that of the true system, we can obtain a state space equation of estimation errors given by

$$\dot{X} = AX + Bu$$

$$\dot{\hat{X}} = A\hat{X} + Bu + L(y - \hat{y})$$

$$(\dot{X} - \dot{\hat{X}}) = A(X - \hat{X}) - L(y - \hat{y}) \tag{4.16}$$

From Equation (4.2), we know that $\hat{y} = C\hat{X}, y = CX$. Therefore, we can write Equation (4.16) as

$$(\dot{X} - \dot{\hat{X}}) = A(X - \hat{X}) - LC(X - \hat{X})$$

$$= (A - LC)(X - \hat{X})$$

$$\Rightarrow \dot{e} = (A - LC)e \tag{4.17}$$

Equation (4.17) is reminiscent of Equation (4.11). Therefore, the condition for the estimation errors to converge to zero is the same as that of Equation (4.12), that is, the eigenvalues of $(A - LC)$ should be negative. In other words, $(A - LC)$ should be a positive definite matrix.

The complete schematic diagram for the observer and the regulator controller is shown in Figure 4.6. You can notice that the observer is based on

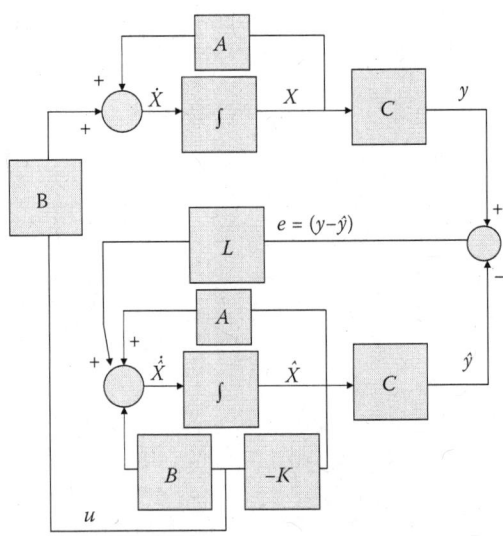

FIGURE 4.6
Schematic diagram of the observer and the regulator.

the model of the system. Since it has no restriction on the sensors, it can estimate all four states. It computes a control command based on the full knowledge of all four estimated states. This control command is given to both the observer and the real system. The error between the sensed output of the real system and the estimated values of the corresponding states is used to make a correction in the observer. According to Equation (4.17), $\hat{X} \to X$ over time.

4.2.3 Simulations on an Inverted Pendulum Control

Let us run a simulation using the MATLAB® code given in Chap 4/pendulum_observer.m. Here, we make small modifications to the system parameters to simulate the errors in estimating a model for the observer: $M = 0.45$, $m = 0.21$, $b = 0.12$, $I = 0.005$, $l = 0.31$.

Based on these parameters, we calculate the A_{obs} and B_{obs} matrices in Equation (4.17), where the subscript *obs* denotes the observer. Then, we calculate an optimum feedback gain matrix K and the eigenvalues of the matrix $(A_{obs} - B_{obs} K)$ in Equation (4.11). In addition to the above errors in the parameters, we introduce an offset to the initial state. The initial state of the system is $(\theta \ \dot{\theta} \ \dot{x} \ x)^T = (0 \ 0 \ 0 \ 1)^T$. The initial state of the observer is $(\hat{\theta} \ \dot{\hat{\theta}} \ \dot{\hat{x}} \ \hat{x})^T = (0 \ 0 \ 0 \ 1.2)^T$.

The feedback gain obtained by implementing the linear quadratic optimizer is given by $K = [-3.6262 \ -19.2997 \ -2.0261 \ -1.0000]$. The resulting eigenvalues of $(A_{obs} - B_{obs} K)$ are given by $\lambda = [-9.1372 \ -3.7533 \ -1.1496 + 0.4255j \ -1.1496 \ -0.4255j]$.

We know that the real part of these eigenvalues decides the rate of convergence of the states of the system to the target state. We want the estimated states to converge to the true states of the system faster than the rate of convergence of states to the target state in order to make the observer provide meaningful feedback to the system. Therefore, we make the eigenvalues of $(A_{obs} - LC)$ larger than those of the system. First, we set $\rho = 2$ in $Eig(A_{obs} - LC) = \rho[Eig(A_{obs} - BK)]$, where $Eig(\bullet)$ denotes the eigenvalues.

Then, we use the MATLAB® command "Place" to calculate the feedback gain L. Now we have all the parameters needed to run the regulator controller backed by the observer. Figure 4.7 shows how it performs for $\rho = 2$. The lines in black color are the true states. The lines in magenta color are those states estimated by the observer. Compare the behavior of the observer for $\rho = 2$ with that for $\rho = 5$ in Figure 4.8. We can see that the estimated states converge to the true states faster in the case of $\rho = 5$. An interesting phenomenon can be seen in the estimation of cart velocity (broken line). In Figure 4.9 ($\rho = 2$), the estimated cart velocity has tried to track the same profile of the true cart velocity. Yet, in Figure 3.8 ($\rho = 2$), the observer has compromised the estimation error of the cart velocity at the initial stage to catch up with the true value later. This is

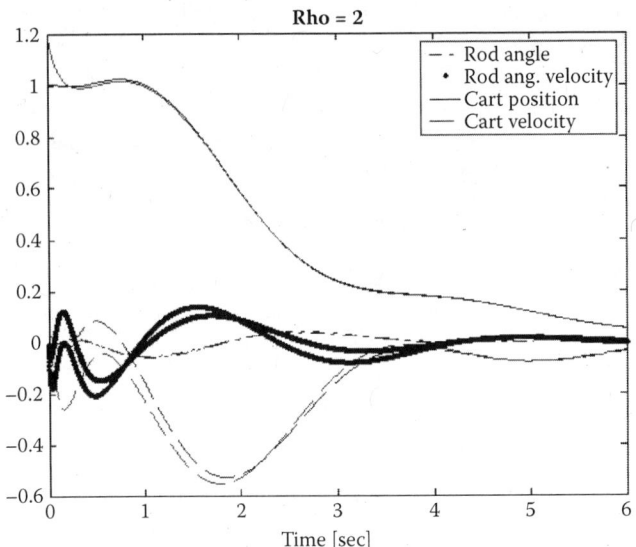

FIGURE 4.7
Tracking performance of the observer for $\rho = 2$.

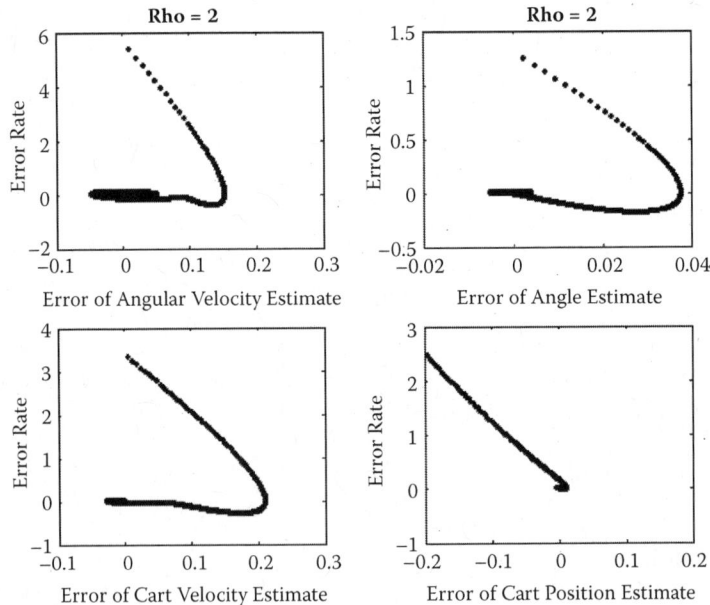

FIGURE 4.8
Phase portrait of observation errors for $\rho = 2$.

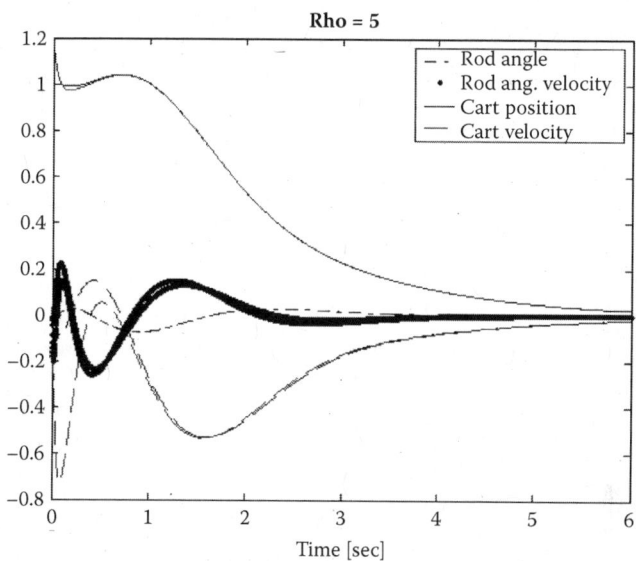

FIGURE 4.9
Tracking performance of the observer for $\rho = 5$.

more evident in the corresponding phase portrait plots in Figures 4.8 and 4.10. In Figure 4.10, we can see that the initial error rates are larger than that of Figure 4.8. Yet, in Figure 4.10, the errors of all four states converge to the origin without much encirclement around the origin. Therefore, the time taken by the estimated states to converge to true states is less in the case of $\rho = 5$.

We have to remember that the above simulations are far from reality in terms of sensor noise in practical situations. Most of the standard industrial sensors have measurement noise that has to be filtered out before sampling. Lets us simulate the performance of the observer by adding noise to the sensed values of the cart position and angle of the rod given by $\theta_{measured} = N(\theta, 0.05)$, $x_{measured} = N(x, 0.05)$, where $N(\mu, \sigma)$ is a normally distributed random number with mean μ and standard deviation σ.

By carefully examining the behavior of the system shown in Figures 4.7 through 4.16 in the two cases for $\rho = 2$ and $\rho = 5$, you can observe that the control performance deteriorates in the case of $\rho = 5$ because the observer tries to overreact to the noisy measured states. Therefore, when the measured states are contaminated with noise, we have to reduce the feedback gains. But how can we know the optimum gains? The answer to this is found in a powerful state filter and observer known as the Kalman filter.

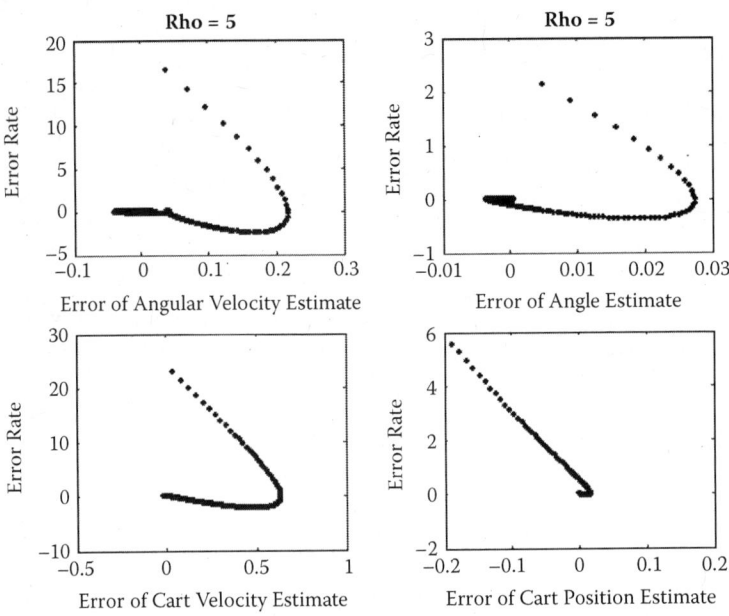

FIGURE 4.10
Phase portrait of observation errors for $\rho = 5$.

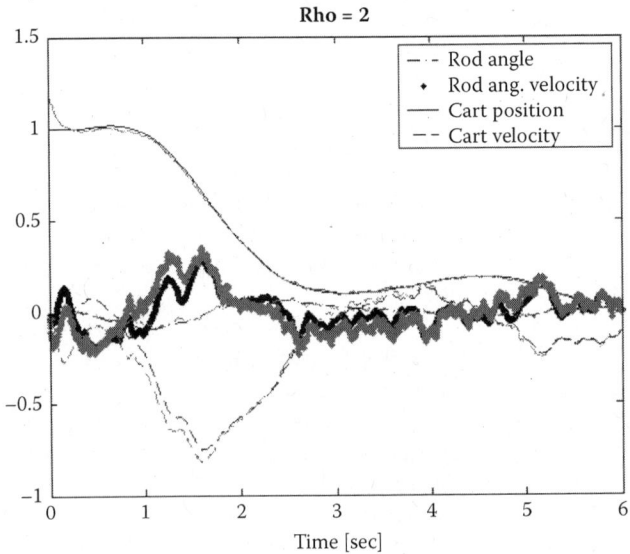

FIGURE 4.11
Tracking performance of the observer with noise for $\rho = 2$.

Observer Design and Kalman Filtering

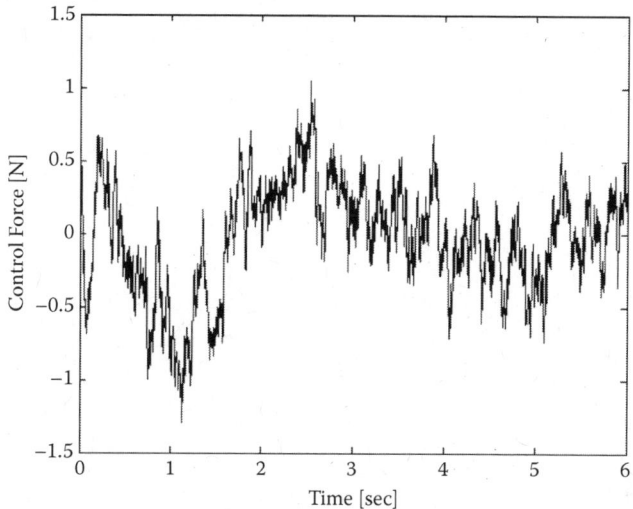

FIGURE 4.12
Control input based on the noisy observed states for $\rho = 2$.

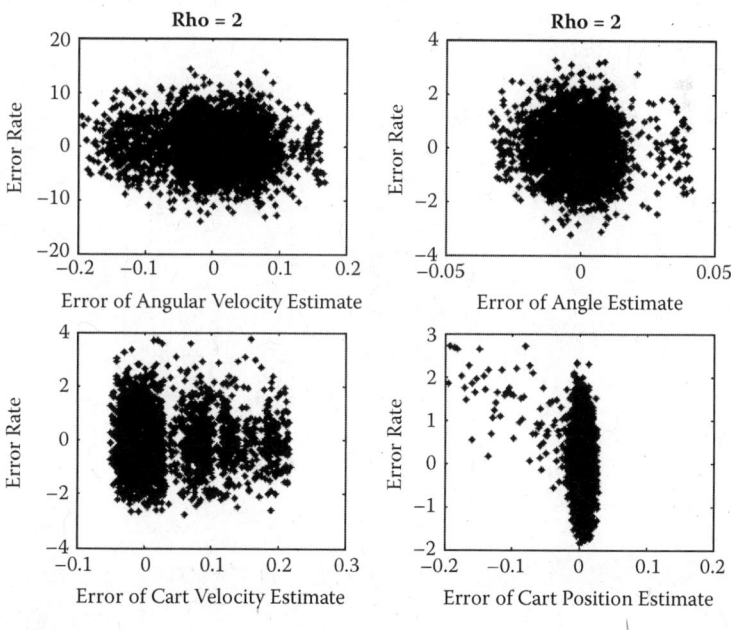

FIGURE 4.13
Phase portrait of observation errors with noisy measurements for $\rho = 2$.

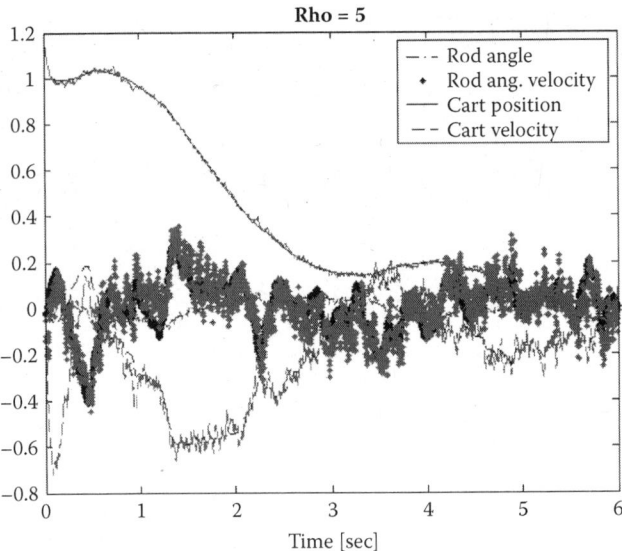

FIGURE 4.14
Tracking performance of the observer with noise for $\rho = 5$.

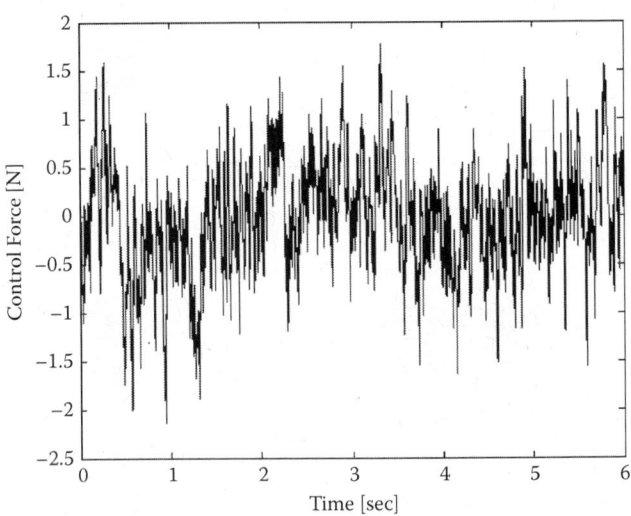

FIGURE 4.15
Control input based on the noisy observed states for $\rho = 5$.

Observer Design and Kalman Filtering

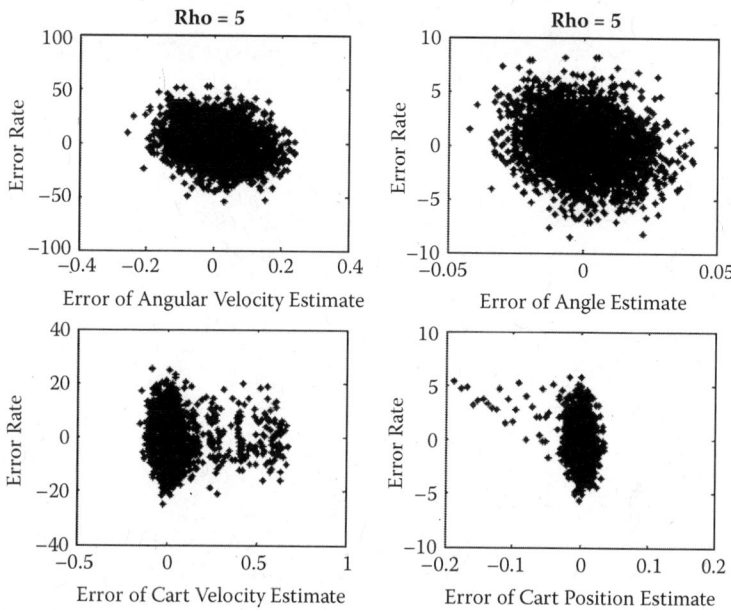

FIGURE 4.16
Phase portrait of observation errors with noisy measurements for $\rho = 5$.

4.3 Derivation of the Discrete Kalman Filter

Before going into the mathematics of Kalman filtering, let us remember an experience almost all of us know. If you have ever tried to ride a bicycle on rough terrain, you have experienced that the best way to stay balanced is to give some compliance to the handle and pedals. If you hold the handle firmly and try to pedal the way you want, the bicycle falls. If you have not experienced the above, please try the following experiment.

Try to climb up or down a staircase fairly fast with a glass of water in one hand. Try one trial with the hand relaxed and try one trial with the hand very stiff. You will see that, when you run up the stairs with your hand stiff, the water tends to spill over. If you do several more trials at different climbing speeds, you will notice that you can afford to have different stiffness levels of your hand to keep the water in the glass stable.

This is the fundamental phenomenon behind the stability of a controller of a dynamic system in the presence of noise and disturbances. The controller should have some mechanism to be compliant to the disturbances injected by the environment and from within the process in order to remain converging toward the general goal.

Now, let us review the mathematics behind the above phenomenon. First, we define the state space equation for a general dynamic system given by

$$x_{k+1} = A_k x_k + B_k u_k + w_k$$
$$z_k = H_k x_k + v_k \qquad (4.18)$$

where x_k is the state of the system, u_k is the control command, w_k is the noise generated from within the process, v_k is the measurement noise, and z_k is the measurement of the states at time t_k.

$$E[w_k w_i^T] = \begin{cases} Q_k & i=k \\ 0 & i \neq k \end{cases}, \quad E[v_k v_i^T] = \begin{cases} R_k & i=k \\ 0 & i \neq k \end{cases}, \quad E[w_k v_i^T] = 0, \forall (i,k) \quad (4.19)$$

We define two forms of the state of the system: one is the estimate of the state with all available information before time t_k and the other is that soon after knowing the measurement of the state at time t_k. The former is called the *a priori* estimate and the latter is called the *a posteriori* estimate of the state. The two forms are shown in Figures 4.17 and 4.18.

The *a priori* estimate is denoted by $\hat{x}_k^-$ and the *a posteriori* estimate is denoted by $\hat{x}_k$. The relationship between the two is given by

$$\hat{x}_k = \hat{x}_k^- + K(z_k - H_k \hat{x}_k^-) \qquad (4.20)$$

The *a priori* estimate is refined by blending it with the weighted error of estimation upon the receipt of the sensor data z_k. Therefore, K is known as the blending parameter.

We define the *a priori* and *a posteriori* errors of estimation given by

$$e_k^- = x_k - \hat{x}_k^-, \quad e_k = x_k - \hat{x}_k \qquad (4.21)$$

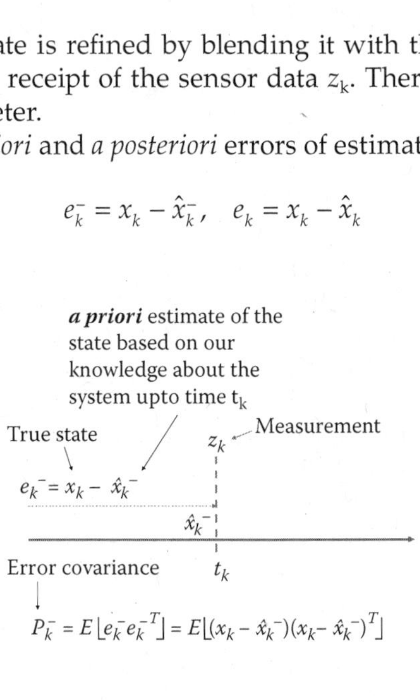

a priori estimate of the state based on our knowledge about the system upto time t_k

True state   Measurement
       z_k
$e_k^- = x_k - \hat{x}_k^-$
       $\hat{x}_k^-$
Error covariance t_k

$$P_k^- = E[e_k^- e_k^{-T}] = E[(x_k - \hat{x}_k^-)(x_k - \hat{x}_k^-)^T]$$

FIGURE 4.17
Relevant events to the *a priori* estimate.

Observer Design and Kalman Filtering

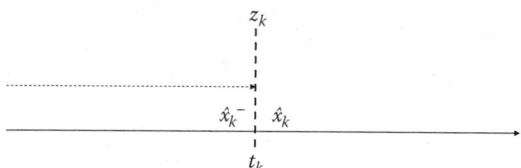

As soon as we get the measurement, we refine the *a priori* estimate to make the *a posteriori* estimate

$$\hat{x}_k = \hat{x}_k^- + K_k(z_k - H_k\hat{x}_k^-)$$

$$P_k = E[e_k e_k^T] = E[(x_k - \hat{x}_k)(x_k - \hat{x}_k)^T]$$

FIGURE 4.18
Relevant events to the *a posteriori* estimate.

The objective of the estimation process is to minimize the estimation error covariance. To derive a recursive equation for the error covariance, we write an equation for the *a posteriori* error in terms of the *a priori* error given by

$$\begin{aligned}
e_k &= x_k - \hat{x}_k \\
&= x_k - \left(\hat{x}_k^- + K_k^T(z_k - H_k\hat{x}_k^-)\right) \\
&= x_k - \hat{x}_k^- - K_k^T\left(H_k x_k + v_k - H_k x_k\right) \\
&= (x_k - \hat{x}_k^-) - K_k^T H_k(x_k - \hat{x}_k^-) - K_k^T v_k \\
&= e_k^- - K_k^T H_k e_k^- - K_k^T v_k
\end{aligned} \quad (4.22)$$

Therefore, we can write an equation for the *a posteriori* error covariance in terms of *a priori* error covariance given by

$$\begin{aligned}
P_k &= E\left[e_k e_k^T\right] \\
&= E\left[(e_k^- - K_k H_k e_k^- - K v_k)(e_k^- - K_k H_k e_k^- - K v_k)^T\right] \\
&= E\left[e_k^- e_k^{-T}\right] - E\left[e_k^- e_k^{-T} H_k^T K_k^T\right] - E\left[e_k^- v_k^T K_k^T\right] - E\left[K_k H_k e_k^- e_k^{-T}\right] \\
&\quad - E\left[K_k H_k e_k^- e_k^{-T} H_k^T K_k^T\right] \\
&\quad - E\left[K_k H_k e_k^- v_k^T K_k^T\right] - E\left[K_k v_k e_k^{-T}\right] + E\left[K_k v_k e_k^{-T} H_k^T K_k^T\right] + E\left[K_k v_k v_k^T K_k^T\right] \\
&= P_k^- - P_k^- H_k^T K_k^T - K_k H_k P_k^- + K_k\left[H_k P_k^- H_k^T + R_k\right]K_k^T
\end{aligned} \quad (4.23)$$

Our goal is to minimize the trace of P_k given by

$$J_k = Tr(P_k) = \sum_{i=1, j=1}^{i=N, j=N} P_k(i, j) \tag{4.24}$$

We try to minimize J_k in terms of the blending parameter K_k. Therefore, we can write the following equation:

$$\frac{\partial(Tr(P_k))}{\partial K_k} = 0 \tag{4.25}$$

$$\frac{\partial(Tr(P_k))}{\partial K_k} = \frac{\partial\left(Tr\left(P_k^-\right)\right)}{\partial K_k} - \frac{\partial\left(Tr\left(P_k^- H_k^T K_k^T\right)\right)}{\partial K_k} - \frac{\partial\left(Tr\left(K_k H_k P_k^-\right)\right)}{\partial K_k}$$

$$+ \frac{\partial\left(Tr\left(K_k \left[H_k P_k^- H_k^T + R_k\right] K_k^T\right)\right)}{\partial K_k} = 0$$

$$= -2 P_k^{-T} H_k^T + 2 K_k \left(H_k P_k^- H_k^T + R_k\right) = 0$$

$$\Rightarrow \quad K_k = P_k^- H_k^T \left(H_k P_k^- H_k^T + R_k\right)^{-1} \tag{4.26}$$

Here, we note that $P_k^{-T} = P_k^-$, $Tr(P_k^- H_k^T K_k^T) = Tr(K_k H_k P_k^-)$ due to the symmetry of the error covariance matrix. By substituting the result in Equation (4.26) in Equation (4.23), we can obtain

$$P_k = (I - K_k H_k) P_k^- \tag{4.27}$$

We want to calculate the error covariance of the *a priori* estimate at time t_{k+1} to complete the recursive algorithm. The estimation error of the *a priori* estimate at time t_{k+1} is given by

$$e_{k+1}^- = x_{k+1} - \hat{x}_{k+1}^-$$
$$= A_k x_k + B_k u_k + w_k - A_k \hat{x}_k - B_k u_k$$
$$= A_k e_k + w_k \tag{4.28}$$

Therefore,

$$P_{k+1}^- = E\left[e_{k+1}^- e_{k+1}^{-T}\right] = E[(A_k e_k + w_k)(A_k e_k + w_k)^T]$$
$$= E\left[A_k e_k e_k^T A_k^T\right] + E\left[w_k w_k^T\right]$$
$$= A_k P_k A_k^T + Q_k \tag{4.29}$$

4.4 Worked Out Project on the Inverted Pendulum

If you carefully observe the above steps, you can notice that the Kalman gain K is not a constant gain vector as in Equation (4.10) and Figure 4.4. It directly depends on the error covariance state estimation and the covariance of measurement noise. The error covariance of estimation, in turn, depends on the covariance of the process noise. The Kalman gain is expected to go down when the measurement noise, process noise, and error or state estimation go up. This is analogous to the water cup example we discussed earlier. We are now in a position to do some simulations. Please refer to Chap4/descrete_kalman_filter.m; it calls several functions in the folder. Each function implements one step in Figure 4.19. We implemented the linear Kalman filter for the inverted pendulum example. Both the process noise w_k and measurement noise v_k were set to be signal-dependent noise given by

$$w_k = \alpha_w N(0, x_k^-), v_k = \alpha_v N(0, H_k x_k^-) \quad (4.30)$$

where $\alpha_w = 0.01$, $\alpha_v = 0.1$, and $N(\mu,\sigma)$ is a normally distributed random vector of the same size of the vector of mean μ and standard deviation σ. Note that the cart velocity shown in Figure 4.20 and the angular velocity of the rod shown in Figure 4.21 are observed without any sensor. In the case of cart position shown in Figure 4.22 and the rod angle shown in Figure 4.23, the noise has been filtered out to observe the true state of the system. Therefore, the Kalman filter works both as a filter and as an observer.

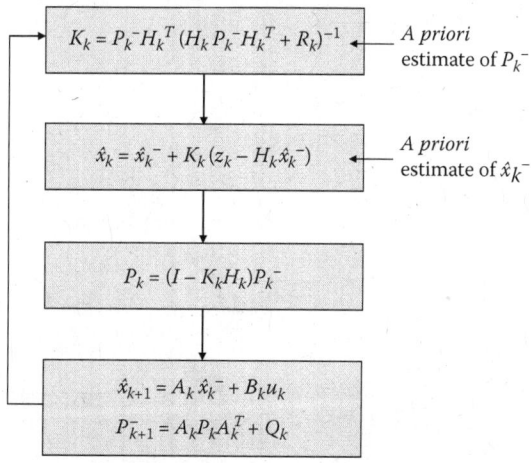

FIGURE 4.19
Steps of the linear Kalman filter implementation.

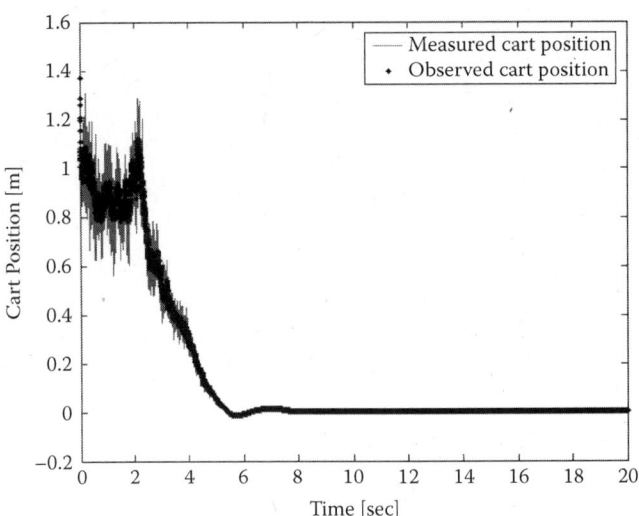

FIGURE 4.20
Observed cart velocity.

Figure 4.24 shows the behavior of each element of the Kalman gain matrix. The first column blends the rod angle estimation error with each state. The second column blends the cart position estimation error with each state. Each row refers to one state variable in the order given by Equation (4.13). Observe

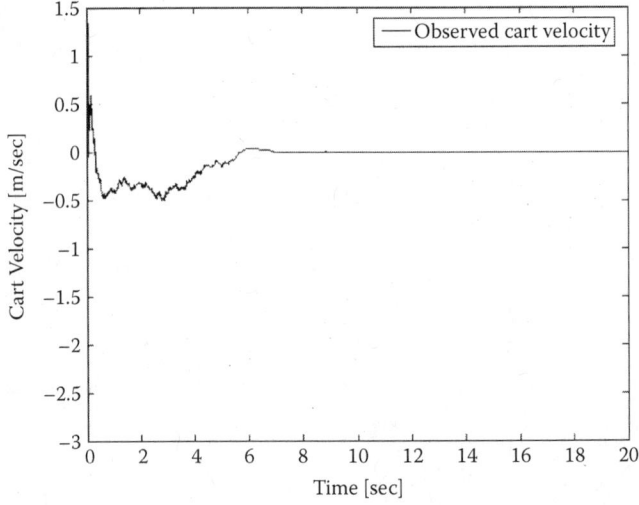

FIGURE 4.21
Observed angular velocity.

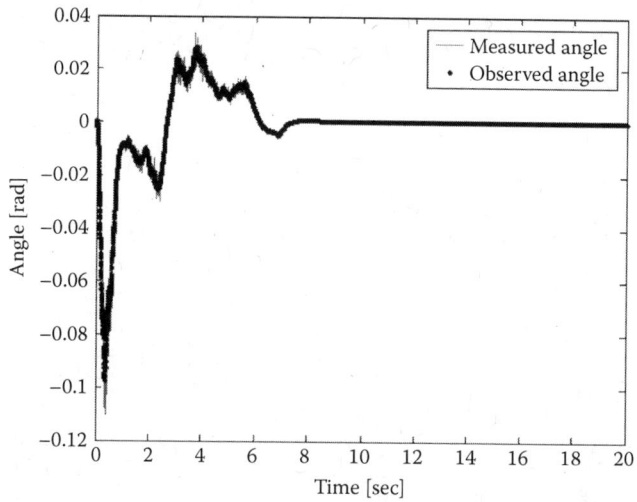

FIGURE 4.22
Measured and observed cart position.

how each element in the Kalman gain matrix changes with the noise level of the measured value of the corresponding states (Figures 4.20 through 4.23).

Case 2: A linear random disturbance $\epsilon = [0,1]$ added to the system at $t_k = 8$ (s). Please observe how the Kalman gains and the observed states respond to this disturbance.

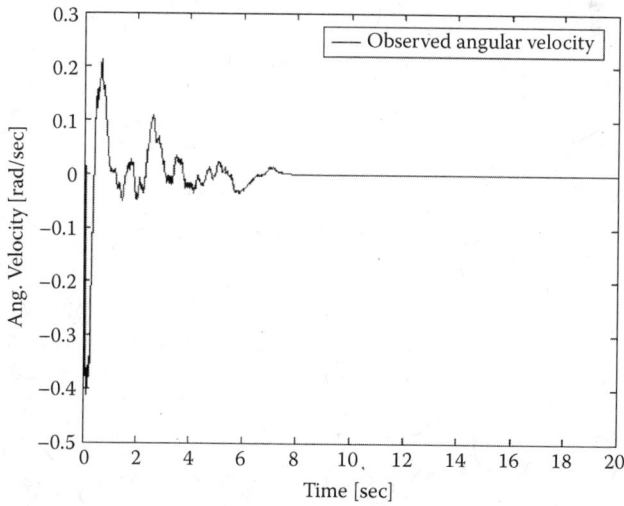

FIGURE 4.23
Measured and observed angle.

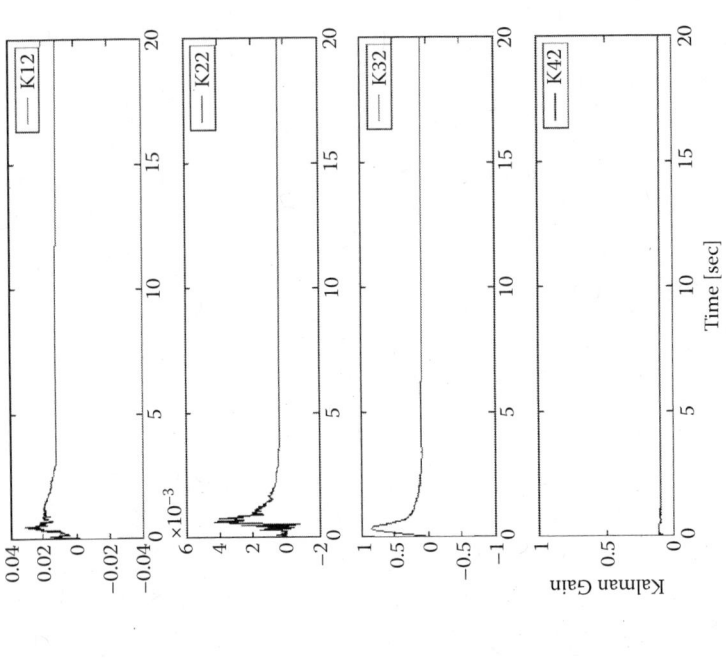
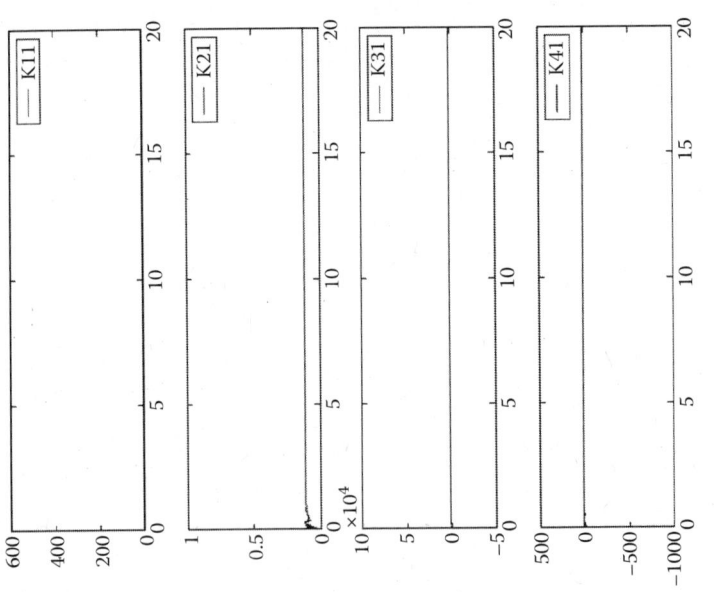

FIGURE 4.24
Behavior of the elements of the Kalman gain matrix $K_k = \begin{bmatrix} K_{11} & K_{12} \\ K_{21} & K_{22} \\ K_{31} & K_{32} \\ K_{41} & K_{42} \end{bmatrix}$.

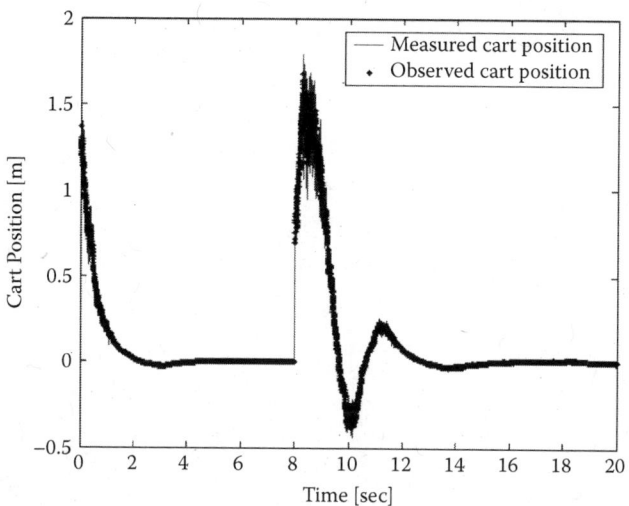

FIGURE 4.25
Measured and observed position.

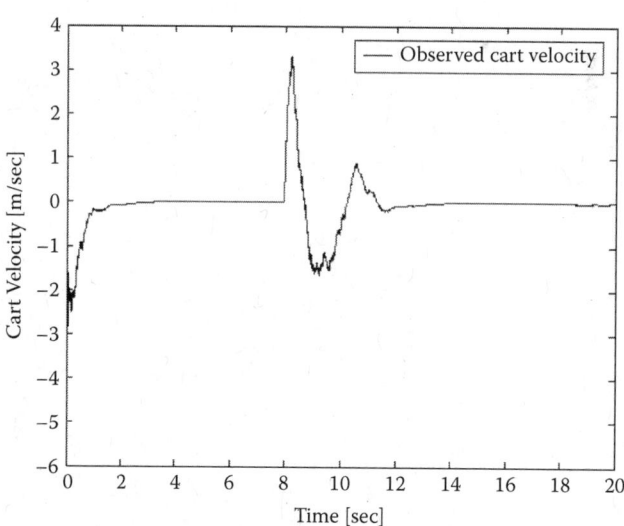

FIGURE 4.26
Observed velocity of the cart.

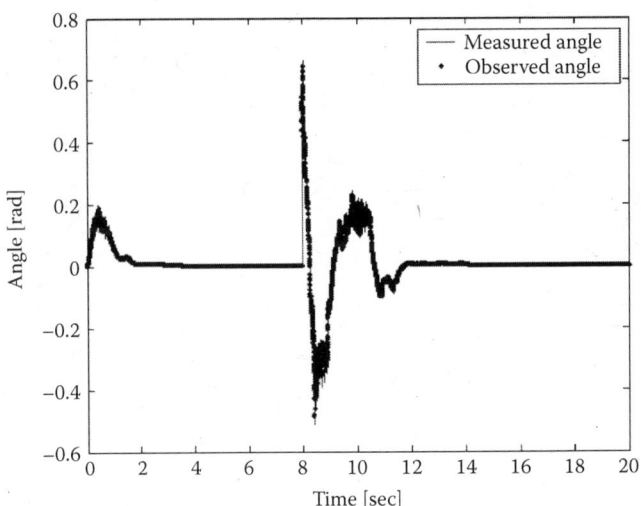

FIGURE 4.27
Measured and observed angle of the rod.

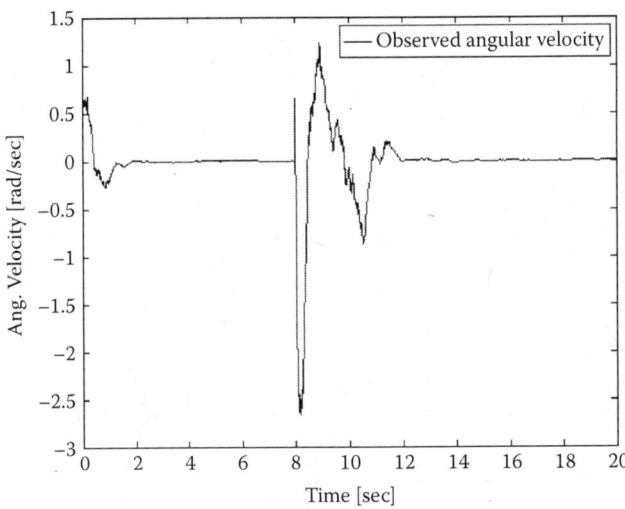

FIGURE 4.28
Observed angular velocity of the rod.

Observer Design and Kalman Filtering 93

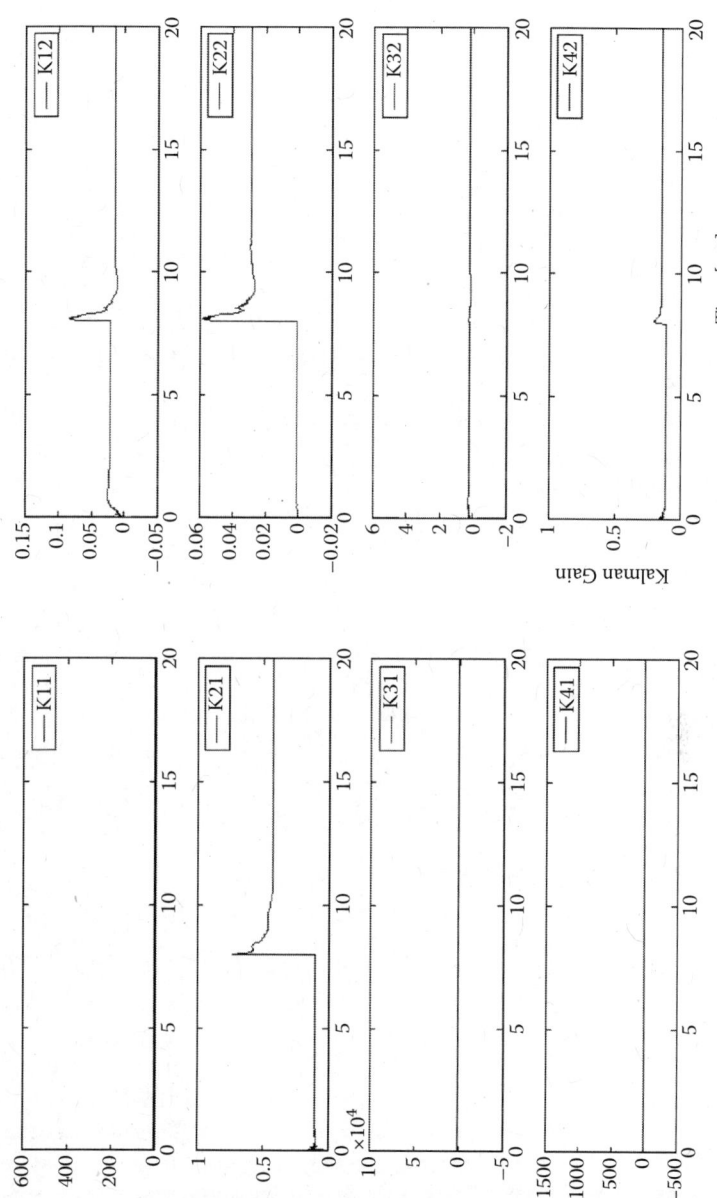

FIGURE 4.29
Behavior of the elements of the Kalman gain matrix.

4.5 Particle Filters

In the above derivation and applications of Kalman filter, one of the most important assumptions was that the process and the measurement noise signals are Gaussian random variables [1]. However, there can be many situations where one or both noise signals are non-Gaussian, or we do not know the exact parameters of the distribution. Particle filtering is an alternative for such applications because it makes no assumptions on the distribution of noise [2]. Rather than worrying about separating noise from the signal, particle filters directly try to increase the probability of choosing a noise-free state out of the bank of candidate states. This is done by keeping a bank of (state, importance) pairs known as "particles." Importance of a state given a measurement is calculated using the sensor model. This can be understood more clearly using Bayesian probability equations.

Let X be the random variable of states and Z be that of the measurements. According to the Bayes rule, we can write the probability of the true state $X = x$ given the measurement $Z = z$, as given by

$$P(X = x | Z = z) = P(Z = z | X = x) P(X = x) \tag{4.31}$$

where $P(X = x)$ is known as the prior distribution of the state, meaning it is the distribution we can think of for the state $X = x$ with whatever information we have before we obtain the sensor reading $Z = z$. For instance, in the inverted pendulum case, one of the state variables was the angle of the pendulum. Before we take any sensor feedback at the start of the pendulum balancing exercise, we know that the pendulum will be held upright. However, the exact angle may not be zero. Therefore, we can have a Gaussian distribution centered about mean zero with a reasonably small variance as the prior distribution of the pendulum angle.

$P(Z = z | X = x)$ is known as the likelihood of measurement, meaning it is the expected uncertainty of the sensor reading $Z = z$ given a state $X = x$. If we go back to the pendulum-balancing exercise, the angle of the pendulum is read using a potentiometer. Suppose that we read an angle reading 0.3 (rad) rather than the value 0 (rad) that we expect. We are not sure as to whether what we read is the correct state of the angle because of the noise in the sensor. If we look at the calibration data of the sensor, or if we do a calibration ourselves, we can see the uncertainty of the voltage signal we get for any given angle. This can come from the uncertainty of the primary voltage source or due to changes in the resistance at the contact point. Good sensors have fixed narrow distributions. Therefore, in the above case, we can expect the true angle to be anywhere around 0.3 (rad) with a variance given by the sensor model.

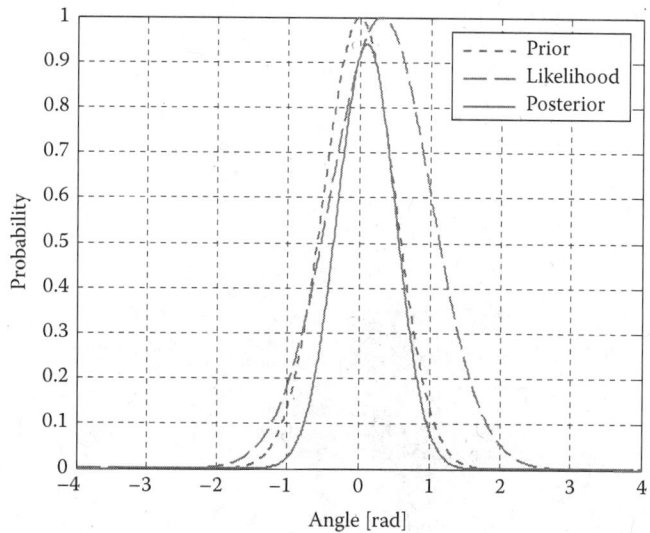

FIGURE 4.30
Posterior distribution combines information from uncertain sensor readings and prior knowledge about the true state.

$P(X = x|Z = z)$ is known as the posterior distribution of the state, meaning it is the updated distribution of the state by combining the information obtained from the sensor readings and the prior information of the state.

Following Figure 4.30 shows how the posterior distribution gets sharply defined when we combine prior information with sensor readings. Note that we use this posterior distribution as prior knowledge for the subsequent state estimations. Now, let us discuss how the particle filter works in this application.

Step 1
Initialize a bank of particles in the given range. For example, $\epsilon = [-4,4]$, $w_i = 0, i = 1, 2, 3, \ldots, N$, where, x_i is the candidate of the state and w_i is the corresponding importance. We set $N = 100$ and choose a uniform distribution to initialize the particles. The result is shown in Figure 4.31.

Step 2
Obtain a sensor reading and assign importance to each particle based on the sensor model $P(Z = z|X = x)$ so that each particle x_i, $i = 1, 2, 3, \ldots, N$, will be associated with importance $w_i = P(Z = z|X = x)$, $i = 1, 2, 3, \ldots, N$.

Assume that the sensor reading was 0.3 (rad), and the non-Gaussian sensor model was given by $P(Z = z|X = x) = \exp[-2(x-z)^2]\exp[-3(x-z)^2]$. In this case, we choose a non-Gaussian distribution for the sensor model to demonstrate that it does not require being Gaussian in particle filters. Figure 4.32 shows the cumulative distribution of the importance assigned to each state.

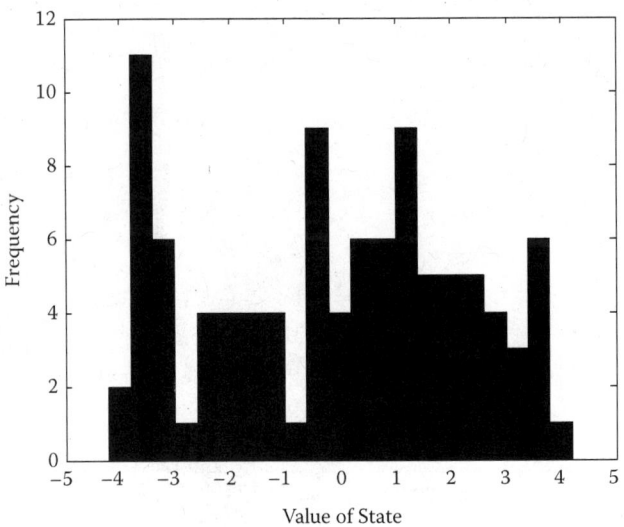

FIGURE 4.31
Initial distribution of particles.

Step 3

Draw N particles according to their importance, so that the probability of choosing $x_i \propto w_i$. Figure 4.33 shows the distribution of states resampled based on their importance. Please find the MATLAB® code used to plot Figures 4.31 to 4.33

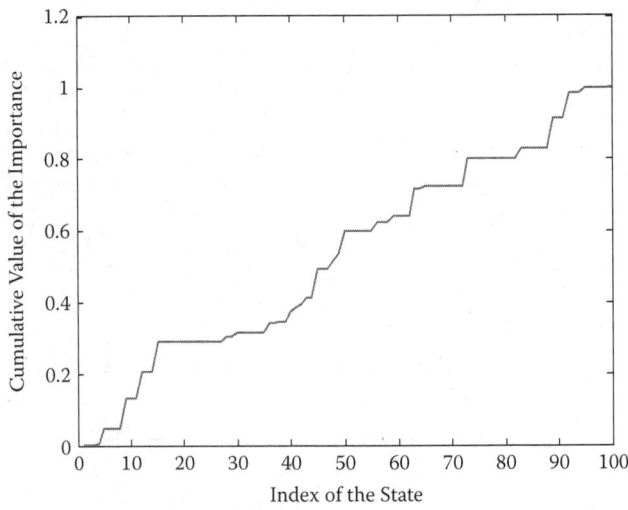

FIGURE 4.32
Cumulative distribution of importance.

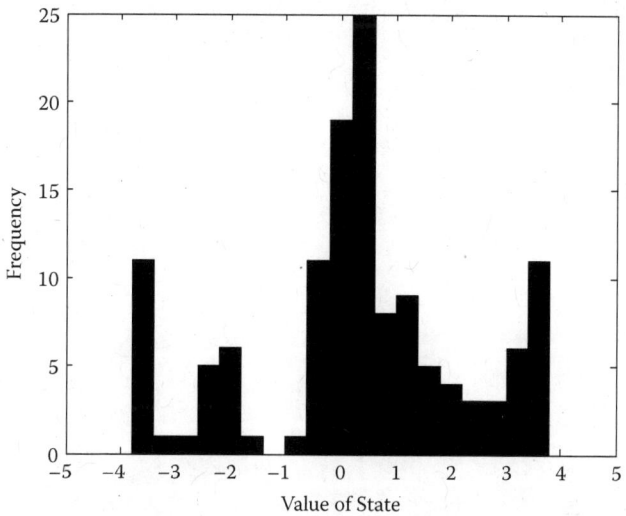

FIGURE 4.33
Distribution of resampled states.

at Chap4/particle_ex.m. Please note in Figure 4.33 that more copies of states around 0.3 have gone into the resamples set of particles.

References

[1] Chui, C. K., and Chen, G., *Kalman Filtering with Real-Time Applications*, 2nd edition, Springer Verlag, New York, 1991.
[2] Ristic, B., Arulampalam, S., and Gordon, N., *Beyond the Kalman Filter—Particle Filters for Tracking Applications*, Artech House, London, 2004.

5

Fuzzy Systems—Sets, Logic, and Control

5.1 Introduction

One of the more popular new technologies is "intelligent control," which is defined as a combination of control theory, operations research, and artificial intelligence (AI). Judging by the billions of dollars worth of sales and thousands of patents issued worldwide, led by Japan since the announcement of the first fuzzy chips in 1987, fuzzy logic is still perhaps the most popular area in AI. Thanks to tremendous technological and commercial advances in fuzzy logic in Japan and other nations, today fuzzy logic continues to enjoy an unprecedented popularity in technological and engineering fields, including manufacturing. Fuzzy logic technology is used in numerous consumer and electronic products and systems, even in the stock market and medical diagnostics. The most important issue facing many industrialized nations in the next several decades will be global competition to an extent that has never before been posed. The arms race is diminishing and the economic race is in full swing. Fuzzy logic is but one such front for global technological, economical, and manufacturing competition.

To understand fuzzy logic, it is important to discuss fuzzy sets. In 1965, Zadeh [1] wrote a seminal paper in which he introduced fuzzy sets, that is, sets with unsharp boundaries. These sets are generally in better agreement with the human mind and reasoning that works with shades of gray, rather than with just black or white. Fuzzy sets are typically able to represent linguistic terms, for example, warm, hot, high, low, close, far, etc. Nearly 10 years later (in 1974), Mamdani [2] succeeded in applying fuzzy logic for control in practice. Today, in Japan, United States, Europe, Asia, and many other parts of the world, fuzzy control is widely accepted and applied. In many consumer products such as washing machines and cameras, fuzzy controllers are used to obtain intelligent machines (Machine Intelligence Quotient, MIQ®) and user-friendly products. The MIQ®-based applications of fuzzy logic were first reported by Jamshidi et al. [3]. A few interesting applications can be cited: control of subway systems, image stabilization of video cameras, image enhancement, and autonomous control of helicopters. Although United States and Europe hesitated in accepting fuzzy logic, they have become more enthusiastic about applying this technology.

Fuzzy set theory is developed comparing the precepts and operations of fuzzy sets with those of classical set theory. Fuzzy sets will be seen to contain the vast majority of the definitions, precepts, and axioms that define classical sets. In fact, very few differences exist between the two set theories. Fuzzy set theory is actually a fundamentally broader theory than the current classical set theory in that it considers an infinite number of "degrees of membership" in a set other than the canonical values of 0 and 1 apparent in classical set theory. In this sense, one could argue that classical sets are a limited form of fuzzy sets. Hence, it will be shown that fuzzy set theory is a comprehensive set theory.

Conceptually, a fuzzy set can be defined as a collection of elements in a universe of information where the boundary of the set contained in the universe is ambiguous, vague, and otherwise fuzzy. It is instructive to introduce fuzzy sets by first reviewing the elements of classical (crisp) set theory.

Fuzzy sets, being introduced in this chapter, are a member of a consortium called "soft computing," first coined by Zadeh (e.g., see the book by Aminzadeh and Jamshidi [4]). Other elements of soft computing are neurocomputing (neural nets), evolutionary computing (genetic algorithms and genetic programming), probabilistic programming, etc. Details of the applications of soft computing tools to control theory can be found in the work of Zilouchian and Jamshidi [5].

A new logic system based on the premises of fuzzy sets is known as fuzzy logic. The need and use of multilevel logic can be traced from the ancient works of Aristotle, who is quoted as saying "There will be a sea battle tomorrow." Such a statement is not yet true or false, but is potentially either. Much later, around AD 1285–1340, William of Occam supported two-valued logic but speculated on what the truth value of "if p then q" might be if one of the two components, p or q, as neither true nor false. During the period of 1878–1956, Lukasiewicz proposed three-level logic as a "true" (1), a "false" (0), and a "neuter" (1/2), which represented half-true or half-false. In subsequent times, logicians in China and other parts of the world continued on the notion of multilevel logic. Zadeh [1], in his seminal 1965 paper, finished the task by following through with the speculation of previous logicians and showing that what he called "fuzzy sets" was the foundation of any logic, regardless of the number of truth levels assumed. He chose the innocent word "fuzz" for the continuum of logical values between 0 (completely false) and 1 (completely true). The theory of fuzzy logic deals with two problems: (1) the fuzzy set theory, which deals with the vagueness found in semantics, and (2) the fuzzy measure theory, which deals with the ambiguous nature of judgments and evaluations.

The primary motivation and "banner" of fuzzy logic is the possibility of exploiting tolerance for some inexactness and imprecision. Precision is often very costly, so if a problem does not warrant great precision, one should not have to pay for it. The traditional example of parking a car is a noteworthy illustration. If the driver is not required to park the car within an exact distance from the curb, why spend any more time than necessary on the task as long as it is a legal parking operation? Fuzzy logic and classical logic differ in the sense that the former can handle both symbolic and numerical manipulation, whereas the latter can handle symbolic manipulation only. In a broad sense, fuzzy logic is a union of fuzzy (fuzzified) crisp logics [7]. To quote Zadeh, "Fuzzy logic's primary aim is to provide a formal, computationally-oriented system of concepts and techniques for dealing with modes of reasoning which are approximate rather than exact." Thus, in fuzzy logic, exact (crisp) reasoning is considered to be the limiting case of approximate reasoning. In fuzzy logic, one can see that everything is a matter of degrees.

This chapter is organized as follows. Section 5.2 briefly describes classical sets, followed by an introduction to classical set operations in Section 5.3. Properties of classical sets are given in Section 5.4. Section 5.5 is a quick introduction to fuzzy sets. Fuzzy set operations and properties are given in Sections 5.6 and 5.7, respectively. Section 5.8 presents fuzzy versus classical relations. In Section 5.9, a brief introduction to predicate logic is given. In Section 5.10, fuzzy logic is presented, followed by approximate reasoning in Section 5.11. Section 5.12 presents fuzzy control systems and the stability of such systems. Finally, a conclusion is given in Section 5.13.

5.2 Classical Sets

In classical set theory, a set is denoted as a so-called *crisp set* and can be described by its characteristic function as follows:

$$\mu_C: U \to \{0, 1\} \tag{5.1}$$

In the above equation, U is called the universe of discourse, that is, a collection of elements that can be continuous or discrete. In a crisp set, each element of the universe of discourse either belongs to the crisp set ($\mu_C = 1$) or does not belong to the crisp set ($\mu_C = 0$).

Consider a characteristic function μ_{Chot} representing the crisp set hot, a set with all "hot" temperatures. Figure 5.1 graphically describes this crisp set,

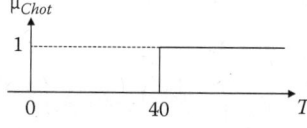

FIGURE 5.1
The characterstic function μ_{Chot}.

considering temperatures higher than 40°C as hot. (Note that, for all temperatures T, we have $T \in U$.)

5.3 Classical Set Operations

Let A and B be two sets in the universe U, and let $\mu_A(x)$ and $\mu_B(x)$ be the characteristic functions of A and B *in* the universe of discourse in sets A and B, respectively. The characteristic function $\mu_A(x)$ is defined as follows:

$$\mu_A(x) = \begin{cases} 1, & x \in A \\ 0, & x \notin A \end{cases} \tag{5.2}$$

and $\mu_B(x)$ is defined as

$$\mu_B(x) = \begin{cases} 1, & x \in B \\ 0, & x \notin B \end{cases} \tag{5.3}$$

Using the above definitions, the following operations are defined [6].

5.3.1 Union

The union between two sets, that is, $C = A \cup B$, where $\cup$ is the union operator, represents all those elements in the universe that reside in either set A, or set B, or both [7] (see Figure 5.2). The characteristic function μ_C is defined in Equation (5.4).

$$\forall x \in U : \mu_C = \max[\mu_A(x), \mu_B(x)] \tag{5.4}$$

The operator in Equation (5.4) is referred to as the max operator.

5.3.2 Intersection

The intersection of two sets, that is, $C = A \cap B$, where $\cap$ is the intersection operator, represents all those elements in the universe U that reside in both

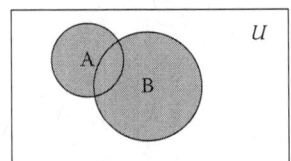

FIGURE 5.2
The characteristic function μ_{Chot}.

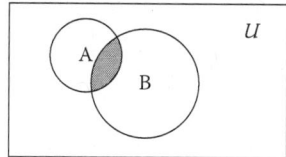

FIGURE 5.3
Union.

sets A and B simultaneously (see Figure 5.3). Equation (5.5) shows how to obtain the characteristic function μ_C.

$$\forall x \in U : \mu_C = \min[\mu_A(x), \mu_B(x)] \tag{5.5}$$

The operator in Equation (5.5) is referred to as the min operator.

5.3.3 Complement

The complement of a set A, denoted $\bar{A}$, is defined as the collection of all elements in the universe that do not reside in the set A (see Figure 5.4). The characteristic function $\mu_{\bar{A}}$ is defined by Equation (5.6).

$$\forall x \in U : \mu_{\bar{A}} = 1 - \mu_A(x) \tag{5.6}$$

5.4 Properties of Classical Set

Properties of classical sets are very important to consider because of their influence on the mathematical manipulation. Some of these properties are listed below [6,7].

Commutativity:

$$A \cup B = B \cup A$$
$$A \cap B = B \cap A \tag{5.7}$$

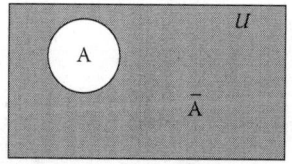

FIGURE 5.4
Intersection.

Associativity:
$$A \cup (B \cup C) = (A \cup B) \cup C$$
$$A \cap (B \cap C) = (A \cap B) \cap C \quad (5.8)$$

Distributivity:
$$A \cup (B \cap C) = (A \cup B) \cap (A \cup C)$$
$$A \cap (B \cup C) = (A \cap B) \cup (A \cap C) \quad (5.9)$$

Idempotency:
$$A \cup A = A$$
$$A \cap A = A \quad (5.10)$$

Identity:
$$A \cup \phi = A \quad (5.11a)$$
$$A \cap X = A \quad (5.11b)$$
$$A \cap \phi = \phi \quad (5.12a)$$
$$A \cup X = X \quad (5.12b)$$

Excluded middle laws are very important since they are the only set operations that are not valid for both classical and fuzzy sets. Excluded middle laws consist of two laws. The first, known as the *law of excluded middle*, deals with the union of a set A and its complement. The second law, known as *law of contradiction*, represents the intersection of a set A and its complement. The following equations describe these laws.

Law of excluded middle
$$A \cup \bar{A} = X \quad (5.13)$$

Law of contradiction
$$A \cap \bar{A} = \phi \quad (5.14)$$

5.5 Fuzzy Sets

The definition of a fuzzy set [1] is given by the characteristic function
$$\mu_F : U \to [0, 1] \quad (5.15)$$

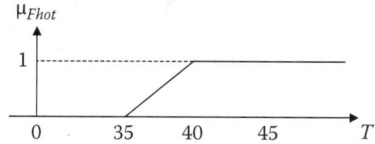

FIGURE 5.5
Complement.

In this case, the elements of the universe of discourse can belong to the fuzzy set with any value between 0 and 1. This value is called the *degree of membership*. If an element has a value close to 1, the degree of membership or truth value is high. The characteristic function of a fuzzy set is called the *membership function*, because it gives the degree of membership for each element of the universe of discourse. If now the characteristic function μ_{Fhot} is considered, one can express the human opinion, for example, that 37°C is still fairly hot, and that 38°C is hot, but not as hot as 40°C and higher. This result is a gradual transition from membership (completely true) to nonmembership (not true at all). Figure 5.5 shows the membership function μ_{Fhot} for the fuzzy set F_{hot}.

The membership functions for fuzzy sets can have many different shapes, depending on definition. Figure 5.6 provides a description of the various features of membership functions. Some of the possible membership functions are shown in Figure 5.7.

Figure 5.7 illustrates some of the possible membership functions: (a) the Γ function, an increasing membership function with straight lines; (b) the L function, a decreasing function with straight lines; (c) the Λ function, a triangular function with straight lines; and (d) the singleton, a membership function with a membership function value 1 for only one value and the rest is zero. There are many other possible functions such as trapezoidal, Gaussian, sigmoidal, or even arbitrary.

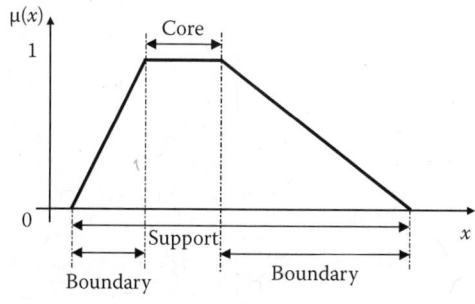

FIGURE 5.6
The membership function μ_{Fhot}. Fuzzy membership functions.

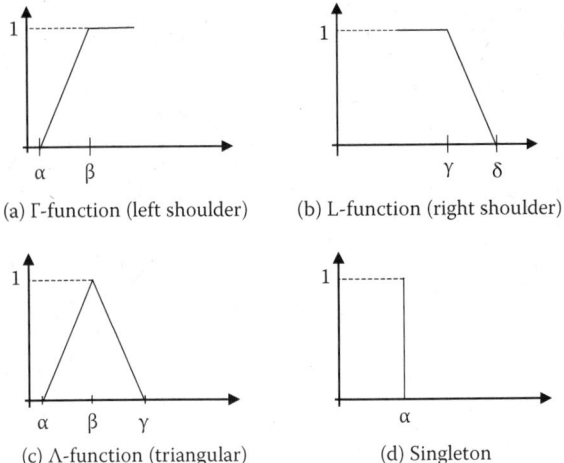

FIGURE 5.7
Description of fuzzy membership functions [7].

A notation convention for fuzzy sets that is popular in the literature when the universe of discourse U is discrete and finite is given below for a fuzzy set A by

$$\underset{\sim}{A} = \frac{\mu_{\underset{\sim}{A}}(x_1)}{x_1} + \frac{\mu_{\underset{\sim}{A}}(x_2)}{x_2} + \cdots = \sum_i \frac{\mu_{\underset{\sim}{A}}(x_i)}{x_i} \qquad (5.16)$$

and, when the universe of discourse U is continuous and infinite, the fuzzy set A is denoted by

$$\underset{\sim}{A} = \int \frac{\mu_{\underset{\sim}{A}}(x)}{x} \qquad (5.17)$$

5.6 Fuzzy Set Operations

As in the traditional crisp sets, logical operations, for example, union, intersection, and complement, can be applied to fuzzy sets [1] (Figures 5.8 to 5.10).

5.6.1 Union

The union operation (and the intersection operation as well) can be defined in many different ways. Here, the definition that is used in most cases is

Fuzzy Systems—Sets, Logic, and Control

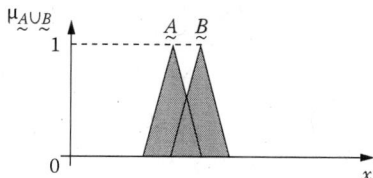

FIGURE 5.8
Examples of membership functions. (a) Γ Function (left shoulder), (b) L function (right shoulder), (c) Λ function (triangular), (d) singleton.

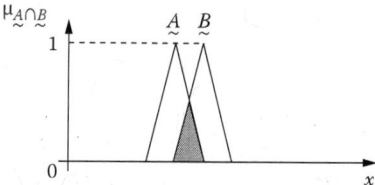

FIGURE 5.9
Union of two fuzzy sets.

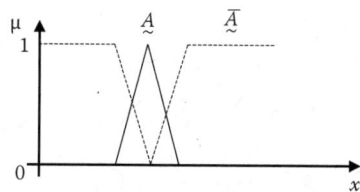

FIGURE 5.10
Intersection of two fuzzy sets.

discussed. The union of two fuzzy sets A and B with the membership functions $\mu_A(x)$ and $\mu_B(x)$ is a fuzzy set C, written as $C = A \cup B$, whose membership function is related to those of A and B as follows:

$$\forall x \in U : \mu_C = \max[\mu_A(x), \mu_B(x)] \tag{5.18}$$

5.6.2 Intersection

According to the min operator, the intersection of two fuzzy sets A and B with the membership functions $\mu_A(x)$ and $\mu_B(x)$ respectively, is a fuzzy set C, written as $C = A \cap B$, whose membership function is related to those of A and B as follows:

$$\forall x \in U : \mu_C = \min[\mu_A(x), \mu_B(x)] \tag{5.19}$$

5.6.3 Complement

The complement of a set A, denoted $\bar{A}$, is defined as the collection of all elements in the universe that do not reside in the set A.

$$\forall x \in U : \mu_{\bar{A}} = 1 - \mu_A(x) \tag{5.20}$$

Keep in mind that, even though the equations of the union, intersection, and complement appear to be the same for classical and fuzzy sets, they differ in the fact that $\mu_A(x)$ and $\mu_B(x)$ can take only a value of 0 or 1 in the case of classical set, whereas in fuzzy sets, they include the whole interval from 0 to 1.

5.7 Properties of Fuzzy Sets

Similar to classical sets, fuzzy sets also have some properties that are important for mathematical manipulations [6,7]. Some of these properties are listed below.

Commutativity:

$$\begin{aligned} A \cup B &= B \cup A \\ A \cap B &= B \cap A \end{aligned} \tag{5.21}$$

Associativity:

$$\begin{aligned} A \cup (B \cup C) &= (A \cup B) \cup C \\ A \cap (B \cap C) &= (A \cap B) \cap C \end{aligned} \tag{5.22}$$

Distributivity:
$$A \cup (B \cap C) = (A \cup B) \cap (A \cup C)$$
$$A \cap (B \cup C) = (A \cap B) \cup (A \cap C) \tag{5.23}$$

Idempotency:
$$A \cup A = A$$
$$A \cap A = A \tag{5.24}$$

Identity:
$$A \cup \phi = A$$
$$A \cap X = A$$
$$A \cap \phi = \phi$$
$$A \cup X = X \tag{5.25}$$

Most of the properties that hold for classical sets (e.g., commutativity, associativity, and idempotency) also hold for fuzzy sets, except for following two properties [6]:

Law of contradiction ($A \cap \bar{A} \neq \phi$): One can easily note that the intersection of a fuzzy set and its complement results in a fuzzy set with membership values of up to ½ and, thus, does not equal the empty set (as in the case of classical sets), as shown in Figure 5.11.

Law of excluded middle ($A \cup \bar{A} \neq \cup$): The union of a fuzzy set and its complement does not give the universe of discourse (see Figure 5.12).

5.7.1 Alpha-Cut Fuzzy Sets

It is in the crisp domain that we perform all computations with today's computers. The conversion from fuzzy to crisp sets can be done via two means, one of which is alpha-cut sets.

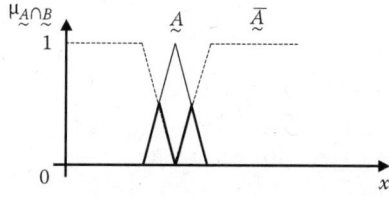

FIGURE 5.11
Complement of a fuzzy set.

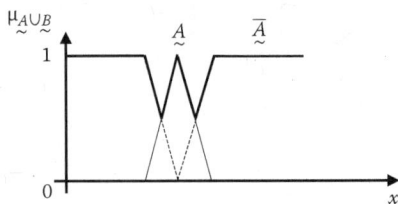

FIGURE 5.12
Law of contradiction.

5.7.1.1 Alpha-Cut Sets

Given a fuzzy set $\underset{\sim}{A}$, the alpha-cut (or lambda cut) set of $\underset{\sim}{A}$ is defined by

$$A_\alpha = \{x \mid \mu_{\underset{\sim}{A}}(x) \geq \alpha\} \quad (5.26)$$

Note that, by virtue of the condition on $\mu_{\underset{\sim}{A}}(x)$ in Equation (5.26), that is, a common property, the set A_α in Equation (5.26) is now a crisp set. In fact, any fuzzy set can be converted to an infinite number of cut sets.

5.7.2 Extension Principle

In fuzzy sets, just as in crisp sets, one needs to find means to extend the domain of a function, that is, given a fuzzy set $\underset{\sim}{A}$ and a function $f(x)$, what is the value of function $f(\underset{\sim}{A})$? This notion is called the *extension principle*, which was first proposed by Zadeh.

Let the function f be defined by

$$f : U \to V \quad (5.27)$$

where U and V are domain and range sets, respectively. Define a fuzzy set $\underset{\sim}{A} \subset U$ as

$$\underset{\sim}{A} = \left\{ \frac{\mu_1}{u_1} + \frac{\mu_2}{u_2} + \cdots + \frac{\mu_n}{u_n} \right\} \quad (5.28)$$

Then, the extension principle asserts that the function f is a fuzzy set, as well, which is defined below:

$$\underset{\sim}{B} = f(\underset{\sim}{A}) = \left\{ \frac{\mu_1}{f(u_1)} + \frac{\mu_2}{f(u_2)} + \cdots + \frac{\mu_n}{f(u_n)} \right\} \quad (5.29)$$

The complexity of the extension principle would increase when more than one member of $u_1 \times u_2$ is mapped to only one member of v; one would take the maximum membership grades of these members in the fuzzy set $\underset{\sim}{A}$.

Example 5.1

Given two universes of discourse $U_1 = U_2 = \{1, 2, 10\}$ and two fuzzy sets (numbers) defined by

"Approximately 2" $= \dfrac{0.5}{1} + \dfrac{1}{2} + \dfrac{0.8}{3}$ and "Approximately 5" $= \dfrac{0.6}{3} + \dfrac{0.8}{4} + \dfrac{1}{5}$

It is desired to find "approximately 10."

Solution

The function $f = u_1 \times u_2 :\to v$ represents the arithmetic product of these two fuzzy numbers and is given by

$$\text{"approximately 10"} = \left(\dfrac{0.5}{1} + \dfrac{1}{2} + \dfrac{0.8}{3}\right) \times \left(\dfrac{0.6}{3} + \dfrac{0.8}{4} + \dfrac{1}{5}\right) = \dfrac{\min(0.5, 0.6)}{3}$$

$$+ \dfrac{\min(0.5, 0.8)}{4} + \dfrac{\min(0.5, 1)}{5} + \dfrac{\min(1, 0.6)}{6} + \dfrac{\min(1, 0.8)}{8}$$

$$+ \dfrac{\min(1, 1)}{10} + \dfrac{\min(0.8, 0.6)}{9} + \dfrac{\min(0.8, 0.8)}{12} + \dfrac{\min(0.8, 1)}{15}$$

$$= \dfrac{0.5}{3} + \dfrac{0.5}{4} + \dfrac{0.5}{5} + \dfrac{0.6}{6} + \dfrac{0.8}{8} + \dfrac{0.6}{9} + \dfrac{1}{10} + \dfrac{0.8}{12} + \dfrac{0.8}{15}$$

The above resulting fuzzy number has its *prototype*, that is, value 10 with a membership function 1 and the other eight pairs are spread around the point (1, 10).

Example 5.2

Consider two fuzzy sets (numbers) defined by

"Approximately 2" $= \dfrac{0.5}{1} + \dfrac{1}{2} + \dfrac{0.5}{3}$ and "Approximately 4" $= \dfrac{0.8}{2} + \dfrac{0.9}{3} + \dfrac{1}{4}$

It is desired to find "approximately 8."

Solution

The function $f = u_1 \times u_2 : \to v$ represents the arithmetic product of these two fuzzy numbers and is given by

$$\text{"approximately 8"} = \left(\dfrac{0.5}{1} + \dfrac{1}{2} + \dfrac{0.5}{3}\right) \times \left(\dfrac{0.8}{2} + \dfrac{0.9}{3} + \dfrac{1}{4}\right) = \dfrac{\min(0.5, 0.8)}{2}$$

$$+ \dfrac{\min(0.5, 0.9)}{3} + \dfrac{\max[\min(0.5, 1), \min(1, 0.8)]}{4}$$

$$+ \dfrac{\max[\min(1, 0.9), \min(0.5, 0.8)]}{6} + \dfrac{\min(1, 1)}{8} + \dfrac{\min(0.5, 0.9)}{9}$$

$$+ \dfrac{\min(0.5, 1)}{12} = \dfrac{0.5}{2} + \dfrac{0.5}{3} + \dfrac{0.8}{4} + \dfrac{0.9}{6} + \dfrac{1}{8} + \dfrac{0.5}{9} + \dfrac{0.5}{12}$$

5.8 Classical Relations versus Fuzzy Relations

Classical relations are structures that represent the presence or absence of correlation or interaction among elements of various sets. There are only two degrees of relationship between elements of the sets in a crisp relation: the "completely related" or "not related" relationships. Fuzzy relations, on the other hand, are developed by allowing the relationship between elements of two or more sets to take an infinite number of degrees of relationship between the extremes of "completely related" and "not related" [6,7].

The classical relation of two universes, U and V, is defined as

$$U \times V = \{(u,v) \mid u \in U, v \in V\} \quad (5.30)$$

which combines $\forall u \in U$ and $\forall v \in V$ in an ordered pair and forms unconstrained matches between u and v, that is, every element in universe U is completely related to every element in universe V. The *strength* of this relationship between ordered pairs of elements in each universe is measured by the characteristic function, where a value of unity is associated with *complete relationship* and a value of zero is associated with *no relationship*, that is, the binary values 1 and 0.

As an example, if $U = \{1,2\}$ and $V = \{a,b,c\}$, then $U \times V = \{(1,a),(1,b),(1,c),(2,a),(2,b),(2,c)\}$. The above product is said to be a *crisp relation*, which can be expressed by either a matrix expression

$$R = U \times V = \begin{matrix} & \begin{matrix} a & b & c \end{matrix} \\ \begin{matrix} 1 \\ 2 \end{matrix} & \begin{bmatrix} 1 & 1 & 1 \\ 1 & 1 & 1 \end{bmatrix} \end{matrix} \quad (5.31)$$

or in a so-called *Sagittal* diagram (see Figure 5.13).

Fuzzy relations map elements of one universe to those of another universe through Cartesian product of the two universes. Unlike crisp relations, the *strength* of the relation between ordered pairs of the two universes is not measured with the characteristic function but, rather, with a membership

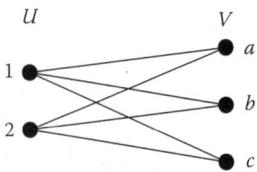

FIGURE 5.13
Law of excluded middle.

function expressing various *degrees* of the strength of the relation on the unit interval [0, 1]. In other words, a fuzzy relation R is a mapping:

$$R : U \times V \to [0,1] \qquad (5.32)$$

The following example illustrates this relationship, that is,

$$\mu_R(u,v) = \mu_{A \times B}(u,v) = \min(\mu_A(u), \mu_B(v)) \qquad (5.33)$$

Example 5.3

Consider two fuzzy sets, $A_1 = \frac{0.2}{x_1} + \frac{0.9}{x_2}$ and $A_2 = \frac{0.3}{y_1} + \frac{0.5}{y_2} + \frac{1}{y_3}$. Determine the fuzzy relation between these sets.

Solution

The fuzzy relation R is

$$R = A_1 \times A_2 = \begin{bmatrix} 0.2 \\ 0.9 \end{bmatrix} \times [0.3 \quad 0.5 \quad 1]$$

$$= \begin{bmatrix} \min(0.2, 0.3) & \min(0.2, 0.5) & \min(0.2, 1) \\ \min(0.9, 0.3) & \min(0.9, 0.5) & \min(0.9, 1) \end{bmatrix} = \begin{bmatrix} 0.2 & 0.2 & 0.2 \\ 0.3 & 0.5 & 0.9 \end{bmatrix}$$

Let R be a relation that relates elements from universe U to universe V, and let S be a relation that relates elements from universe V to universe W. Is it possible to find the relation T that relates the same elements in universe U that R contains to elements in universe W that S contains? The answer is yes, using an operation known as *composition*.

In crisp or fuzzy relations, the composition of two relations, using the max-min rule, is given below. Given two fuzzy relations $R(u,v)$ and $S(v,w)$, then the composition of these is

$$T = R \circ S = \max_{v \in V} \{ \min(\mu_R(u,v), \mu_S(v,w)) \} \qquad (5.34a)$$

and using the max-product rule, the characteristic function is given by

$$T = R \circ S = \max\{(\mu_R(u,v), \mu_S(v,w))\} \qquad (5.34b)$$

The same composition rules hold for crisp relations.

Example 5.4

Consider two fuzzy relations $R = \begin{bmatrix} 0.6 & 0.8 \\ 0.7 & 0.9 \end{bmatrix}$ and $S = \begin{bmatrix} 0.3 & 0.1 \\ 0.2 & 0.8 \end{bmatrix}$. It is desired to evaluate $R \circ S$ and $S \circ R$.

Solution

Using the max-min composition for $R \circ S$, we have

$$R \circ S = \begin{bmatrix} 0.3 & 0.8 \\ 0.3 & 0.8 \end{bmatrix}$$

where, for example, the element (1,1) is obtained by max{min(0.6,0.3),min (0.8,0.2)} = 0.3.
For $S \circ R$, we obtain the following result:

$$S \circ R = \begin{bmatrix} 0.3 & 0.3 \\ 0.7 & 0.8 \end{bmatrix} \neq R \circ S$$

Using the max-product rule, we have

$$R \circ S = \begin{bmatrix} 0.18 & 0.64 \\ 0.21 & 0.72 \end{bmatrix}$$

where, for example, the element (2,2) is obtained by max{(0.7), (0.1), (0.9)(0.8)} = 0.72.
For $S \circ R$, we obtain the following result

$$S \circ R = \begin{bmatrix} 0.18 & 0.24 \\ 0.56 & 0.72 \end{bmatrix} \neq R \circ S$$

5.9 Predicate Logic

Let a predicate logic proposition P be a linguistic statement contained within a universe of propositions that are either completely true or false. The truth value of the proposition P can be assigned a binary truth value, called $T(P)$, just as an element in a universe is assigned a binary quantity to measure its membership in a particular set. For binary (Boolean) predicate logic, $T(P)$ is assigned a value of 1 (truth) or 0 (false). If U is the universe of all propositions, then T is a mapping of these propositions to the binary quantities (0,1), or

$$T : U \rightarrow \{0,1\}$$

Now, let P and Q be two simple propositions on the same universe of discourse that can be combined using the following five logical connectives:

Disjunction ($\vee$)
Conjunction ($\wedge$)
Negation (−)
Implication ($\rightarrow$)
Equality ($\leftrightarrow$ or $\equiv$)

to form logical expressions involving two simple propositions. These connectives can be used to form new propositions from simple propositions.

Now, define sets A and B from universe X where these sets might represent linguistic ideas or thoughts. Then, a propositional calculus will exist for the case where proposition P measures the truth of the statement that an element, x, from the universe X is contained in set A and the truth of the statement that this element, x, is contained in set B, or more conventionally:

- P: truth that $x \in A$.
- Q: truth that $x \in B$, where truth is measured in terms of the truth value, that is,
  - If $x \in A$, $T(P) = 1$; otherwise, $T(P) = 0$.
  - If $x \in B$, $T(Q) = 1$; otherwise, $T(P) = 0$, or using the characteristic function to represent true (1) and false (0):

$$\chi_A(x) = \begin{cases} 1, & x \in A \\ 0, & x \notin A \end{cases}$$

The above five logical connectives can be used to create compound propositions, where a compound proposition is defined as a logical proposition formed by logically connecting two or more simple propositions. Just as one is interested in the truth of a simple proposition, predicate logic also involves the assessment of the truth of compound propositions. Given a proposition $P : x \in A, \bar{P} : x \notin A$, the resulting compound propositions are defined below in terms of their binary truth values:

Disjunction:

$$P \vee Q \Rightarrow x \in A \text{ or } B$$

Hence, $T(P \vee Q) = \max(T(P), T(Q))$

Conjunction:

$$P \wedge Q \Rightarrow x \in A \text{ and } B$$

Hence, $T(P \wedge Q) = \min(T(P), T(Q))$

Negation:

If $T(P) = 1$, then $T(\bar{P}) = 0$; If $T(P) = 0$, then $T(\bar{P}) = 1$

Equivalence:

$$P \leftrightarrow Q \Rightarrow x \in A, B$$

Hence, $T(P \leftrightarrow Q) \Rightarrow T(P) = T(Q)$

Implication:

$$P \rightarrow Q \Rightarrow x \notin A \text{ or } x \in B$$

Hence, $T(P \rightarrow Q) = T(\bar{P} \cup Q)$

The logical connective implication presented here is also known as the classical implication, to distinguish it from an alternative form attributed to Lukasiewicz, a Polish mathematician in the 1930s, who was first credited with exploring logic other than Aristotelian (classical or binary) logic. This classical form of the implication operation requires some explanation.

For a proposition P defined on set A and a proposition Q defined on set B, the implication "P implies Q" is equivalent to taking the union of elements in the complement of set A with the elements in the set B. That is, the logical implication is analogous to the set-theoretic form.

$$P \rightarrow Q \equiv \bar{A} \cup B \text{ is true} \equiv \text{either "not in } A\text{" or "in } B\text{"}$$

So that $(P \rightarrow Q) \leftrightarrow (\bar{P} \vee Q)$

$$T(P \rightarrow Q) = T(\bar{P} \vee Q) = \max(T(\bar{P}), T(Q))$$

This is linguistically equivalent to the statement "P implies Q is true" when either "not A" or "B" is true [11]. Graphically, this implication and the analogous set operation are represented by the Venn diagram in Figure 5.14. As noted, the region represented by the difference $A \backslash B$ is the set region where the implication "P implies Q" is false (the implication *fails*). The shaded region in Figure 5.14 represents the collection of elements in the universe where the implication is true, that is, the shaded area is the set presented in Figure 5.1.

Now, with two propositions (P and Q), each being able to take on one of two truth values (*true* or *false*, 1 or 0), there will be a total of $2^2 = 4$ propositional situations. These situations are illustrated in Table 5.1, along with the appropriate truth values for the propositions P and Q and the various logical connectives between them in the truth table.

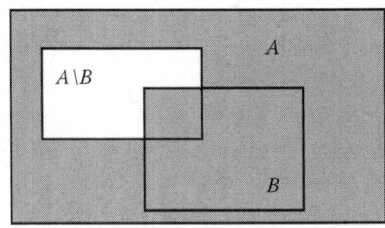

FIGURE 5.14
Sagittal diagram.

TABLE 5.1
Truth Table for Propositions P and Q

P	Q	$\bar{P}$	$P \vee Q$	$P \wedge Q$	$P \to Q$	$P \leftrightarrow Q$
True	True	False	True	True	True	True
True	False	False	True	False	False	False
False	True	True	True	False	True	False
False	False	True	False	False	True	True

To help you understand this concept, assume you have two propositions, P and Q: P, you are a graduate student; Q, you are a university student. Let us examine the implication "P implies Q." If you are a student in general and a graduate student in particular, then the implication is true. On the other hand, the implication would be false if you are a graduate student without being a student. Now, let us assume that you are an undergraduate student; regardless of whether you are a graduate or not, then the implication is true (since the case where you are not a graduate student does not negate the fact that you are an undergraduate). Then, we come to the final case: you are neither a graduate nor undergraduate student. In this case, the implication is true because the fact that you are not a graduate or undergraduate student does not negate the implication that, for you to be a graduate student, you have to be a student at the university.

Suppose the implication operation involves two different universes of discourse, P is a proposition described by set A, which is defined on universe X, and Q is a proposition described by set B, which is defined on universe Y. Then, the implication "P implies Q" can be represented in set theory terms by the relation R, where R is defined by

$$R = (A \times B) \cup (\bar{A} \times Y) \equiv \text{IF } A, \text{THEN } B$$

If $x \in A$ (where $x \in X$, $A \subset X$)

Then $y \in B$ (where $y \in Y$, $B \subset Y$)

where $A \times B$ and $A \times Y$ are Cartesian products [8].

This implication is also equivalent to the linguistic rule form: IF A, THEN B. The graphic shown in Figure 5.15 represents the Cartesian space of the product $X \times Y$, showing typical sets A and B, and superimposed on this space is the set theory equivalent of the implication. That is,

$$P \to Q \Rightarrow \text{IF } x \in A, \text{ then } y \in B, \text{ or } P \to Q \equiv \bar{A} \cup B$$

The shaded regions of the compound Venn diagram in Figure 5.15 represent the truth domain of the implication, IF A, THEN B (P implies Q).

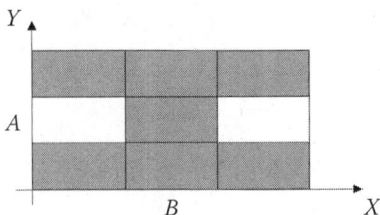

FIGURE 5.15
Classical implication operation (shaded area is where implication holds) [2].

5.9.1 Tautologies

In predicate logic, it is useful to consider compound propositions that are always true, regardless of the truth values of the individual simple propositions. Classical logic compound propositions with this property are called *tautologies*. Tautologies are useful for deductive reasoning and for making deductive inferences. So, if a compound proposition can be expressed in the form of a tautology, the truth value of that compound proposition is known to be true. Inference schemes in expert systems often use tautologies. The reason for this is that tautologies are logical formulas that are true on logical grounds alone [8].

One of these, known as *modus ponens* deduction, is a very common inference scheme used in forward chaining rule-based expert systems. It is an operation whose task is to find the truth value of a consequent in a production rule, given the truth value of the antecedent in the rule. Modus ponens deduction concludes that, given two propositions, a and a-implies-b, both of which are true, and then the truth of the simple proposition b is automatically inferred. Another useful tautology is the *modus tollens* inference, which is used in backward-chaining expert systems. In *modus tollens*, an implication between two propositions is combined with a second proposition, and both are used to imply a third proposition. Some common tautologies are listed below.

$$\bar{B} \cup B \leftrightarrow X$$

$$A \cup X \leftrightarrow X$$

$$\bar{A} \cup X \leftrightarrow X$$

$$(A \wedge (A \rightarrow B)) \rightarrow B \quad \text{(modus ponens)}$$

$$(\bar{B} \wedge (A \rightarrow B)) \rightarrow \bar{A} \quad \text{(modus tollens)}$$

5.9.2 Contradictions

Compound propositions that are always false, regardless of the truth value of the individual simple propositions comprising the compound proposition, are called contradictions. Some simple contradictions are listed below.

$$\bar{B} \cap B \leftrightarrow \phi$$
$$A \cap \phi \leftrightarrow \phi$$
$$\bar{A} \cap \phi \leftrightarrow \phi$$

5.9.3 Deductive Inferences

The *modus ponens* deduction is used as a tool for inferencing in rule-based systems. A typical IF-THEN rule is used to determine whether an antecedent (cause or action) infers a consequent (effect or action). Suppose we have a rule of the form

IF A, THEN B

This rule could be translated into a relation using the Cartesian product sets A and B, that is,

$$R = (A \times B) \cup (\bar{A} \times Y) \tag{5.35}$$

Now, suppose a new antecedent, say A', is known. Can we use *modus ponens* deduction to infer that a new consequent, say B', is resulting from the new antecedent? That is, in rule form,

IF A', THEN B'?

The answer, of course, is yes, through the use of the composition relation. Since "A implies B" is defined on the Cartesian space $X \times Y$, B' can be found through the following set-theoretic formulation:

$$B' = A' \circ R = A' \circ ((A \times B) \cup (\bar{A} \times Y)) \tag{5.36}$$

Modus ponens deduction can also be used for the compound rule,

IF A, THEN B, ELSE C

Using the relation defined as

$$R = (A \times B) \cup (\bar{A} \times C), \tag{5.37}$$

and hence $B' = A' \circ R$.

Example 5.5

Let two universes of discourse be described by $X = \{1, 2, 3, 4, 5, 6\}$ and $Y = \{1, 2, 3, 4\}$ and define the crisp set $A = \{2, 3\}$ on X and $B = \{3, 4\}$ on Y. Determine the deductive inference IF A, THEN B.

Solution

Expressing the crisp sets in Zadeh's notation,

$$A = \frac{0}{1} + \frac{1}{2} + \frac{1}{3} + \frac{0}{4}$$

$$B = \frac{0}{1} + \frac{0}{2} + \frac{1}{3} + \frac{1}{4} + \frac{0}{5} + \frac{0}{6}$$

Taking the Cartesian product $A \times B$, which involves taking the pairwise min of each pair from sets A and B [8],

$$A \times B = \begin{array}{c} \\ 1 \\ 2 \\ 3 \\ 4 \end{array} \begin{array}{cccccc} 1 & 2 & 3 & 4 & 5 & 6 \\ \left[\begin{array}{cccccc} 0 & 0 & 0 & 0 & 0 & 0 \\ 0 & 0 & 1 & 1 & 0 & 0 \\ 0 & 0 & 1 & 1 & 0 & 0 \\ 0 & 0 & 0 & 0 & 0 & 0 \end{array}\right] \end{array}$$

Then, computing $\bar{A} \times Y$

$$\bar{A} = \frac{1}{1} + \frac{0}{2} + \frac{0}{3} + \frac{1}{4}$$

$$Y = \frac{1}{1} + \frac{1}{2} + \frac{1}{3} + \frac{1}{4} + \frac{1}{5} + \frac{1}{6}$$

$$\bar{A} \times Y = \begin{array}{c} \\ 1 \\ 2 \\ 3 \\ 4 \end{array} \begin{array}{cccccc} 1 & 2 & 3 & 4 & 5 & 6 \\ \left[\begin{array}{cccccc} 1 & 1 & 1 & 1 & 1 & 1 \\ 0 & 0 & 0 & 0 & 0 & 0 \\ 0 & 0 & 0 & 0 & 0 & 0 \\ 1 & 1 & 1 & 1 & 1 & 1 \end{array}\right] \end{array}$$

again, using pairwise min for the Cartesian product.

The deductive inference yields the following characteristic function in matrix form, following the relation

$$R = (A \times B) \cup (\bar{A} \times Y) = \begin{array}{c} \\ 1 \\ 2 \\ 3 \\ 4 \end{array} \begin{array}{cccccc} 1 & 2 & 3 & 4 & 5 & 6 \\ \left[\begin{array}{cccccc} 1 & 1 & 1 & 1 & 1 & 1 \\ 0 & 0 & 1 & 1 & 0 & 0 \\ 0 & 0 & 1 & 1 & 0 & 0 \\ 1 & 1 & 1 & 1 & 1 & 1 \end{array}\right] \end{array}$$

5.10 Fuzzy Logic

The third primary topic in this chapter is fuzzy logic, that is, how does the classical (predicate) logic extend to fuzzy domain? The extension of the above discussions to fuzzy deductive inference is straightforward. The fuzzy proposition $\underset{\sim}{P}$ has a value on the closed interval [0,1]. The truth value of a proposition $\underset{\sim}{P}$ is given by

$$T(\underset{\sim}{P}) = \mu_{\underset{\sim}{A}}(x) \text{ where } 0 \leq \mu_{\underset{\sim}{A}} \leq 1$$

Thus, the degree of truth for $\underset{\sim}{P} : x \in \underset{\sim}{A}$ is the membership grade of x in A. The logical connectives of negation, disjunction, conjunction, and implication are similarly defined for fuzzy logic, for example, disjunction.

Negation:

$$T(\overline{\underset{\sim}{P}}) = 1 - T(\underset{\sim}{P})$$

Disjunction:

$$\underset{\sim}{P} \vee \underset{\sim}{Q} \Rightarrow x \in \underset{\sim}{A} \text{ or } \underset{\sim}{B}$$

$$\text{Hence, } T(\underset{\sim}{P} \vee \underset{\sim}{Q}) = \max(T(\underset{\sim}{P}), T(\underset{\sim}{Q}))$$

Conjunction:

$$\underset{\sim}{P} \wedge \underset{\sim}{Q} \Rightarrow x \in \underset{\sim}{A} \text{ and } \underset{\sim}{B}$$

$$\text{Hence, } T(\underset{\sim}{P} \wedge \underset{\sim}{Q}) = \min(T(\underset{\sim}{P}), T(\underset{\sim}{Q}))$$

Implication:

$$\underset{\sim}{P} \to \underset{\sim}{Q} \Rightarrow x \text{ is } \underset{\sim}{A}, \text{ then } x \text{ is } \underset{\sim}{B}$$

$$T(\underset{\sim}{P} \to \underset{\sim}{Q}) = T(\overline{\underset{\sim}{P}} \vee \underset{\sim}{Q}) = \max(T(\overline{\underset{\sim}{P}}), T(\underset{\sim}{Q}))$$

Thus, a fuzzy logic implication would result in a fuzzy rule

$$\underset{\sim}{P} \to \underset{\sim}{Q} \Rightarrow \text{If } x \text{ is } \underset{\sim}{A}, \text{ then } y \text{ is } \underset{\sim}{B}$$

and the equivalent to the following fuzzy relation

$$\underset{\sim}{R} = (\underset{\sim}{A} \times \underset{\sim}{B}) \cup (\overline{\underset{\sim}{A}} \times Y)$$

with a grade membership function,

$$\mu_R = \max\{(\mu_A(x) \wedge \mu_B(y)), (1-\mu_A(x))\}$$

Example 5.6

Consider two universes of discourse described by $X = \{1, 2, 3, 4\}$ and $Y = \{1, 2, 3, 4, 5, 6\}$. Let two fuzzy sets $\underset{\sim}{A}$ and $\underset{\sim}{B}$ be given by

$$\underset{\sim}{A} = \frac{0.8}{2} + \frac{1}{3} + \frac{0.3}{4}$$

$$\underset{\sim}{B} = \frac{0.4}{2} + \frac{1}{3} + \frac{0.6}{4} + \frac{0.2}{5}$$

It is desired to find a fuzzy relation $\underset{\sim}{R}$ corresponding to IF $\underset{\sim}{A}'$, THEN $\underset{\sim}{B}'$.

Solution
Using the relation in Equation (5.37) would give

$$\underset{\sim}{A} \times \underset{\sim}{B} = \begin{array}{c} \\ 1 \\ 2 \\ 3 \\ 4 \end{array} \begin{array}{cccccc} 1 & 2 & 3 & 4 & 5 & 6 \\ \left[\begin{array}{cccccc} 0 & 0 & 0 & 0 & 0 & 0 \\ 0 & 0.4 & 0.8 & 0.6 & 0.2 & 0 \\ 0 & 0.4 & 1 & 0.6 & 0.2 & 0 \\ 0 & 0.3 & 0.3 & 0.3 & 0.2 & 0 \end{array}\right] \end{array}$$

$$\overline{\underset{\sim}{A}} \times Y = \begin{array}{c} \\ 1 \\ 2 \\ 3 \\ 4 \end{array} \begin{array}{cccccc} 1 & 2 & 3 & 4 & 5 & 6 \\ \left[\begin{array}{cccccc} 1 & 1 & 1 & 1 & 1 & 1 \\ 0.2 & 0.2 & 0.2 & 0.2 & 0.2 & 0.2 \\ 0 & 0 & 0 & 0 & 0 & 0 \\ 0.7 & 0.7 & 0.7 & 0.7 & 0.7 & 0.7 \end{array}\right] \end{array}$$

and, hence $\underset{\sim}{R} = \max\{\underset{\sim}{A} \times \underset{\sim}{B}, \overline{\underset{\sim}{A}} \times Y\}$

$$\underset{\sim}{R} = \begin{array}{c} \\ 1 \\ 2 \\ 3 \\ 4 \end{array} \begin{array}{cccccc} 1 & 2 & 3 & 4 & 5 & 6 \\ \left[\begin{array}{cccccc} 1 & 1 & 1 & 1 & 1 & 1 \\ 0.2 & 0.4 & 0.8 & 0.6 & 0.2 & 0.2 \\ 0 & 0.4 & 1 & 0.6 & 0.2 & 0 \\ 0.7 & 0.7 & 0.7 & 0.7 & 0.7 & 0.7 \end{array}\right] \end{array}$$

5.11 Approximate Reasoning

The primary goal of fuzzy systems is to formulate a theoretical foundation for reasoning about imprecise propositions, which is termed *approximate reasoning* in fuzzy logic technological systems [4,5].

Let us have a rule-based format to represent fuzzy information. These rules are expressed in conventional antecedent-consequent form, such as

$$\text{Rule 1: IF } x \text{ is } \underset{\sim}{A}, \text{ THEN } y \text{ is } \underset{\sim}{B}$$

where $\underset{\sim}{A}$ and $\underset{\sim}{B}$ represent fuzzy propositions (sets).

Now, let us introduce a new antecedent, say $\underset{\sim}{A}'$, and we consider the following rule:

$$\text{Rule 2: IF } x \text{ is } \underset{\sim}{A}', \text{ THEN } y \text{ is } \underset{\sim}{B}'$$

From the information derived from rule 1, is it possible to derive the consequent rule 2, $\underset{\sim}{B}'$? The answer is yes, and the procedure is a fuzzy composition. The consequent $\underset{\sim}{B}'$ can be found from the composition operation

$$\underset{\sim}{B}' = \underset{\sim}{A}' \circ \underset{\sim}{R}$$

Example 5.7

Reconsider the fuzzy system of Example 5.6. Let a new fuzzy set $\underset{\sim}{A}'$ be given by $\underset{\sim}{A}' = \frac{0.5}{1} + \frac{1}{2} + \frac{0.2}{3}$. It is desired to find an approximate reason (consequent) for the rule IF $\underset{\sim}{A}'$, THEN $\underset{\sim}{B}'$.

Solution

Relations (5.35) and (5.36) are used to determine $\underset{\sim}{B}'$.

$$\underset{\sim}{B}' = \underset{\sim}{A}' \circ \underset{\sim}{R} = [0.5 \; 0.5 \; 0.8 \; 0.6 \; 0.5 \; 0.5]$$

or

$$\underset{\sim}{B}' = \frac{0.5}{1} + \frac{0.5}{2} + \frac{0.8}{3} + \frac{0.6}{4} + \frac{0.5}{5} + \frac{0.5}{6}$$

where the composition is of the max-min form.

Note the inverse relation between fuzzy antecedents and fuzzy consequences arising from the composition operation. More accurately, if we have a fuzzy relation $\underset{\sim}{R} : \underset{\sim}{A} \to \underset{\sim}{B}$, then will the value of the composition $\underset{\sim}{A} \circ \underset{\sim}{R} = \underset{\sim}{B}$? The answer is no, and one should not expect an inverse to exist for fuzzy composition. This is not, however, the case in crisp logic, that is, $B' = A' \circ R = A \circ R = B$, where all these latter sets and relations are crisp [5,6]. The following example illustrates the nonexistence of the inverse.

Example 5.8

Let us reconsider the fuzzy system of Examples 5.6 and 5.7. Let $\underset{\sim}{A'} = \underset{\sim}{A}$ and evaluate $\underset{\sim}{B'}$.

Solution

We have

$$\underset{\sim}{B'} = \underset{\sim}{A'} \circ \underset{\sim}{R} = \underset{\sim}{A} \circ \underset{\sim}{R} = \frac{0.3}{1} + \frac{0.4}{2} + \frac{0.8}{3} + \frac{0.6}{4} + \frac{0.3}{5} + \frac{0.3}{6} \neq \underset{\sim}{B}$$

which yields a new consequent, since the inverse is not guaranteed. The reason for this situation is the fact that fuzzy inference is imprecise but approximate. The inference, in this situation, represents approximate linguistic characteristics of the relation between two universes of discourse.

5.12 Fuzzy Control

A common definition of a fuzzy control system is that it is a system that emulates a human expert. In this situation, the knowledge of the human operator would be put in the form of a set of fuzzy linguistic rules. These rules would produce an approximate decision, just as a human would. Consider Figure 5.16, where a block diagram of this definition is shown. As shown, the human operator observes quantities by observing the inputs, that is, reading a meter or measuring a chart, and performs a definite action (e.g., pushes a knob, turns on a switch, closes a gate, or replaces a fuse), thus leading to a crisp action, shown here by the output variable $y(t)$. The human operator can be replaced by a combination of a fuzzy rule-based system (FRBS) and a block called a *defuzzifier*. The input sensory (crisp or numerical) data are fed

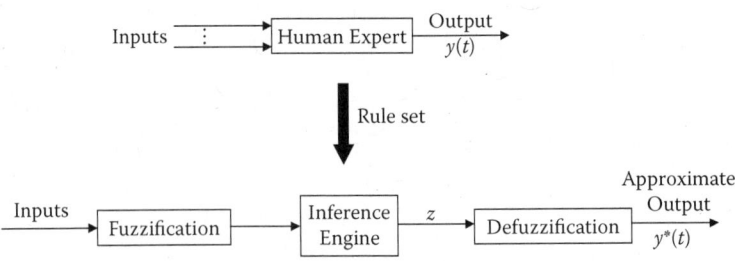

FIGURE 5.16
Conceptual definition of a fuzzy control system.

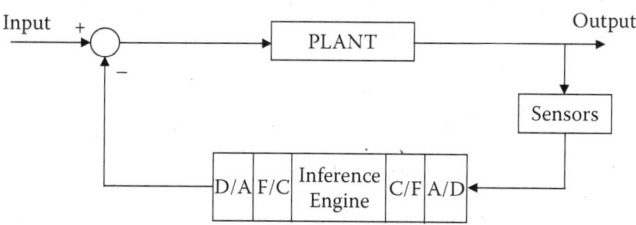

FIGURE 5.17
Block diagram for a laboratory implementation of a fuzzy controller.

into the FRBS, where physical quantities are represented or compressed into linguistic variables with appropriate membership functions. These linguistic variables are then used in the *antecedents* (IF part) of a set of fuzzy rules within an inference engine to result in a new set of fuzzy linguistic variables, or the *consequent* (THEN part). Variables are then denoted in this figure by z and are combined and changed to a crisp (numerical) output $y^*(t)$, which represents an approximation to actual output $y(t)$. It is therefore noted that a fuzzy controller consists of three operations: (1) fuzzification, (2) inference engine, and (3) defuzzification.

Before a formal description of fuzzification and defuzzification processes is made, let us consider a typical structure of a fuzzy control system, which is presented in Figure 5.17. As shown, the sensory data go through two levels of interface, that is, the analog to digital and the crisp to fuzzy, and at the other end in reverse order, that is, fuzzy to crisp and digital to analog.

Another structure for a fuzzy control system is a fuzzy inference, connected to a knowledge base, in a supervisory or adaptive mode. The structure is shown in Figure 5.18. As shown, a classical crisp controller (often an existing one) is left unchanged, but through a fuzzy inference engine or a fuzzy adaptation algorithm, the crisp controller is altered to cope with the system's unmodeled dynamics, disturbances, or plant parameter changes much like a standard adaptive control system. Here, the function $h(x)$ represents the unknown nonlinear controller or mapping function $h: e \circledR u$, which, along with any two input components, e_1 and e_2, of e, represents a nonlinear surface, sometimes known as the *control surface* [16].

The fuzzification operation, or the *fuzzifier* unit, represents a mapping from a crisp point $x = (x_1 x_2 \ldots x_n)^T \in X$ into a fuzzy set $A \in X$, where X is the universe of discourse and T denotes vector or matrix transposition.* There are normally two categories of fuzzifiers in use: singleton and nonsingleton.

---

* For convenience, in this chapter, the tilde (~) symbol that was used earlier to express fuzzy sets will be omitted.

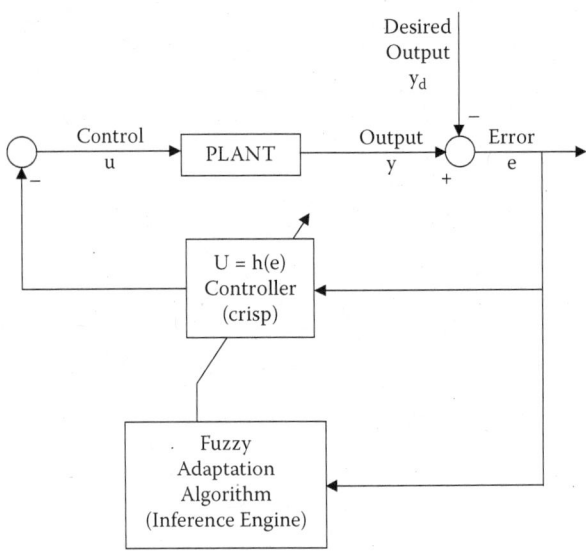

FIGURE 5.18
An adaptive (tuner) fuzzy control system, fuzzification.

A singleton fuzzifier has one point (value) x_p as its fuzzy set support, that is, the membership function is governed by the following relation:

$$\mu_A(x) = \begin{cases} 1, & x = x_p \in X \\ 0, & x \neq x_p \in X \end{cases}$$

The nonsingleton fuzzifiers are those in which the support is more than a point. Examples of these fuzzifiers are triangular, trapezoidal, Gaussian, etc. In these fuzzifiers, $\mu_A(x) = 1$ at $x = x_p$, where x_p may be one or more than one point, and then $\mu_A(x)$ decreases from 1 as x moves away from x_p or the "core" region to which x_p belongs, such that $\mu_A(x_p)$ remains 1 (see Section 5.12.5). For example, the following relation represents a Gaussian-type fuzzifier:

$$\mu_A(x) = \exp\left\{-\frac{(x-x_p)^T(x-x_p)}{\sigma^2}\right\}$$

where the variance, σ^2, is a parameter characterizing the shape of $\mu_A(x)$.

5.12.1 Inference Engine

The cornerstone of any expert controller is its inference engine, which consists of a set of expert rules, which reflect the knowledge base and reasoning structure of the solution of any problem. A fuzzy (expert) control system is no exception, and its rule base is the heart of the nonlinear fuzzy controller. A typical fuzzy rule can be composed as [8]

$$\text{IF } A \text{ If } A \text{ is } A_1 \text{ AND } B \text{ is } B_1 \text{ OR } C \text{ is } C_1 \text{ THEN } U \text{ is } U_1 \qquad (5.38)$$

where A, B, C, and U are fuzzy variables; A_1, B_1, C_1, and U_1 are fuzzy linquistic values (membership functions or fuzzy linguistic labels); and "AND," "OR," and "NOT" are connectives of the rule. The rule in Equation (5.38) has three antecedents and one consequent. Typical fuzzy variables may, in fact, represent physical or system quantities such as "temperature," "pressure," "output," "elevation," etc., and typical fuzzy linguistic values (labels) may be "hot," "very high," "low," etc. The portion "very" in a label "very high" is called a *linquistic hedge*. Other examples of a hedge are "much," "slightly," "more," "less," etc. The above rule is known as Mamdani-type rule. Under Mamdani rules, the antecedents and the consequent parts of the rule are expressed using linguistic labels. In general, in fuzzy system theory, there are many forms and variations of fuzzy rules, some of which will be introduced here and throughout the chapter. Another form is *Takagi-Sugeno* rules, under which the consequent part is expressed as an analytical expression or equation.

Two cases will be used here to illustrate the process of inferencing graphically. In the first case, the inputs to the system are crisp values, and we use the max-min inference method. In the second case, the inputs to the system are also crisp, but we use the max-product inference method. Please keep in mind that there could also be cases where the inputs are fuzzy variables.

Consider the following rule, whose consequent is not a fuzzy implication

$$\text{IF } x_1 \text{ is } A_1^i \text{ AND } x_2 \text{ is } A_2^i \text{ THEN } y^i \text{ is } B^i, \text{ for } i = 1, 2, \ldots, l, \qquad (5.39)$$

where A_1^i and A_2^i are the fuzzy sets representing the ith-antecedent pairs, and B^i are the fuzzy sets representing the ith consequent, and l is the number of rules.

Case 5.1: Inputs x_1 and x_2 are crisp values and the max-min inference method is used. Based on the Mamdani implication method of inference, and for a set of *disjunctive rules*, that is, rules connected by the OR connective, the aggregated output for the l rules presented in Equation (5.4) will be given by

$$\mu_{B^i}(y) = \max_i [\min [\mu_{A_1^i}(x_1), \mu_{A_2^i}(x_2)]], \text{ for } i = 1, 2, \ldots, l$$

Figure 5.19 shows a graphical illustration of Mamdani-type rules using max-min inference, for $l = 2$, where A_1^1 and A_2^1 refer to the first and second fuzzy antecedents of the first rule, respectively, and B^1 refers to the fuzzy consequent of the first rule. Similarly, A_1^2 and A_2^2 refer to the first and second fuzzy antecedents of the second rule, respectively, and B^2 refers to the fuzzy consequent of the second rule. Because the antecedent pairs used in general form

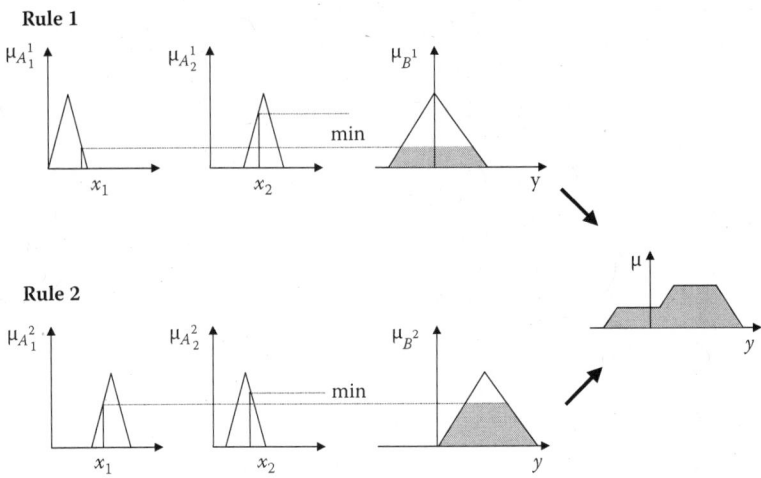

FIGURE 5.19
Graphical execution of a min-max inference in a Mamdani rule.

presented in Equation (5.39) are connected by a logical *AND*, the minimum function is used. For each rule, the minimum value of the antecedent propagates through and truncates the membership function for the consequent. This is done graphically for each rule. Assuming that the rules are disjunctive, the aggregation operation *max* results in an aggregated membership function consisting of the outer envelope of the individual truncated membership forms from each rule. To compute the final crisp value of the aggregated output, defuzzification is used, which will be explained in the next section.

Case 5.2: Inputs x_1 and x_2 are crisp values, and the max-product inference method is used. Based on the Mamdani implication method of inference, and for a set of *disjunctive rules*, the aggregated output for the l rules presented in Equation (10.4) will be given by

$$\mu_{B^i}(y) = \max_i \; [\mu_{A_1^i}(x_1) \cdot \mu_{A_2^i}(x_2)], \text{ for } i = 1, 2, \ldots, l \qquad (5.40)$$

Figure 5.20 is a graphical illustration of Equation (5.40), for $l = 2$, where A_1^1 and A_2^1 refer to the first and second fuzzy antecedents of the first rule, respectively, and B^1 refers to the fuzzy consequent of the first rule. Similarly, A_1^2 and A_2^2 refer to the first and second fuzzy antecedents of the second rule, respectively, and B^2 refers to the fuzzy consequent of the second rule. Since the antecedent pairs used in general form presented in Equation (5.39) are connected by a logical *AND*, the minimum function is used again. For each rule, the minimum value of the antecedent propagates through and scales the membership function for the consequent. This is done graphically for each rule. Similar to the first case, the aggregation operation *max* results in

Fuzzy Systems—Sets, Logic, and Control

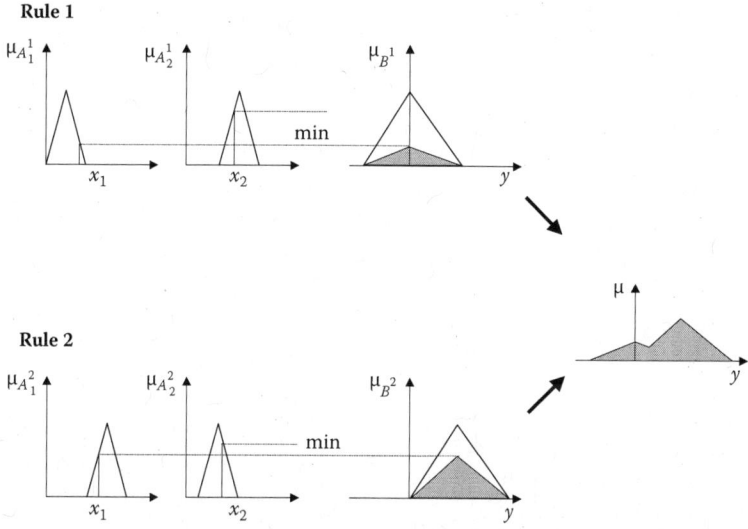

FIGURE 5.20
Graphical execution of a min-product inference in a Mamdani rule.

an aggregated membership function consisting of the outer envelope of the individual truncated membership forms from each rule. To compute the final crisp value of the aggregated output, defuzzification is used (Figure 5.20).

5.12.2 Defuzzification

Defuzzification is the third important element of any fuzzy controller. In this section, only the *center of gravity defuzzifier*, which is the most common one, is discussed. In this method, the weighted average of the membership function or the center of gravity of the area bounded by the membership function curve is computed as the most typical crisp value of the union of all output fuzzy sets:

$$y_c = \frac{\int y \cdot \mu_A(y) dy}{\int \mu_A(y) dy}$$

5.12.3 Fuzzy Control Design

One of the first steps in the design of any fuzzy controller is to develop a knowledge base for the system to eventually lead to an initial set of rules. There are at least five different methods to generate a fuzzy rule base [4]:

1. Simulate the closed-loop system through its mathematical model.
2. Interview an operator who has had many years of experience controlling the system.

3. Generate rules through an algorithm using numerical input/output data of the system.
4. Use learning or optimization methods such as neural networks or genetic algorithms to create the rules.
5. In the absence of all of the above, if a system does exist, experiment with it in the laboratory or factory setting and gradually gain enough experience to create the initial set of rules.

Example 5.9

Consider the linearized model of an inverted pendulum Figure 5.21, described by the equation given below:

$$\dot{x} = \begin{pmatrix} 0 & 1 \\ 15.79 & 0 \end{pmatrix} x + \begin{pmatrix} 0 \\ 1.46 \end{pmatrix} u$$

with $l = 0.5$ m, $m = 100$ g, and initial conditions $x^T(0) = [\theta(0) \; \dot{\theta}(0)]^T = [1 \; 0]^T$. It is desired to stabilize the system using fuzzy rules.

Clearly, this system is unstable and a controller is needed to stabilize it. To generate the rules for this problem, only common sense is needed, that is, if the pole is falling in one direction, then push the cart in the same direction to counter the movement of the pole. To put this into rules of the form Equation (5.39), we obtain the following:

IF θ is θ_Positive AND $\dot{\theta}$ is $\dot{\theta}$_Positive THEN u is u_Negative

IF θ is θ_Negative AND $\dot{\theta}$ is $\dot{\theta}$_Negative THEN u is u_Positive

where the membership functions described above are defined in Figure 5.22.

As shown in Figure 5.22, the membership functions for the inputs are half-triangular, whereas the membership function of the output is

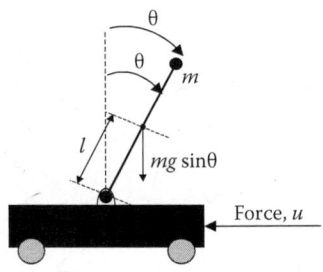

FIGURE 5.21
Inverted pendulum.

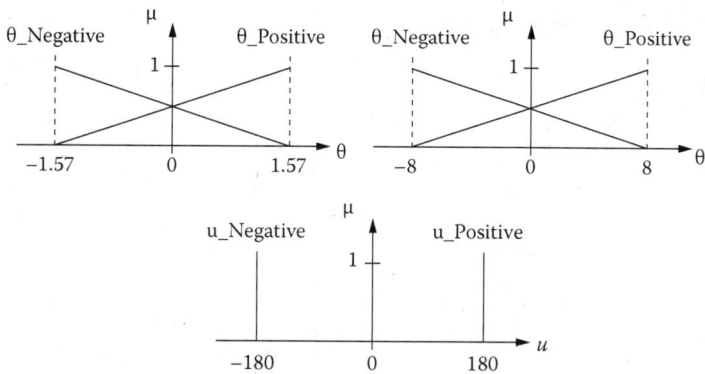

FIGURE 5.22
Membership functions for the inverted pendulum problem.

singleton. By simulating the system with the fuzzy controller, we obtain the response shown in Figure 5.23. It is clear that the system is stable. In this example, only two rules were used, but more rules could be added to obtain a better response, that is, less undershoot.

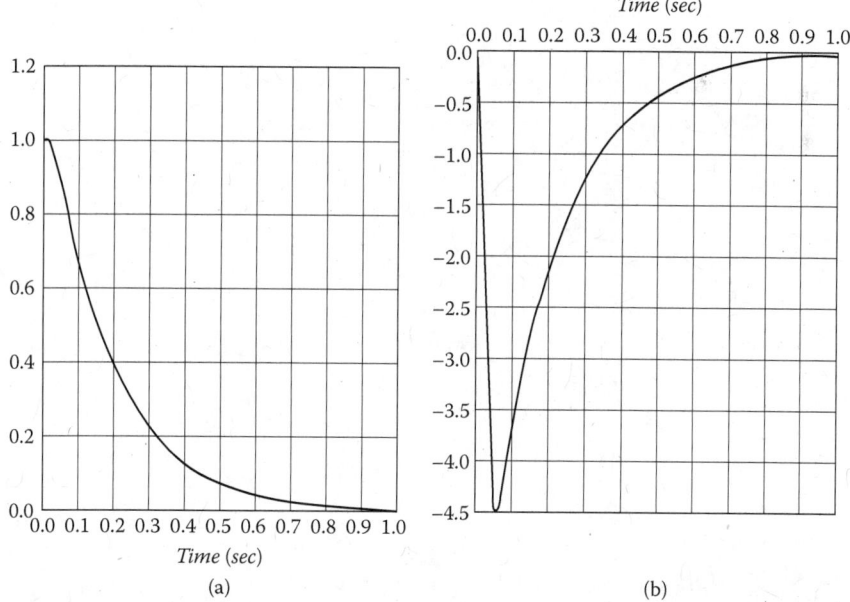

FIGURE 5.23
Simulation results for Example 5.9: (a) $\theta(t)$ and (b) $\dot{\theta}(t)$.

5.12.4 Analysis of Fuzzy Control Systems

In this section, some results of Tanaka and Sugeno [14] with respect to analysis of feedback fuzzy control systems will be briefly discussed. This section would use Takagi-Sugeno models to develop fuzzy block diagrams and fuzzy closed-loop models.

Consider a typical Takagi-Sugeno fuzzy plant model represented by implication P^i in Figure 5.23.

$$P^i: \text{IF } x(k) \text{ is } A_1^i \text{ AND } \ldots x(k-n+1) \text{ is } A_n^i \text{ AND}$$

$$u(k) \text{ is } B_1^i \text{ AND } \ldots \text{ AND } u(k-m+1) \text{ is } B_n^i$$

$$\text{THEN } x^i(k+1) = a_0^i + a_1^i x(k) + \cdots + a_n^i x(k-n+1)$$

$$+ b_1^i u(k) + \cdots + b_n^i u(k-m+1)$$

where P_i ($i = 1, 2, \ldots, l$) is the ith implication; l is the total number of implications; a_p^i ($p = 1, 2, \ldots, n$) and b_q^i ($q = 1, 2, \ldots, m$) are constant consequent parameters; k is time sample, $x(k), \ldots, x(k-n+1)$ are input variables; and n and m are the number of antecedents for states and inputs, respectively. The terms A_p^i and B_p^i are fuzzy sets with piecewise-continuous polynomial (PCP) membership functions. PCP is defined as follows.

DEFINITION 5.1:
A fuzzy set A satisfying the following properties is said to be a PCP membership function $A(x)$ [13]:

$$A(x) = \begin{cases} \mu_1(x), & x \in [p_0, p_1] \\ \vdots \\ \mu_s(x), & x \in [p_{s-1}, p_s] \end{cases}$$

where $\mu_i(x) \in [0,1]$ for $x \in [p_{i-1}, p_i]$, $i = 1, 2, \ldots, s$, and $-p_0 p_1 p_{s-1}, p_s$

$$\mu_i(x) = \sum_{j=0}^{n_i} c_j^i x^j$$

where c_j^i are known parameters of polynomials $\mu_i(x)$.

Given the inputs

$$\begin{aligned} \mathbf{x}(k) &\equiv [x(k) \quad x(k-1) \ldots x(k-n+1)]^T \\ \mathbf{u}(k) &\equiv [u(k) \quad u(k-1) \ldots u(k-m+1)]^T \end{aligned} \quad (5.41)$$

Fuzzy Systems—Sets, Logic, and Control

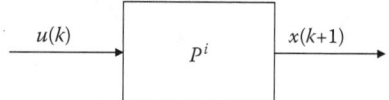

FIGURE 5.24
Single-input, single-output fuzzy block represented by ith implication P^i.

Using the above vector notation, Equation (5.41) can be represented in the following form:

$$P^i: \text{IF } \mathbf{x}(k) \text{ is } \mathbf{A}^i \text{ AND } \mathbf{u}(k) \text{ is } \mathbf{B}^i$$

$$\text{THEN } x^i(k+1) = a_0^i + \sum_{p=1}^{n} a_p^i x(k-p+1) + \sum_{q=1}^{m} b_q^i u(k-q+1) \quad (5.42)$$

where $\mathbf{A}^i \equiv [A_1^i \ A_2^i \ \ldots \ A_n^i]^T$, $\mathbf{B}^i \equiv [B_1^i \ B_2^i \ \ldots \ B_m^i]^T$, and "$\mathbf{x}(k)$ is $\mathbf{A}^i$" are equivalent to antecedent "$x(k)$ is A_1^i AND $\ldots$ $x(k-n+1)$ is A_n^i."

The final defuzzified output of the inference is given by a weighted average of $x^i(k+1)$ values:

$$x(k+1) = \frac{\sum_{i=1}^{l} w^i x^i(k+1)}{\sum_{i=1}^{l} w^i} \quad (5.43)$$

where it is assumed that the denominator of Equation (5.43) is positive, and $x^i(k+1)$ is calculated from the ith implication, and the weight w^i refers to the overall truth value of the ith implication premise for the inputs in Equation (5.42).

Since the product of two PCP fuzzy sets can be considered a series connection of two fuzzy blocks of the type in Figure 5.24, it is concluded that the convexity of fuzzy sets in succession is not preserved in general. Now, let us consider a fuzzy control system whose plant model and controller are represented by fuzzy implications as depicted in Figure 5.25. In

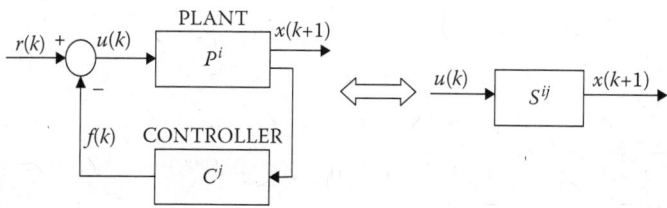

FIGURE 5.25
A fuzzy control system depicted by two implications and its equivalent implication [13].

this figure, $r(k)$ represents a reference input. The plant implication P^i is already defined by Equation (5.42), whereas the controller's jth implication is given by

$$C^j: \text{IF } \mathbf{x}(k) \text{ is } \mathbf{D}^j \text{ AND } \mathbf{u}(k) \text{ is } \mathbf{F}^j$$

$$\text{THEN } f^j(k+1) = c_0^j + \sum_{p=1}^{n} c_p^j x(k-p+1)$$

where $\mathbf{D}^j \equiv [D_1^j \; D_2^j \; \ldots \; D_n^j]^T; \mathbf{F}^i \equiv [F_1^i \; F_2^i \; \ldots \; F_m^i]^T$; and, of course, $u(k) = r(k) - f(k)$. The equivalent implication S^{ij} is given by

$$S^{ij}: \text{IF } \mathbf{x}(k) \text{ is } (\mathbf{A}^i \text{ AND } \mathbf{D}^j) \text{ AND } \mathbf{v}^*(k) \text{ is } (\mathbf{B}^i \text{ AND } \mathbf{F}^j)$$

$$\text{THEN } x^{ij}(k+1) = a_0^i - b^i c_0^j + b^i r(k) +$$

$$\sum_{p=1}^{n} (a_p^i - b^i c_p^j) x(k-p+1)$$

where $i = 1, \ldots, l_1, j = 1, \ldots, l_2$, and l_1 and l_2 are the total number of implications for the plant and the controller, respectively. The term $v^*(k)$ is defined by

$$v^*(k) = \big[r(k) - e^*(x(k)), \; r(k-1) - e^*(x(k-1))$$

$$\ldots, r(k-m+1) - e^*(x(k-m+1)) \big]^T$$

where $e^*(x)$ is the input-output mapping function of block C^j in Figure 5.25, that is, $f(k) = e^*(x(k))$.

Example 5.10

Consider a fuzzy feedback control system of the type shown in Figure 5.25 with the following implications:

P^1: IF $x(k)$ is A^1 THEN $x^1(k+1) = 1.85x(k) - 0.65x(k-1) + 0.35u(k)$

P^2: IF $x(k)$ is A^2 THEN $x^2(k+1) = 2.56x(k) - 0.135x(k-1) + 2.22u(k)$

C^1: IF $x(k)$ is D^1 THEN $f^1(k+1) = k_1^1 x(k) - k_2^1 x(k-1)$

C^2: IF $x(k)$ is D^2 THEN $f^2(k+1) = k_1^2 x(k) - k_2^2 x(k-1)$

It is desired to find the closed-loop implications S^{ij}, $i = 1, 2$ and $j = 1, 2$.

Solution

Noting that $u(k) = r(k) - f(k)$ in Figure 5.25 and the implications in Equation (5.15), we have

S^{11}: IF $x(k)$ is $(A^1$ AND $D^1)$ THEN $x^{11}(k+1) = \left(1.85 - 0.35k_1^1\right)x(k)$

$\quad + \left(-0.65 - 0.35k_2^1\right)x(k-1) + 0.35r(k)$

S^{12}: IF $x(k)$ is $(A^1$ AND $D^2)$ THEN $x^{12}(k+1) = \left(1.85 - 0.35k_1^2\right)x(k)$

$\quad + \left(-0.65 - 0.35k_2^2\right)x(k-1) + 0.35r(k)$

S^{21}: IF $x(k)$ is $(A^2$ AND $D^1)$ THEN $x^{21}(k+1) = \left(2.56 - 2.22k_1^1\right)x(k)$

$\quad + \left(-0.135 - 2.22k_2^1\right)x(k-1) + 2.22r(k)$

S^{22}: IF $x(k)$ is $(A^2$ AND $D^2)$ THEN $x^{22}(k+1) = \left(2.56 - 2.22k_1^2\right)x(k)$

$\quad + (-0.135 - 2.22k_2^2)x(k-1) + 2.22r(k)$

5.12.5 Stability of Fuzzy Control Systems

One of the most important issues in any control system fuzzy or otherwise is stability. Briefly, a system is said to be *stable* if it would come to its equilibrium state after any external input, initial conditions, and/or disturbances have impressed the system. The issue of stability is of even greater relevance when questions of safety, lives, and environment are at stake, as in such systems as nuclear reactors, traffic systems, and airplane autopilots. The stability test for fuzzy control systems, or lack of it, has been a subject of criticism by many control engineers in some control engineering literature [6].

Almost any linear or nonlinear system under the influence of a closed-loop crisp controller has one type of stability test or another. For example, the stability of a linear time-invariant system can be tested by a wide variety of methods such as Routh-Hurwitz, root locus, Bode plots, Nyquist criterion, and even via traditionally nonlinear systems methods of Lyapunov, Popov, and circle criterion. The common requirement in all these tests is the availability of a mathematical model, either in time or frequency domain. A reliable mathematical model for a very complex and large-scale system may, in practice, be unavailable or not feasible. In such cases, a fuzzy controller may be designed based on expert knowledge or experimental practice. However, the issue of the stability of a fuzzy control system remains and must be addressed. The aim of this section is to present an up-to-date survey of available techniques and tests for fuzzy control system stability.

From the viewpoint of stability, a fuzzy controller can be either acting as a conventional (low-level) controller or as a supervisory (high-level) controller. Depending on the existence and nature of a system's mathematical model and the level in which fuzzy rules are being utilized for control and

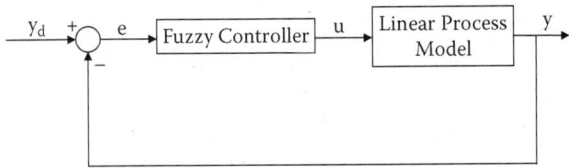

FIGURE 5.26
Class 1 of fuzzy control system stability problem.

robustness, four classes of fuzzy control stability problems can be distinguished. These four classes are:

- Class 1: Process model is crisp and linear and fuzzy controller is low level.
- Class 2: Process model is crisp and nonlinear and the fuzzy controller is low level.
- Class 3: Process model (linear or nonlinear) is crisp and a fuzzy tuner or an adaptive fuzzy controller is present at high level.
- Class 4: Process model is fuzzy and fuzzy controller is low level.

Figures 5.26 through 5.29 show all four classes of fuzzy control systems whose stability is of concern. Here, we are concerned mainly with the first three classes. For the last class, traditional nonlinear control theory could fail and is beyond the scope of this section. It will be discussed very briefly. The techniques for testing the stability of the first two classes of systems (Figures 5.26 and 5.27) are divided into two main groups: time and frequency.

5.12.5.1 Time-Domain Methods

The state-space approach has been considered by many authors [16–24]. The basic approach here is to subdivide the state space into a finite number of cells based on the definitions of the membership functions. Now, if a separate rule is defined for every cell, a cell-to-cell trajectory can be constructed

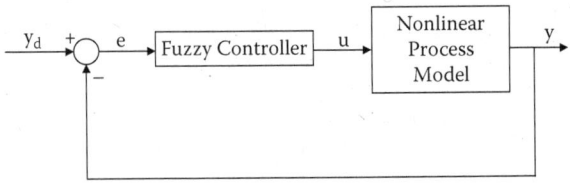

FIGURE 5.27
Class 2 of fuzzy control system stability problem.

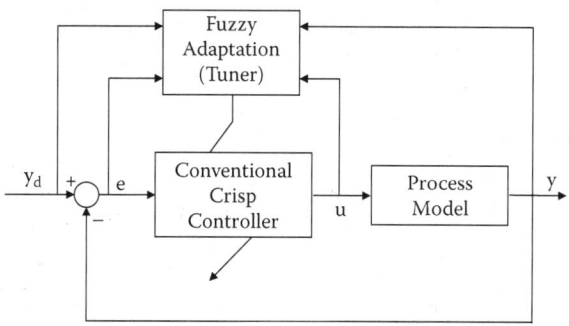

FIGURE 5.28
Class 3 of fuzzy control system stability problem.

from the system's output induced by the new outputs of the fuzzy controller. If every cell of the modified state space is checked, one can identify all the equilibrium points, including the system's stable region. This method should be used with some care since the inaccuracies in the modified description could cause oscillatory phenomenon around the equilibrium points.

The second class of methods is based on Lyapunov's method. Several authors [19,20,22,23,25–32] have used this theory to come up with the criterion for stability of fuzzy control systems. The approach shows that the time derivative of the Lyapunov function at the equilibrium point is negative semi-definite. Many approaches have been proposed. One approach is to define a Lyapunov function and then derive the fuzzy controller's architecture out of the stability conditions. Another approach uses Aiserman's method [16] to find an adopted Lyapunov function, while representing the fuzzy controller by a nonlinear algebraic function $u = f(y)$, when y is the system's output. A third method calls for the use of so-called *facet functions*, where the fuzzy controller is realized by box-wise multilinear facet functions with the system being described by a state space model. To test stability, a numerical parameter optimization scheme is needed.

The *hyperstability* approach, considered by other authors [42–44], has been used to check stability of systems depicted in Figure 5.26. The basic approach here is to restrict the input-output behavior of the nonlinear fuzzy controller

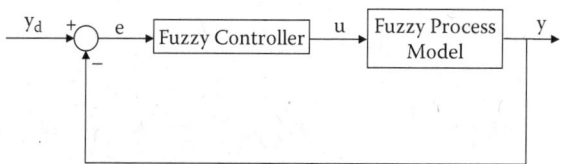

FIGURE 5.29
Class 4 of fuzzy control system stability problem.

by inequality and to derive conditions for the linear part of the closed-loop system to be satisfied for stability.

Bifurcation theory [20] can be used to check stability of fuzzy control systems of the class described in Figure 5.27. This approach represents a tool in deriving stability conditions and robustness indices for stability from small gain theory. The fuzzy controller, in this case, is described by a nonlinear vector function. The stability in this scheme could only be lost if one of the following conditions becomes true: (1) the origin becomes unstable if a pole crosses the imaginary axis into the right half-plane, static bifurcation; (2) the origin becomes unstable if a pair of poles would cross over the imaginary axis and assumes positive real parts, Hopf bifurcation; or (3) new additional equilibrium points are produced. The last time-domain method is the use of graph theory [13]. In this approach, conditions for special nonlinearities are derived to test the bounded input-bounded output stability.

5.12.5.2 Frequency-Domain Methods

There are three primary groups of methods that have been considered here. The harmonic balance approach, considered in references [36–38], among others, has been used to check the stability of the first two classes of fuzzy control systems (see Figures 5.26 and 5.27). The main idea is to check if permanent oscillations occur in the system and whether these oscillations with known amplitude or frequency are stable. The nonlinearity (fuzzy controller) is described by a complex-valued describing function, and the condition of harmonic balance is tested. If this condition is satisfied, then a permanent oscillation exists. This approach is equally applicable to multiple input-multiple output systems.

The circle criterion [17,34,39,40] and Popov criterion [41,42] have been used to check stability of the first class of systems (Figure 5.26). In both criteria, certain conditions on the linear process model and static nonlinearity (controller) must be satisfied. It is assumed that the characteristic value of the nonlinearity remains within certain bounds, and the linear process model must be open-loop stable with proper transfer function. Both criteria can be graphically evaluated in simple manners. A summary of many stability approaches for fuzzy control systems has been presented by Jamshidi [13].

5.12.6 Lyapunov Stability

One of the most fundamental criteria of any control system is to ensure stability as part of the design process. In this section, some theoretical results on this important topic are detailed.

We begin with the ith Takagi-Sugeno implication of a fuzzy system:

$$P^i: \text{IF } x(k) \text{ is } A_1^i \text{ AND } \ldots x(k-n+1) \text{ is } A_n^i \tag{5.44}$$

$$\text{THEN } x^i(k+1) = a_0^i + a_1^i x(k) + \cdots + a_n^i x(k-n+1)$$

with $i = 1, \ldots, l$. It is noted that this implication is similar to Equation (5.42) except that, since we are dealing with Lyapunov stability, the inputs $u(k)$ are absent. The stability of a fuzzy control system with the presence of the inputs will be considered shortly. The consequent part of Equation (5.44) represents a set of linear subsystems and can be rewritten as [14]

$$P^i: \text{IF } x(k) \text{ is } A_1^i \text{ AND } \ldots x(k-n+1) \text{ is } A_n^i$$
$$\text{THEN } \mathbf{x}(k+1) = \mathbf{A}_i \mathbf{x}(k) \tag{5.45}$$

where $\mathbf{x}(k)$ is defined by Equation (5.41) and $n \times n$ $\mathbf{A}_i$ is

$$\mathbf{A}_i = \begin{bmatrix} a_1^i & a_2^i & \cdots & a_{n-1}^i & a_n^i \\ 1 & 0 & \cdots & 0 & 0 \\ 0 & 1 & \cdots & 0 & 0 \\ \vdots & \vdots & \ddots & \vdots & \vdots \\ 0 & 0 & \cdots & 1 & 0 \end{bmatrix} \tag{5.46}$$

The output of the fuzzy system described by Equations (5.45) and (5.46) is given by

$$\mathbf{x}(k+1) = \frac{\sum_{i=1}^{l} w^i \mathbf{A}_i \mathbf{x}(k)}{\sum_{i=1}^{l} w^i} \tag{5.47}$$

where w^i is the overall truth value of the ith implication and l is the total number of implications. Using this notation, we then present the first stability result of fuzzy control systems [14].

THEOREM 5.1

The equilibrium point of the fuzzy system Equation (5.47) is globally asymptotically stable if there exists a common positive definite matrix $\mathbf{P}$ for all subsystems such that

$$\mathbf{A}_i^T \mathbf{P} \mathbf{A}_i - \mathbf{P} < 0 \quad \text{for} \quad i = 1, \ldots, l \tag{5.48}$$

It is noted that this theorem can be applied to any nonlinear system that can be approximated by a piecewise linear function if the stability condition (5.48) is satisfied. Moreover, if there exists a common positive definite matrix $\mathbf{P}$, then all $\mathbf{A}_i$ matrices are stable. Since Theorem 5.1 is a sufficient condition for stability, it is possible not to find a $\mathbf{P} > 0$ even if all the $\mathbf{A}_i$ matrices are stable. In other words, a fuzzy system may be globally asymptotically stable even if a $\mathbf{P} > 0$ is not found. The fuzzy system is not always stable even if all $\mathbf{A}_i$ matrices are stable.

THEOREM 5.2

Let $\mathbf{A}_i$ be stable and nonsingular matrices for $i = 1, \ldots, l$. Then, $\mathbf{A}_i \mathbf{A}_j$ are stable matrices for $i, j, = 1, \ldots, l$, if there exists a common positive definite matrix $\mathbf{P}$ such that

$$\mathbf{A}_i^T \mathbf{P} \mathbf{A}_i - \mathbf{P} < 0 \quad \text{for} \quad i = 1, \ldots, l$$

Example 5.11

Consider the following fuzzy system

P^1: IF $x(k)$ is A^1 THEN $x^1(k+1) = 1.2x(k) - 0.6x(k-1)$

P^2: IF $x(k)$ is A^2 THEN $x^2(k+1) = x(k) - 0.4x(k-1)$

where A^i are fuzzy sets shown in Figure 5.30. It is desired to check the stability of this system.

Solution

The two subsystems' matrices are

$$\mathbf{A}_1 = \begin{pmatrix} 1.2 & -0.6 \\ 1 & 0 \end{pmatrix}, \quad \mathbf{A}_2 = \begin{pmatrix} 1 & -0.4 \\ 1 & 0 \end{pmatrix}$$

The product of matrix $\mathbf{A}_1 \mathbf{A}_2$ is

$$\mathbf{A}_1 \mathbf{A}_2 = \begin{pmatrix} 0.6 & -0.48 \\ 1 & -0.4 \end{pmatrix}$$

whose eigenvalues are $\lambda_{1,2} = 0.1 \pm j0.48$, which indicates that $\mathbf{A}_1 \mathbf{A}_2$ is a stable matrix. Thus, by Theorem 5.2, a common $\mathbf{P}$ exists, and if we use $\mathbf{P}$ with the following

$$\mathbf{P} = \begin{pmatrix} 2 & -1.2 \\ -1.2 & 1 \end{pmatrix}$$

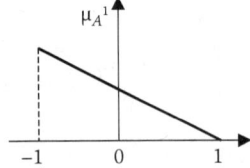
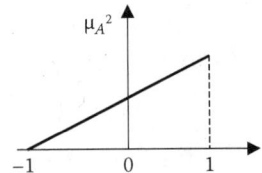

FIGURE 5.30
Fuzzy sets for Example 5.11.

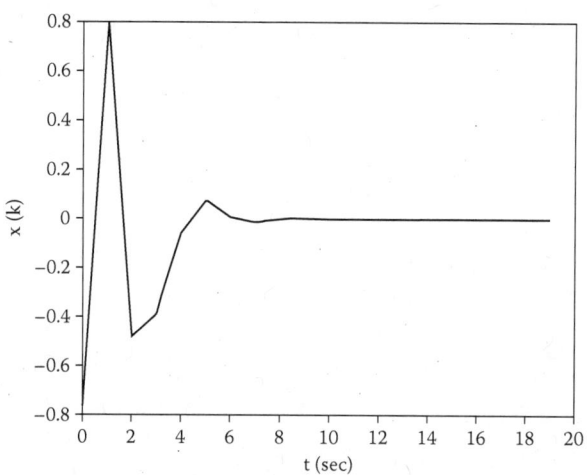

FIGURE 5.31
Simulation result for Example 5.11.

then both equations $\mathbf{A}_i^T \mathbf{P} \mathbf{A}_i - \mathbf{P} < 0$ for $i = 1, 2$ are simultaneously satisfied. This result was also verified using simulation. Figure 5.31 shows the simulation result, which is clearly stable.

Thus far, the criteria that have been presented treat autonomous (either closed-loop or no-input) systems. Consider the following nonautonomous fuzzy system:

P^i: IF $x(k)$ is A_1^i AND ... AND $x(k-n+1)$ is A_n^i AND

$u(k)$ is B_1^i AND ... AND $u(k-m+1)$ is B_m^i

THEN $x^i(k+1) = a_0^i + a_1^i x(k) + \cdots + a_n^i x(k-n+1) +$

$b_1^i u(k) + \cdots + b_m^i x(k-m+1)$

Here, we use some results from Sheikholeslam [43] and Tahani and Sheikholeslam [32] to test the stability of the above system. We begin with a definition.

DEFINITION 5.2
The nonlinear system

$$\mathbf{x}(k+1) = \mathbf{f}[\mathbf{x}(k), \mathbf{u}(k), k], \quad \mathbf{y} = \mathbf{g}[\mathbf{x}(k), \mathbf{u}(k), k]$$

is *totally stable* if and only if, for any bounded input $\mathbf{u}(k)$ and bounded initial state $\mathbf{x}_0$, the state $\mathbf{x}(k)$ and the output $\mathbf{y}(k)$ of the system are bounded, that is, we have

For all $\|\mathbf{x}_0\| < \infty$ and for all $\|\mathbf{u}(k)\| < \infty \Rightarrow \|\mathbf{x}(k)\| < \infty$ and $\|\mathbf{y}(k)\| < \infty$

Now, we consider the following theorem.

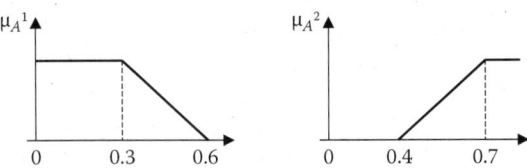

FIGURE 5.32
Fuzzy sets for Example 10.4.

THEOREM 5.3
The fuzzy system Equation (5.23) is totally stable if there exists a common positive definite matrix **P** such that the following inequalities hold:

$$\mathbf{A}_i^T \mathbf{P} \mathbf{A}_i - \mathbf{P} < 0 \quad \text{for } i = 1, \ldots, l$$

where $\mathbf{A}_i$ is defined by Equation (5.19). The proof of this theorem can be found in the work of Sheikholeslam [43].

Example 5.12

Consider the following fuzzy system:

P^1: IF $x(k)$ is A^1 THEN $x^1(k+1) = 0.85x(k) - 0.25x(k-1) + 0.35u(k)$

P^2: IF $x(k)$ is A^2 THEN $x^2(k+1) = 0.56x(k) - 0.25x(k-1) + 2.22u(k)$

where A_i are fuzzy sets shown in Figure 5.32. It is desired to check the stability of this system. Assume that the input $u(k)$ is bounded.

Solution
The two subsystems' matrices are

$$\mathbf{A}_1 = \begin{pmatrix} 0.85 & -0.25 \\ 1 & 0 \end{pmatrix}, \quad \mathbf{A}_2 = \begin{pmatrix} 0.56 & -0.25 \\ 1 & 0 \end{pmatrix}$$

If we choose the positive definite matrix **P**

$$\mathbf{P} = \begin{pmatrix} 3 & -1 \\ -1 & 1 \end{pmatrix}$$

then it can be easily verified that the system is totally stable. The product of matrix $\mathbf{A}_1 \mathbf{A}_2$ is

$$\mathbf{A}_1 \mathbf{A}_2 = \begin{pmatrix} 0.23 & -0.21 \\ 0.56 & -0.25 \end{pmatrix}$$

The eigenvalues of product of matrix $\mathbf{A}_1 \mathbf{A}_2$ eigenvalues are $\lambda_{1,2} = 0.012 \pm j\, 0.25$, which indicates that $\mathbf{A}_1 \mathbf{A}_2$ is a stable matrix.

5.12.7 Stability via Interval Matrix Method

Some results on the stability of time-varying discrete interval matrices by Han and Lee [44] can lead us to some more conservative, but computationally more convenient, stability criteria for fuzzy systems of the Takagi-Sugeno type shown by Equation (5.47). Before we can state these new criteria, some preliminary discussion will be necessary.

Consider a linear discrete time system described by a difference equation in state form:

$$x(k+1) = (A + G(k))x(k), \quad x(0) = x_0 \quad (5.48)$$

where A is an $n \times n$ constant asymptotically stable matrix, x is the $n \times 1$ state vector, and $G(k)$ is an unknown $n \times n$ time-varying matrix on the perturbation matrix's maximum modulus, that is,

$$|G(k)| \le G_m, \quad \text{for all } k \quad (5.49)$$

where the $|\cdot|$ represents the matrix with modulus elements and the inequality holds element-wise. Now, consider the following theorem.

THEOREM 5.4
The time-varying discrete time system Equation (10.27) is asymptotically stable if

$$\rho(|A| + G_m) < 1 \quad (5.50)$$

where $\rho(x)$ stands for the spectral radius of the matrix. The proof of this theorem is straightforward, based on the evaluation of the spectral norm $\|x(k)\|$ or $x(k)$ and showing that, if condition Equation (5.50) holds, then $\lim_{k \to \infty} \|x(k)\| = 0$. The proof can be found in the work of Han and Lee [44].

DEFINITION 5.3
An interval matrix $A_I(k)$ is an $n \times n$ matrix whose elements consist of intervals $[b_{ij}, c_{ij}]$ for $i, j = 1, \ldots, n$, that is,

$$A_I(k) = \begin{bmatrix} [b_{11}, c_{11}] & \cdots & [b_{1n}, c_{1n}] \\ \vdots & [b_{ij}, c_{ij}] & \vdots \\ [b_{n1}, c_{n1}] & \cdots & [b_{nn}, c_{nn}] \end{bmatrix} \quad (5.51)$$

DEFINITION 5.4
The center matrix, A_c, and the maximum difference matrix, A_m, of $A_I(k)$ in Equation (5.51) are defined by

$$A_c = \frac{B+C}{2}, \quad A_m = \frac{C-B}{2}$$

where $B = \{b_{ij}\}$ and $C = \{c_{ij}\}$. Thus, the interval matrix $\mathbf{A}_I(k)$ in Equation (5.51) can also be rewritten as

$$\mathbf{A}_I(k) = [\mathbf{A}_c - \mathbf{A}_m, \mathbf{A}_c + \mathbf{A}_m] = \mathbf{A}_c + \Delta\mathbf{A}(k)$$

with $|\Delta\mathbf{A}(k)| \leq \mathbf{A}_m$.

LEMMA 5.1
The interval matrix $\mathbf{A}_I(k)$ is asymptotically stable if matrix $\mathbf{A}_c$ is stable and

$$\rho(|\mathbf{A}_c| + \mathbf{A}_m) < 1$$

The proof can be found in the work of Han and Lee [44]. Lemma 5.1 can be used to check the sufficient condition for the stability of fuzzy systems of Takagi-Sugeno type given in Equation (5.45). Consider a set of m fuzzy rules such as Equation (5.45):

IF $x(k)$ is A_1^1 AND ... $x(k-n+1)$ is A_n^1

THEN $x(k+1) = \mathbf{A}_1 x(k)$

$\vdots$

IF $x(k)$ is A_1^m AND ... $x(k-n+1)$ is A_n^m

THEN $x(k+1) = \mathbf{A}_m x(k)$,

(5.52)

where $\mathbf{A}_i$ matrices for $i = 1, \ldots, m$ are defined by Equation (5.46). One can now formulate all the m matrices $\mathbf{A}_i$, $i = 1, \ldots, m$ as an interval matrix of the form of Equation (5.51) by simply finding the minimum and the maximum of all elements at the top row of all the $\mathbf{A}_i$ matrices. In other words, we have

$$\mathbf{A}_I(k) = \begin{bmatrix} [\underline{a}_1, \overline{a}_1] & [\underline{a}_2, \overline{a}_2] & \cdots & [\underline{a}_{n-1}, \overline{a}_{n-1}] & [\underline{a}_n, \overline{a}_n] \\ 1 & 0 & \cdots & 0 & 0 \\ 0 & 1 & \cdots & 0 & 0 \\ \vdots & \vdots & \ddots & \vdots & \vdots \\ 0 & 0 & \cdots & 1 & 0 \end{bmatrix}$$

(5.53)

where $\underline{a}_i$ and $\overline{a}_i$, for $i = 1, \ldots, n$ are the minimum and maximum of the respective element of the first rows of $\mathbf{A}_i$ in Equation (5.19), taken element by element.

Using the above definitions and observations, the fuzzy system Equation (5.52) can be rewritten by

$$\text{IF } x(k) \text{ is } A_1^i \text{ AND } \ldots x(k-n+1) \text{ is } A_n^i$$

$$\text{THEN } \mathbf{x}(k+1) = \mathbf{A}_I^i \mathbf{x}(k)$$

where $i = 1, \ldots, m$ and A_I^i is an interval matrix of the form Equation (5.53), except that $\underline{a}_i = \overline{a}_i = a_i$. Now, finding the weighted average, one has

$$\mathbf{x}(k+1) = \frac{\sum_{i=1}^l w^i \mathbf{A}_I^i \mathbf{x}(k)}{\sum_{i=1}^l w^i} \tag{5.54}$$

THEOREM 5.5
The fuzzy system Equation (5.54) is asymptotically stable if the interval matrix $\mathbf{A}_I(k)$ is asymptotically stable, that is, the conditions in Lemma 5.1 are satisfied.

Example 5.13

Reconsider Example 5.11. It is desired to check its stability via the matrix interval approach.

Solution

The system's two canonical matrices are written in the form of an interval matrix (10.30) as

$$\mathbf{A}_I(k) = \begin{pmatrix} [1, 1.2] & [-0.6, -0.4] \\ 1 & 0 \end{pmatrix}$$

The center and maximum difference matrices are

$$\mathbf{A}_c = \begin{pmatrix} 1.1 & -0.5 \\ 1 & 0 \end{pmatrix}, \quad \mathbf{A}_m = \begin{pmatrix} 0.1 & 0.1 \\ 0 & 0 \end{pmatrix}$$

Then, the condition of Lemma 5.1 would become

$$\rho(|\mathbf{A}_c| + \mathbf{A}_m) = \rho\begin{pmatrix} 1.2 & 0.6 \\ 1 & 0 \end{pmatrix} = 1.58 > 1$$

Thus, the stability of the fuzzy system under consideration is inconclusive. In fact, it was shown to be stable.

Example 5.14

Consider the following fuzzy system:

$$P^1: \text{IF } x(k) \text{ is } A^1 \text{ THEN } x^1(k+1) = 0.3x(k) + 0.5x(k-1)$$

$$P^2: \text{IF } x(k) \text{ is } A^2 \text{ THEN } x^2(k+1) = 0.2x(k) + 0.2x(k-1)$$

where A^i are fuzzy sets shown in Figure 5.30. It is desired to check the stability of this system using the matrix interval method.

Solution

The two subsystems' matrices are

$$\mathbf{A}_1 = \begin{pmatrix} 0.3 & 0.5 \\ 1 & 0 \end{pmatrix}, \quad \mathbf{A}_2 = \begin{pmatrix} 0.2 & 0.2 \\ 1 & 0 \end{pmatrix}$$

The systems' two canonical matrices are written in the form of an interval matrix (5.51) as

$$\mathbf{A}_I(k) = \begin{pmatrix} [0.2, 0.3] & [0.2, 0.5] \\ 1 & 0 \end{pmatrix}$$

The center and maximum difference matrices are

$$\mathbf{A}_c = \begin{pmatrix} 0.25 & 0.35 \\ 1 & 0 \end{pmatrix}, \quad \mathbf{A}_m = \begin{pmatrix} 0.05 & 0.15 \\ 0 & 0 \end{pmatrix}$$

Then, the condition in Lemma 5.1 would become,

$$\rho(|\mathbf{A}_c| + \mathbf{A}_m) = \rho \begin{pmatrix} 0.3 & 0.5 \\ 1 & 0 \end{pmatrix} = 0.873 < 1$$

Thus, the system is stable. This result was also verified by simulation (see Figure 5.33).

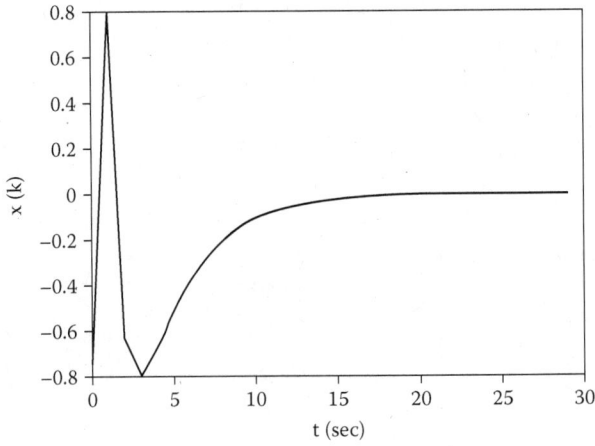

FIGURE 5.33
Simulation result for Example 5.14.

5.13 Conclusions

In this chapter, a quick overviews of classical and fuzzy sets, classical and fuzzy logic, and fuzzy control systems and their stability were given. Main similarities and differences between classical and fuzzy sets were introduced. In general, set operations are the same for classical and fuzzy sets. The exceptions were excluded middle laws. Alpha-cut sets and extension principle were presented, followed by a brief introduction to classical versus fuzzy relations. This chapter presented issues that are important in understanding fuzzy sets and their advantages over classical sets. A set of problems at the end of the book will further enhance the reader's understanding of these concepts.

The second topic was fuzzy logic, which covered classical and fuzzy logic. Most of the tools needed to form an idea about fuzzy logic and its operation have been introduced. These tools are essential in understanding the next chapter addressing fuzzy control and stability. More details can be found in a companion book by Zilouchian and Jamshidi [5].

The third main topic of the chapter was fuzzy control. Both Mamdani rules and Takagai-Sugeno rules were presented. Stability analysis of Takagi-Sugeno-type fuzzy systems was addressed. Fuzzy control systems are very desirable in situations where precise mathematical models are not available and human involvement is necessary. In such cases, fuzzy rules could be used to mimic human behavior and actions. Stability of fuzzy control systems was also discussed, including the Lyapunov stability.

References

[1] Zadeh, L. A., "Fuzzy sets," *Information and Control* 8, 338–353, 1965.
[2] Mamdani, E. H., "Applications of fuzzy algorithms for simple dynamic plant," *Proceedings of the IEEE* 121(12), 1585–1588, 1974.
[3] Jamshidi, M., Titli, A., Zadeh, L. A., and Bverie, S., "Applications of fuzzy logic—Toward high machine intelligence quotient systems," *Prentice Hall Series on Environmental and Intelligent Manufacturing Systems*, Vol. 9 (M. Jamshidi, Ed.). Prentice-Hall, Upper Saddle River, NJ, 1997.
[4] Aminzadeh, F., and Jamshidi, M., "Soft computing," *Prentice Hall Series on Environmental and Intelligent Manufacturing Systems*, Vol. 4 (M. Jamshidi, Ed.). Prentice-Hall, Englewood Cliffs, NJ, 1993.
[5] Zilouchian, A., and Jamshidi, M. (Eds.), *Intelligent Control Systems with Soft Computing Methodologies*, CRC Publishers, Boca Raton, FL, 2001.
[6] Ross, T. J., *Fuzzy Logic with Engineering Application*, McGraw-Hill, New York, 1995.

[7] Jamshidi, M., Vadiee, N., and Ross, T. J., "Fuzzy logic and control: Software and hardware applications," *Prentice Hall Series on Environmental and Intelligent Manufacturing Systems*, Vol. 2 (M. Jamshidi, Ed.). Prentice-Hall, Englewood Cliffs, NJ, 1993.

[8] Ross, T. J., *Fuzzy Logic with Engineering Application*, McGraw-Hill, New York, 1995.

[9] Zadeh, L. A., "A theory of approximate reasoning," *Machine Intelligence* (J. Hayes, D. Michie, and L. Mikulich, Eds.). Halstead Press, New York, pp. 149–194, 1979.

[10] Gaines, B., "Foundation of fuzzy reasoning," *International Journal of Man-Machine Studies* 8, 623–688, 1976.

[11] Yager, R. R., "On the implication operator in fuzzy logic," *Information Sciences* 31, 141–164, 1983.

[12] Wang, L.-X., *Adaptive Fuzzy Systems and Control*, Prentice-Hall, Englewood Cliffs, NJ, 1994.

[13] Jamshidi, M., "Large-scale systems—Modeling, control and fuzzy logic," *Prentice Hall Series on Environmental and Intelligent Manufacturing Systems*, Vol. 8 (M. Jamshidi, Ed.). Prentice-Hall, Saddle River, NJ, 1996.

[14] Tanaka, K., and Sugeno, M., "Stability analysis and design of fuzzy control systems," *Fuzzy Sets and Systems* 45, 135–156, 1992.

[15] *IEEE Control Systems Magazine*, Letters to the Editor, IEEE, Vol. 13, 1993.

[16] Bretthauer, G., and Opitz, H.-P., "Stability of fuzzy systems," in *Proceedings of EUFIT'94*, Aachen, Germany, pp. 283–290, September 1994.

[17] Aracil, J., Garcia-Cezero, A., Barreiro, A., and Ollero, M., "Stability analysis of fuzzy control systems: A geometrical approach," *AI, Expert Systems and Languages in Modeling and Simulation* (C. A. Kulikowski and R. M. Huber, Eds.). North-Holland, Amsterdam, pp. 323–330, 1988.

[18] Chen, Y. Y., and Tsao, T. C., "A description of the dynamical behavior of fuzzy systems," *IEEE Transactions on Systems, Man and Cybernetics* 19, 745–755, 1989.

[19] Wang, P.-Z., Zhang, H.-M., and Xu, W., "Pad-analysis of fuzzy control stability," *Fuzzy Sets and Systems* 38, 27–42, 1990.

[20] Hojo, T., Terano, T., and Masui, S., "Stability analysis of fuzzy control systems," in *Proceedings of IFSA '91, Engineering*, Brussels, pp. 44–49, 1991.

[21] Hwang, G.-C., and Liu, S. C., "A stability approach to fuzzy control design for nonlinear systems," *Fuzzy Sets and Systems* 48, 279–287, 1992.

[22] Driankov, D., Hellendoorn, H., and Reinfrank, M., *An Introduction to Fuzzy Control*, Springer-Verlag, Berlin, 1993.

[23] Kang, H., "Stability and control of fuzzy dynamic systems via cell-state transitions in fuzzy hypercubes," *IEEE Transations on Fuzzy Systems* 1, 267–279, 1993.

[24] Demaya, B., Boverie, S., and Titli, A., "Stability analysis of fuzzy controllers via cell-to-cell root locus analysis," in *Proceedings of EVFIT '94*, Aachen, Germany, pp. 1168–1174, 1994.

[25] Langari, G., and Tomizuka, M., "Stability of fuzzy linguistic control systems," in *Proceedings of the IEEE Conference on Decision and Control*, Hawaii, pp. 2185–2190, 1990.

[26] Bouslama, F., and Ichikawa, A., "Application to limit fuzzy controllers to stability analysis," *Fuzzy Sets and Systems* 49, 103–120, 1992.

[27] Chen, C.-L., Chen, P.-C., and Chen, C.-K., "Analysis and design of a fuzzy control system," *Fuzzy Sets and Systems* 57, 125–140, 1993.

[28] Chen, Y. Y., "Stability analysis of fuzzy control—A Lyapunov approach," *IEEE Annals of Conferences on Systems, Man, and Cybernetics* 19, 1027–1031, 1987.
[29] Franke, D., "Fuzzy control with Lyapunov stability," in *Proceedings of the European Control Conference;* Groningen, 1993.
[30] Gelter, J., and Chang, H. W., "An instability indicator for expert control," *IEEE Transactions on Control Systems* 31, 14–17, 1986.
[31] Kiszka, J. B., Gupta, M. M., and Nikiforuk, P. N., "Energistic stability of fuzzy dynamic systems," *IEEE Transactions on Systems, Man and Cybernetics* 15, 783–792, 1985.
[32] Tahani, V., and Sheikholeslam, F., "Extension of new results on nonlinear systems stability of fuzzy systems," in *Proceedings of EUFIT'94*, Aachen, Germany, pp. 638–686, 1994.
[33] Barreiro, A., and Aracil, J., "Stability of uncertain dynamical systems," in *Proceedings, IFAC Symposium on AI in Real-Time Control*, Delft, pp. 177–182, 1992.
[34] Opitz, H. P., "Fuzzy control, Teil 6: Stabilitat von Fuzzy-Regelungen," *Automatisierungstechnik* 41, A21–A24, 1993.
[35] Opitz, H. P., "Stability analysis and fuzzy control," in *Proceedings of Fuzzy Duisburg '94, Int. Workshop on Fuzzy Technologies in Automation and Intelligent Systems*, Duisburg, 1994.
[36] Braee, M., and Rutherford, D. A., "Selection of parameters for a fuzzy logic controller," *Fuzzy Set Systems* 49, 103–120, 1978.
[37] Braee, M., and Rutherford, D. A., "Theoretical and linguistic aspects of the fuzzy logic controller," *Automatica* 15, 553–577, 1979.
[38] Kickert, W. J., and Mamdani, E. H., "Analysis of fuzzy logic controller," *Fuzzy Sets and Systems* 1, 29–44, 1978.
[39] Ray, K. S., and Majumder, D. D., "Application of circle criteria for stability analysis associated with fuzzy logic controller," *IEEE Transactions on Systems, Man and Cybernetics* 14, 345–349, 1984.
[40] Ray, K. S., Ananda, S. G., and Majumder, D. D., "L-stability and the related design concept for SISO linear systems associated with fuzzy logic controller," *IEEE Transaction on Systems, Man and Cybernetics* 14, 932–939, 1984.
[41] Böhm, R., "Ein Ansatz Zur Stabilitätasalyse von Fuzzy-Reglern," in *Forschungsberichte Universitäte Dortmund, Fakultät fur Elektrotechnik, Band Nr. 3,2. Workshop Fuzzy Control des GMA-UA 1.4.2.* am 19/20.11.1992, pp. 24–35, 1992.
[42] Bühler, H., "Stabilitatsuntersuchung von Fuzzy-Regelungssystemem," in *Proc., 3, Workshop Fuzzy Control des GMA-UA 1.4.1*, Dortmund, pp. 1–12, 1993.
[43] Sheikholeslam, F., "Stability Analysis of Nonlinear and Fuzzy Systems," MSc thesis, Department of EECS Isfahan University of Technology, Isfahan, Iran, 1994.
[44] Han, H. S., and Lee, J. G, "Necessary and sufficient conditions for stability of time-varying discrete interval matrices," *International Journal of Control* 59, 1021–1029, 1994.

6

Neural Network-Based Control

6.1 Introduction to Function Approximation

Dynamic systems follow a set of rules except for some degree of uncertainty. This allows an engineer to design a controller based on a model of the system. Yet, the symbolic representations and the mathematical axioms we use sometimes make the derivation of differential equations that govern the dynamics of a system a daunting task. Even if we could derive the dynamics of a system, sometimes the mathematical knowledge we have is not enough to explain certain dynamic properties of complex systems. Yet, if we carefully look at what we are doing by deriving mathematical equations to model a system, we notice that we are merely looking for a reasonable graph that helps us to map a given set of independent variables to a set of dependent variables. Mathematically, this mapping task is denoted by $f: x(t) \rightarrow y(t)$, where function f maps variable $x(t)$ to variable $y(t)$ in the respective number spaces at time t. For instance, if we look at Figure 6.1, the graph shows how the independent variable x is mapped to the dependent variable y by the function $y = (x^3 + \sin^2 x) e^{-|x|}$. If you look at the mathematical expression, it looks fairly complex. Yet, if you look at the landscape of the graph it makes, it looks very simple. Therefore, if the requirement is to do what this function does in terms of mapping x to y, it is good enough to have some function that gives us the same landscape.

Now let us look at how we can approximate the landscape created by the above equation in Figure 6.1 (Chap6/general_graph.m) using a linear combination of a set of localized primitive shapes. Figure 6.2 shows how the landscape of $y = [x^3 + \sin^2(x)] e^{-|x|}$ can be approximated by a linear combination of two Gaussian functions given by $y_4 = y_2 + 2y_3, y_2 = [e^{-(x+0.5)^2} - 1], y_3 = e^{-0.4(x-2)^2}$ (Chap6/general_graph_primitives.m).

Therefore, the fundamental principle behind function approximation is to use a combination of a set of primitive localized functions to construct the mapping landscape created by a given complex mathematical function.

6.1.1 Biological Inspirations

Amazingly, this is how biological brains seem to approximate relationships. Given a dynamic system, the brain does not know how to derive the dynamics

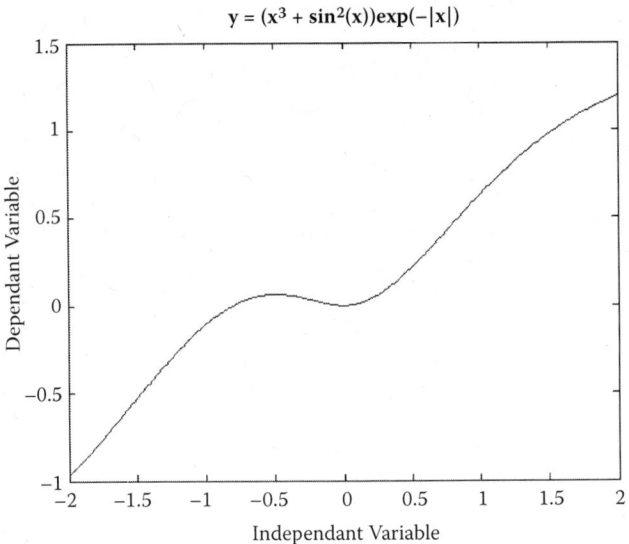

FIGURE 6.1
Graph of $y = [x^3 + \sin^2(x)]e^{-|x|}$.

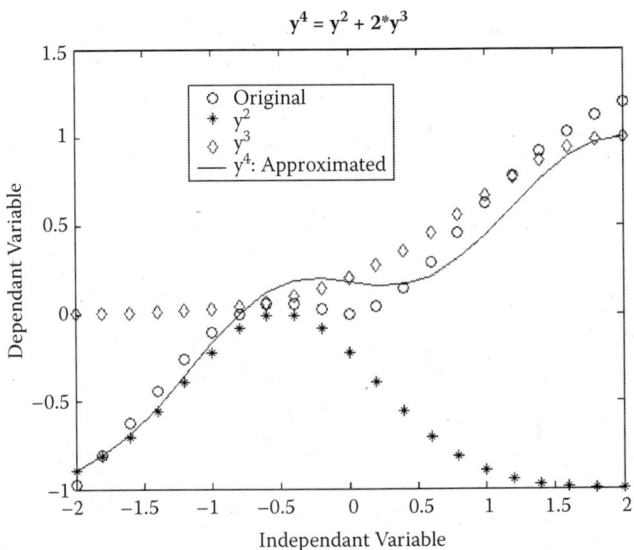

FIGURE 6.2
"Original function" $y = [x^3 + \sin^2(x)]e^{-|x|}$ approximated by $y_4 = y_2 + 2y_3$, $y_2 = [e^{-(x+0.5)^2} - 1]$, $y_3 = e^{-0.4(x-2)^2}$.

using man-made mathematical theories or axioms. Instead, it knows that the natural phenomena are governed by various rules that can be characterized by variables and parameters. In other words, how a physical system maps input variables to output variables given a set of conditions is more or less repeatable and can be generalized around an operating point, just like what we discussed in Figures 6.1 and 6.2. The brain capitalizes on this principle by keeping a set of cells that are active in various local subspaces in the sensory input space. These cells are called neurons [1]. They have a set of input terminals called dendrites that bring sensory inputs to the cell body called the somatic part as shown in Figure 6.3. All communication is done through electric voltage pulses called action potentials. When the cell body receives these action potentials through dendrites, the cell body calculates a single output that is transmitted through an output terminal called the axon. An axon can make connections with a dendrite of another cell by establishing a synaptic connection. There is a lot of research conducted on the shape of the signal mapping behavior of neurons. Research conducted on the human motor system has revealed a few important properties of how motor neurons map inputs to outputs. The motor system is responsible for movement control. Therefore, variables such as desired position, velocity, acceleration, force, and their proprioceptive feedback information are the inputs to the system. The output is

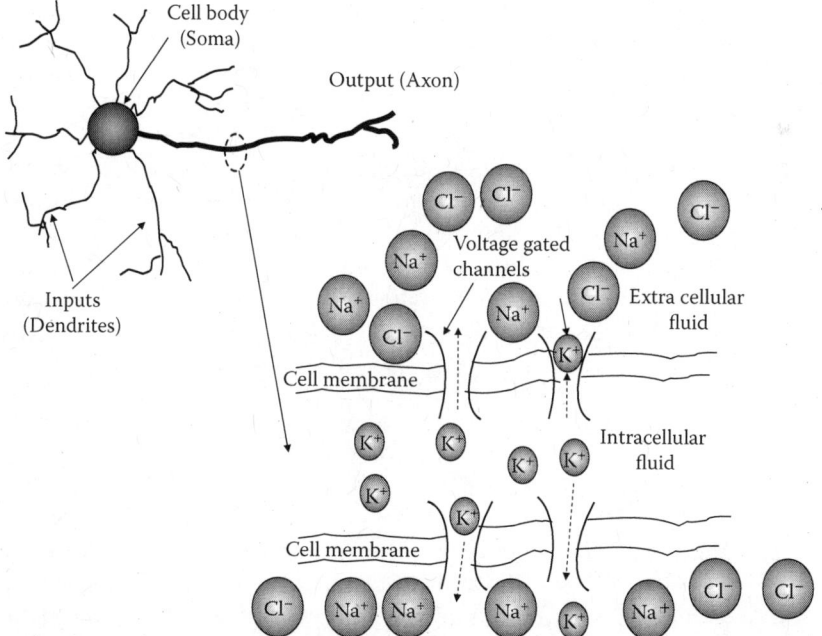

FIGURE 6.3
Voltage-regulated Na^+ and K^+ ion gates helping a neuron to propagate action potentials along its branches.

the motor command sent to the muscles. Therefore, given a movement to be made, the motor system has to calculate the correct motor commands to realize the desired kinematic and dynamic variables mentioned above.

Take, for example, a case where a child is trying to learn how to ride a bicycle. At first, the child does not know what to do to balance it. All what he can do is be seated and pedal the bicycle. Obviously, he wants be able to balance it first, but he will observe what happens to the bicycle when he pedals. He will observe how the bicycle moves when he pedals. This can be the first piece of information to form a simple neural structure to map the force applied on the pedal to the initial acceleration in various directions when he holds the handles in different orientations on different slopes. In this case, the input variable to his neural network (NN) is force applied on the pedal, orientation of the handle, and his perception of the slope and terrain roughness. The output of the network will be the initial acceleration of the bicycle in various directions that can involve angular acceleration around the axis connecting the two ground contact points of the tires, and the linear acceleration along that axis. Even this simple map is a very complex mathematical problem if we were to derive the complete dynamics of the system. However, the brain will try to evolve an approximate map by adjusting parameters known as synaptic weights that scale the signals transmitted from one neuron to the other. In addition, the brain will learn the best way to connect neurons to each other. In other words, the brain will learn a structure for an NN built inside itself. Amazing! The child will master evolving this network across repeated trials. This will help his brain to combine various primitive functions it learned on this task to build a very robust, comprehensive network that can predict a wide range of behaviors of the bicycle.

Then, the child will proceed to experiment with various speeds to see on which conditions the bicycle balances. He will find this magical setting of variables where the bicycle moves without falling. He will approximate the relationships among the above-mentioned variables in dynamically balanced conditions with no idea of Newton's equations governing the motion of physical bodies and how gravity works, or Einstein's equations of warping of space-time to make things move. The child's brain will approximate all these equations by carefully combining and tuning a set of neurons inside it. Therefore, an approximation of a graph that relates inputs to outputs is enough to predict how a dynamical system would behave in a locality.

There has been a tremendous number of breakthroughs made in the recent past on how these NNs work, the time course of the evolution of networks, and how they estimate complex relationships over time [1–3]. In experiments on velocity and position control tasks of the human hand, scientists have found that different motor neurons become active in different postures of the hand, suggesting that a neuron is active only in a local region in the input space. Another important property found in these experiments is that a given neuron is most active in a given movement direction of the hand and less active in other directions within the region it is active. It suggests that a

Neural Network-Based Control

neuron has a preferred movement direction. The most important property we should notice is that a neuron can change the region in which it is active, the activity level given a region, and the preferred direction in order to improve the accuracy of the movements. This last property is called the plasticity of the motor neurons. The plasticity is associated with the generalization ability of the neurons. In other words, when a neuron changes its local behavior in response to an error experienced in one direction, it also updates the activity in other directions. This suggests that neurons have smooth functions.

Now, let us have a closer look at an approximate model of a biological neuron known as the Hodgkin-Huxley model. It tries to simulate the biological mechanism of propagating neural information using voltage pulses known as action potentials.

Neurons are surrounded by a fluid (extracellular fluid) with a chemical composition very similar to that of seawater. Therefore, it has an abundance of Na^+ and Cl^- ions. In contrast, the fluid inside the cell (intracellular fluid) is dominated by K^+ ions. Therefore, the cell can control the potential difference across the membrane by pumping K^+ and Na^+ ions across the membrane. The cell can make use of K^+ and Na^+ ion channels to do this, as shown in Figure 6.4.

The Nernst equation for the potential difference across the cell membrane is given by

$$v(k) = \lambda \log\left(\frac{n_{out}(k)}{n_{in}(k)}\right)$$

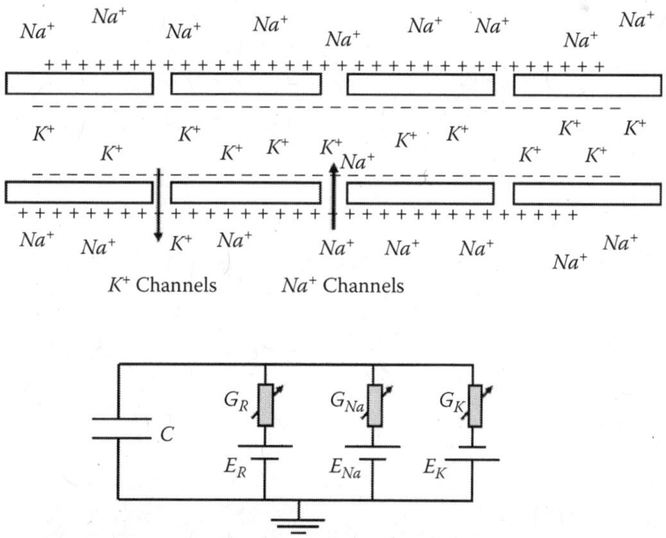

FIGURE 6.4
A simplified model of K^+ and Na^+ ion channels in the cell membrane known as the Hodgkin-Huxley model.

where $v(k)$ is the membrane potential at time k, η_{out} and η_{in} are net ion concentration outside and inside the cell, respectively, and λ is a constant. Therefore, it is clear that the K⁺ and Na⁺ ion pumps can control the membrane potential by pumping ions across the membrane to change the relative ion concentrations in the extracellular and intracellular fluids. The dynamics of how this happens is given by the following equations. When ion pumps move ions across the membrane, currents are formed. The current is defined as the speed of a charge. For instance, if a charge of 1 Coulomb is moved at 1 m/s, it is equal to a current of 1 A. Therefore, there are three types of currents across the membrane: i_K, i_{Na}, and i_R, where i_K is the current made by K⁺ ions, i_{Na} is the current made by Na⁺ ions, and i_R is the current created by leakage currents that cannot be accounted for by K⁺ and Na⁺ ion channels.

Therefore, the membrane potential is related to the currents across various ion pumps as given by $v(k+1) = v(k) + T(I - (i_K + i_{Na} + i_R))/C$, where I is the current injected to the cell from outside, C is the membrane capacitance, and T is the sampling time interval. Currents i_K, i_{Na}, and i_R are related to other internal variables as given by

$$i_{Na} = G_{Na} m^3 h(v(k) - E_{Na})$$

$$i_K = G_K n^4 (v(k) - E_K)$$

$$i_R = G_R (v(k) - E_R)$$

$$m(k+1) = m(k) + T(\alpha_m(k)(1 - m(k))) - \beta_m(k)m(k)$$

$$n(k+1) = n(k) + T(\alpha_n(k)(1 - n(k))) - \beta_n(k)n(k)$$

$$h(k+1) = h(k) + T(\alpha_h(k)(1 - h(k))) - \beta_h(k)h(k)$$

$$\alpha_n(k) = \frac{(0.1 - 0.01v(k))}{(\exp(1 - 0.1v(k)) - 1)}$$

$$\alpha_m(k) = \frac{(2.5 - 0.1v(k))}{(\exp(2.5 - 0.1v(k)) - 1)}$$

$$\alpha_h(k) = 0.07 \exp(-v(k)/20)$$

$$\beta_n(k) = 0.125 \exp(-v(k)/80)$$

$$\beta_m(k) = 4 \exp(-v(k)/18)$$

$$\beta_h(k) = \frac{1}{(\exp(3 - 0.1v(k)) + 1)}$$

where

$$C = 1, G_K = 36 \text{ mS/cm}^2, G_{Na} = 120 \text{ mS/cm}^2, G_R = 0.3 \text{ mS/cm}^2$$

$$E_{Na} = 115 \text{ mV}, E_K = -12 \text{ mV}, E_R = 10.6 \text{ mV}$$

Figures 6.5 through 6.9 show the behavior of each of these variables when the input current is given by $I = 20 + 10 \sin(t/10)$ $\mu A/cm^2$ where t is time.

Please use chap6/hodgkin_huxley.m MATLAB® code to experiment with the Hodgkin-Huxley model and extend it to do more advanced encoding.

Now, you might wonder how these action potentials encode relationships. One of the earliest models, known as the McCulloch-Pitts model, explains this phenomenon in a simple manner.

In Figure 6.10 $x_1, x_2, \ldots, x_N$ are inputs and $w_1, w_2, \ldots, w_N$ are corresponding weights. The variable X can be considered as the input current to the neuron given in the Hodgkin-Huxley model. The activation function $f(X)$ was originally proposed to be a step function. It implies that the activation function of a neuron is one of integrate-and-fire nature as depicted in the behavior of state variables in Figures 6.5 to 6.9. However, please note that there is no fixed model for the shape of the activation function $f(X)$. The shape can be

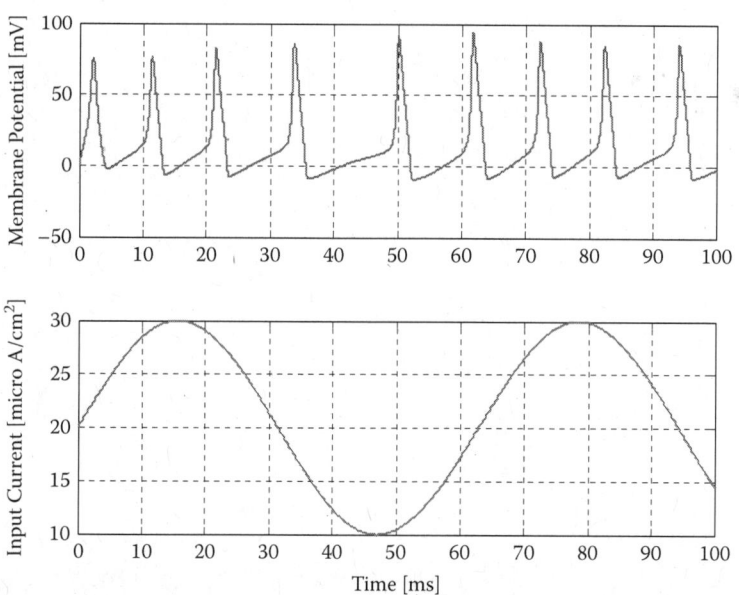

FIGURE 6.5
How the action potentials are generated when the input current behaves by $I = 20 + 10 \sin(t/10) \mu A/cm^2$, where t is time.

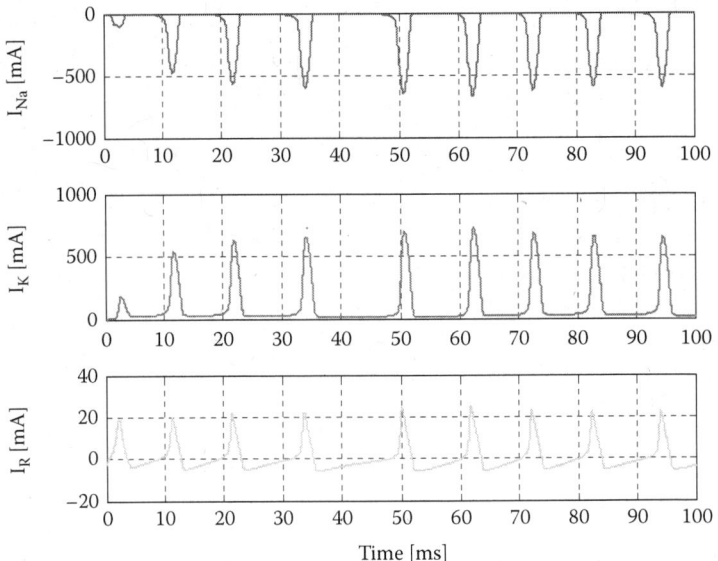

FIGURE 6.6
Behavior of currents across the membrane.

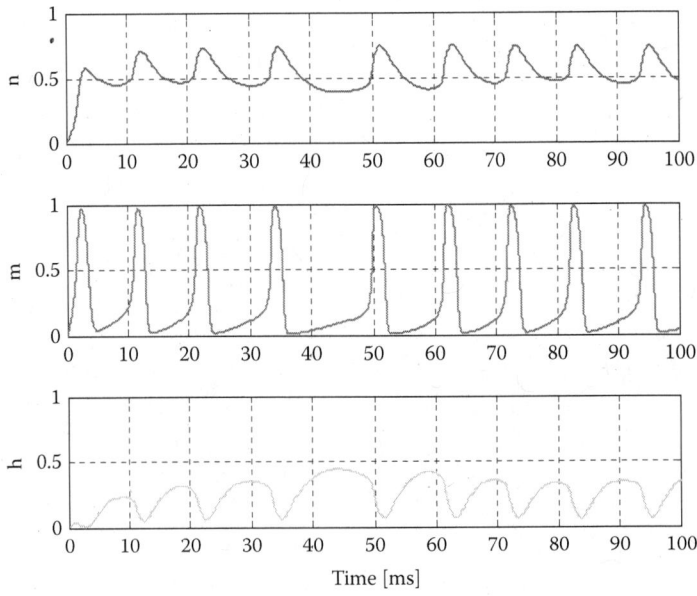

FIGURE 6.7
Behavior of gating variables.

Neural Network-Based Control 159

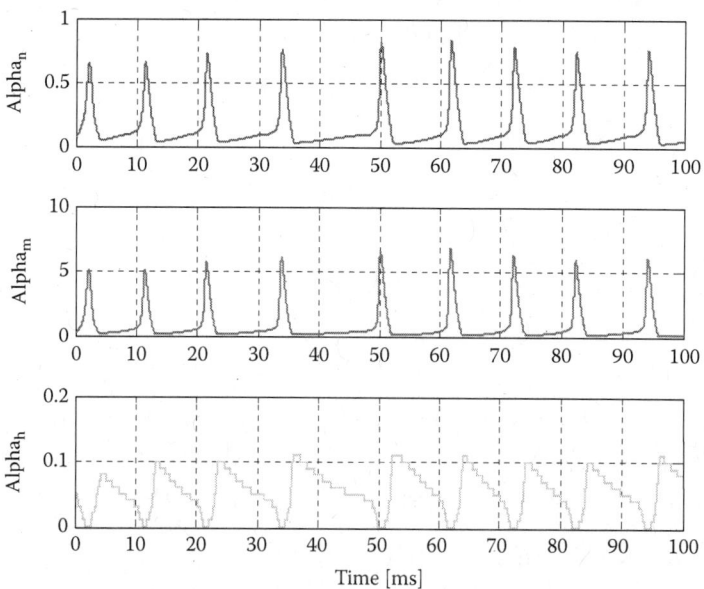

FIGURE 6.8
Behavior of empirical functions α_n, α_m, α_h.

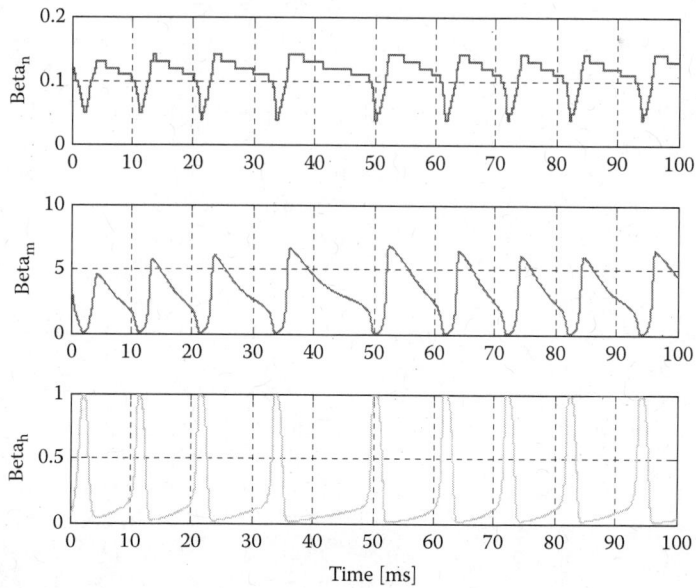

FIGURE 6.9
Behavior of empirical functions β_n, β_m, β_h.

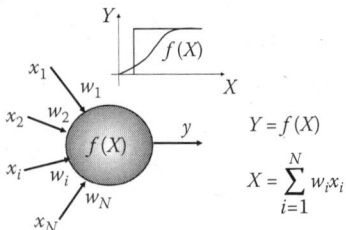

FIGURE 6.10
The McCulloch-Pitts model.

decided based on the prior knowledge about the particular function that the neuron(s) is (are) going to map.

6.1.2 Construction of Complex Functions by Adaptive Combination of Primitives

The most important phenomenon we should observe is that these localized neurons that look very simple perform complex tasks by working together in networks of neurons. Similarly, we should be able to model complex systems by allowing a network of artificial neurons to receive inputs from the physical system and change their local behaviors and the manner in which they communicate among each other to minimize the error of estimating the output of the physical system. The next problem is how to create artificial neurons and how to update their parameters to model real physical systems.

Artificial neurons are mathematical functions with the properties we noticed in biological neurons, that is, an artificial neuron can receive a vector of inputs, it has a localized activation landscape, it can be moved in the input space, and it can change its shape and activity level by changing some set of parameters. Figures 6.11 and 6.12 show how the shape of a simple Gaussian function can be changed locally without causing a global effect (Chap6/Gaussian_neuron.m). We can move the landscape created by the function by changing the center values. We can control the scope of the effect made by the function by changing the width parameter.

Figure 6.13 shows how a complex landscape can be constructed by combining two primitive Gaussian shapes (Chap6/Gausian_nn.m). In this case, we can modulate the height of an individual primitive by introducing weighting parameters.

Figures 6.14 through 6.16 show how the primitive landscape of a Mexican hat function given by $y^M = \frac{\sigma^3}{\sqrt{2\pi}}(1 - \frac{(x-c)^T(x-c)}{\sigma^2})e^{-(x-c)^T(x-c)/2\sigma^2}$ can be controlled (Chap6/Mexican_hat_neuron.m).

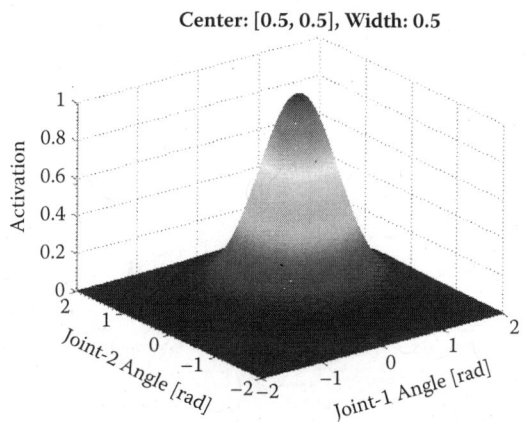

FIGURE 6.11
Graph of Gaussian function $y_1^G = e^{-(x-c)^T(x-c)/2\sigma^2}$ for $c = [0.5, 0.5]$, $\sigma = 0.5$.

Figures 6.17 and 6.18 show two fairly complex landscapes constructed using a combination of primitive shapes shown above (Chap6/Mexican_hat_nn.m).

6.1.3 Concept of Radial Basis Functions

The concept of radial basis functions (RBFs) is perhaps the most significant advancement made in the field of NN structures after Rosenblatt introduced the concept of a perceptron in 1957, where a collection of activation functions was organized to form a layered network to approximate a nonlinear function.

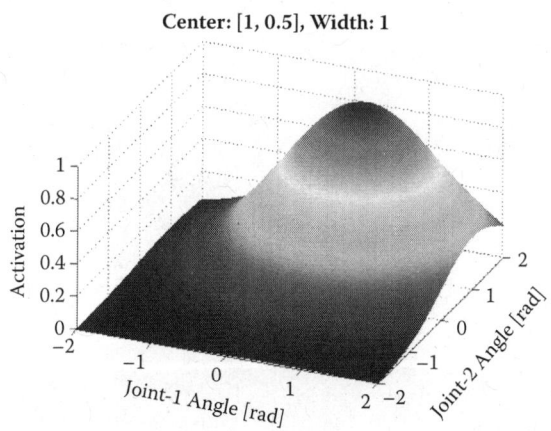

FIGURE 6.12
Graph of Gaussian function $y_2^G = e^{-(x-c)^T(x-c)/2\sigma^2}$ for $c = [1, 0.5]$, $\sigma = 1$.

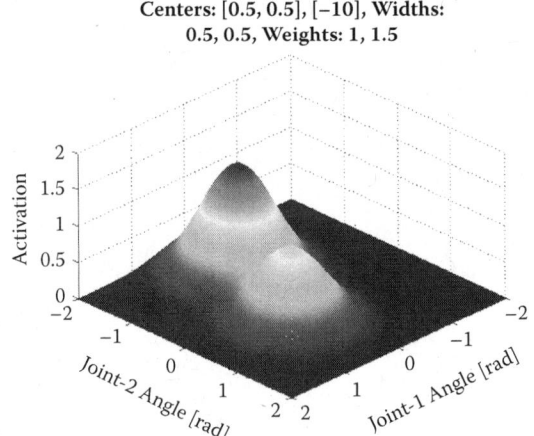

FIGURE 6.13
The landscape created by $w_1 y_1^G + w_2 y_2^G$ for $w_1 = 1$, $w_2 = 1.5$.

The basic way of thinking behind RBFs as primitive activation functions is to look at a primitive as a function centered at a particular point in the input space with different slopes along each axis that defines the input space.

A general function can be written in the form given by

$$y = \varphi\left((X-C)^T \sum (X-C)\right)$$

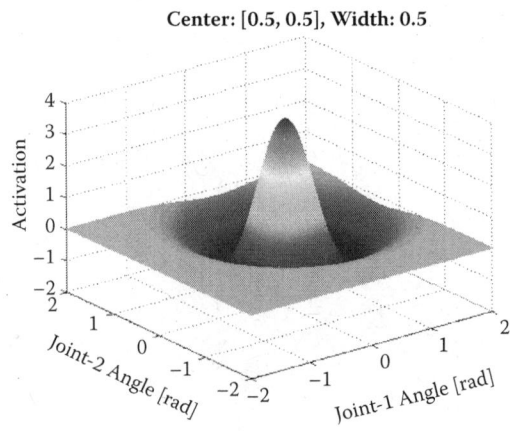

FIGURE 6.14
Graph of the Mexican hat wavelet function $y_1^M = \frac{\sigma^3}{\sqrt{2\pi}}(1 - \frac{(x-c)^T(x-c)}{\sigma^2}) e^{-(x-c)^T(x-c)/2\sigma^2}$ for $c = [0.5, 0.5]$, $\sigma = 0.5$.

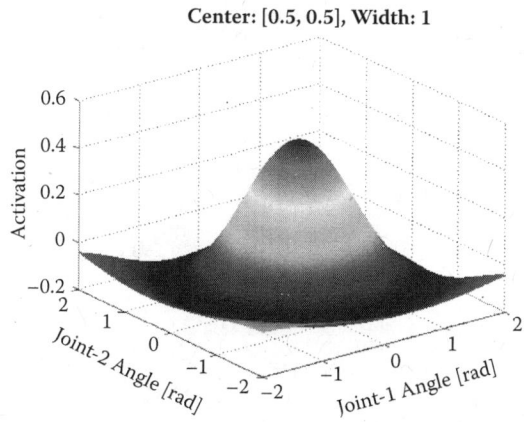

FIGURE 6.15
The graph of the Mexican hat wavelet function $y_2^M = \frac{\sigma^3}{\sqrt{2\pi}}(1-\frac{(x-c)^T(x-c)}{\sigma^2})e^{-(x-c)^T(x-c)/2\sigma^2}$ for $c = [0.5, 0.5]$, $\sigma = 1$.

where $X \in \Re^n$ is the vector of inputs, $C \in \Re^n$ is the vector of centers defined in the input space, and $\Sigma \in \Re^{n \times n}$ is the covariance matrix defining the activation landscape of the function in the input space. Usually, Σ is a diagonal matrix. If we revisit the functions we discussed above, they are RBFs. For instance, $y_3^M = \frac{\sigma^3}{\sqrt{2\pi}}(1-\frac{(x-c)^T(x-c)}{\sigma^2})e^{-(x-c)^T(x-c)/2\sigma^2}$ is an RBF where $\varphi = \frac{\sigma^3}{\sqrt{2\pi}}(1-\frac{z}{\sigma^2})e^{-z/2\sigma^2}$, $z = (x-c)^T\Sigma(x-c), \Sigma = I$.

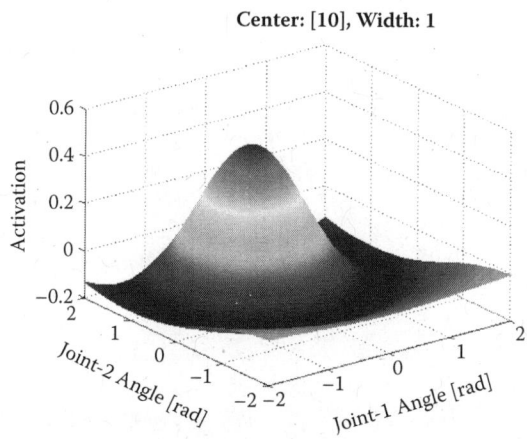

FIGURE 6.16
The graph of the Mexican hat wavelet function $y_3^M = \frac{\sigma^3}{\sqrt{2\pi}}(1-\frac{(x-c)^T(x-c)}{\sigma^2})e^{-(x-c)^T(x-c)/2\sigma^2}$ for $c = [1,0]$, $\sigma = 1$.

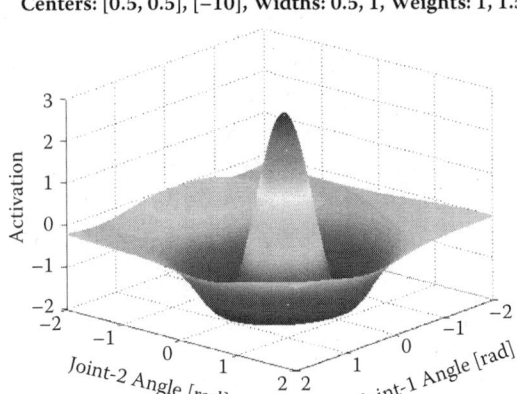

FIGURE 6.17
The landscape created by $w_1 y_1^M + w_2 y_3^M$ for $w_1 = 1$, $w_2 = 1.5$.

An RBF NN is essentially a linear combination of such primitive functions to calculate each output of a layered *NN* given by

$$NN_k = \sum_{l=1}^{N} W_{lk} \varphi((X - C_l)^T \Sigma_l (X - C_l))$$

where NN_k is the *k*th output of the *NN*, W_{lk} is the scalar weight connecting the *l*th RBF with the *k*th output of the network as shown in Figure 6.19.

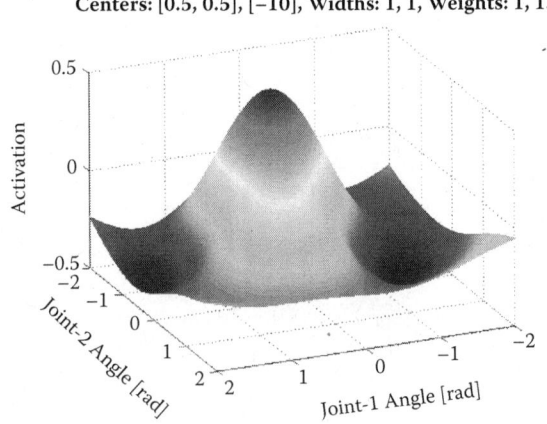

FIGURE 6.18
The landscape created by $w_1 y_2^M + w_2 y_3^M$ for $w_1 = 1$, $w_2 = 1.5$.

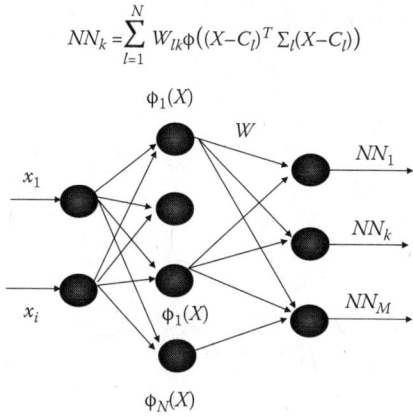

FIGURE 6.19
Structure of an RBF neural network.

6.2 NN-Based Identification of Dynamics of a Robot Manipulator

With this background, we will perform some experiments on modeling the dynamics of the two-link robot manipulator discussed in Chapter 2. The dynamics of the robot manipulator is governed by the equation $\tau = M(\theta)\ddot{\theta} + V(\theta,\dot{\theta}) + G(\theta)$, where the inertia matrix, the Coriolis and centrifugal matrices, and the gravity matrix are given in Equations (6.1) through (6.3). Let us rewrite them as follows:

$$M(\theta) = \begin{bmatrix} I_1 + I_2 + \frac{1}{4}M_1 l_1^2 + m_2\left(\frac{1}{4}l_2^2 + l_1^2 + l_1 l_2 \cos\theta_2\right) & I_2 + \frac{1}{2}m_2 l_2\left(\frac{1}{2}l_2 + l_1 \cos\theta_2\right) \\ I_2 + \frac{1}{2}m_2 l_2\left(\frac{1}{2}l_2 + l_1 \cos\theta_2\right) & I_2 + \frac{1}{4}m_2 l_2^2 \end{bmatrix}$$

(6.1)

is the positive definite inertia matrix,

$$V(\theta,\dot{\theta}) = \begin{bmatrix} -\frac{1}{2}m_2 l_1 l_2 \sin\theta_2 (2\dot{\theta}_1 \dot{\theta}_2 + \dot{\theta}_2^2) \\ \left(\frac{1}{2}m_2 l_1 l_2 \sin\theta_2\right) \dot{\theta}_1^2 \end{bmatrix}$$

(6.2)

is the Coriolis and centrifugal matrix, and

$$G(\theta) = \begin{bmatrix} \frac{1}{2} m_2 l_2 g \sin(\theta_1 + \theta_2) + \left(\frac{1}{2} m_1 l_1 + m_2 l_1 \right) g \sin \theta_1 \\ \frac{1}{2} m_2 l_2 g \sin(\theta_1 + \theta_2) \end{bmatrix} \quad (6.3)$$

is the gravity matrix.

Here, we discuss a few approaches to model the dynamics of the robot manipulator by estimating the landscapes created by each element of the above matrices. For instance, if we take the inertia matrix given in Equation (6.1), Coriolis and centrifugal matrices in Equation (6.2), and gravity matrix given in Equation (6.3), we notice that each element makes a landscape in the joint angle space as shown in Figures 6.20 through 6.22 (Chap6/element_landscapes.m). Therefore, if the requirement is to do what each of these landscapes does in terms of mapping joint angles to a value in a matrix, each element can be approximated by a linear combination of a set of primitive functions. We call this arrangement an artificial NN. Do not be intimidated by the term "artificial neural network." It is just a combination of a set of primitive functions to approximate the landscape of a given element.

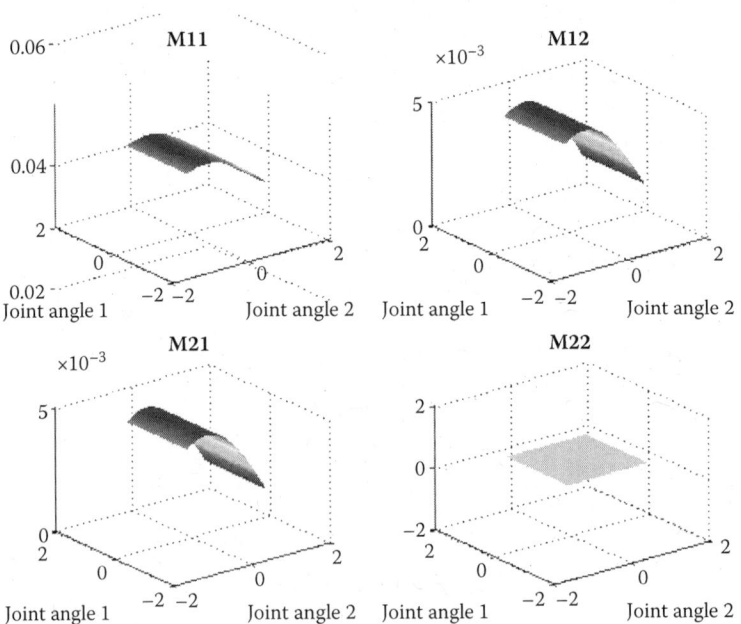

FIGURE 6.20
Landscapes made by each element of the inertia matrix in the joint angle space.

Neural Network-Based Control

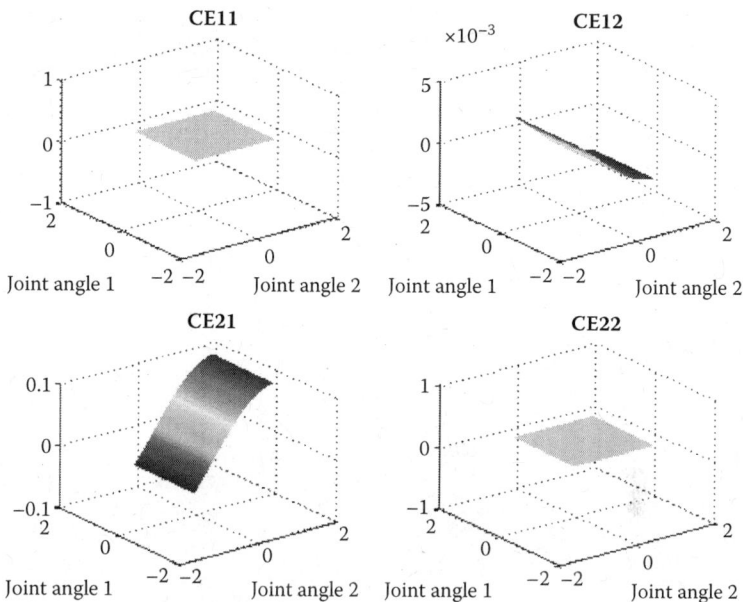

FIGURE 6.21
The landscapes made by each element of the centrifugal matrix in the joint angle space.

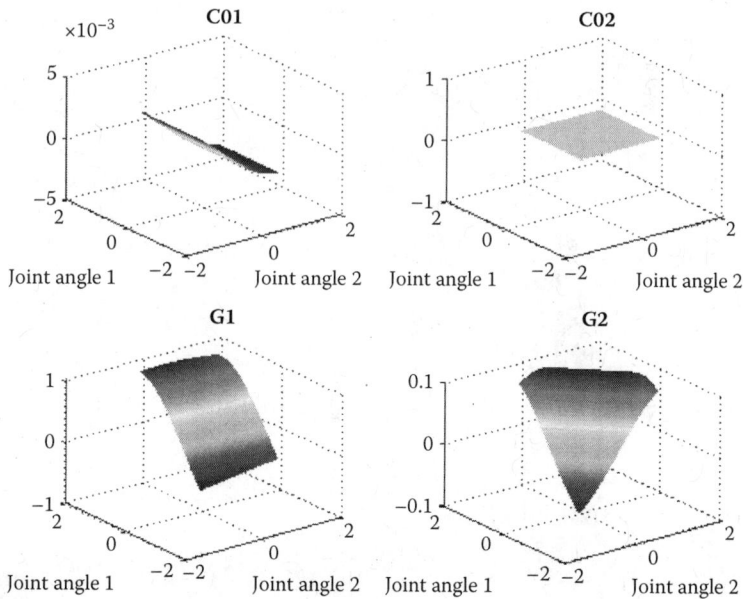

FIGURE 6.22
Landscapes made by each element of the Coriolis and gravity matrices in the joint angle space.

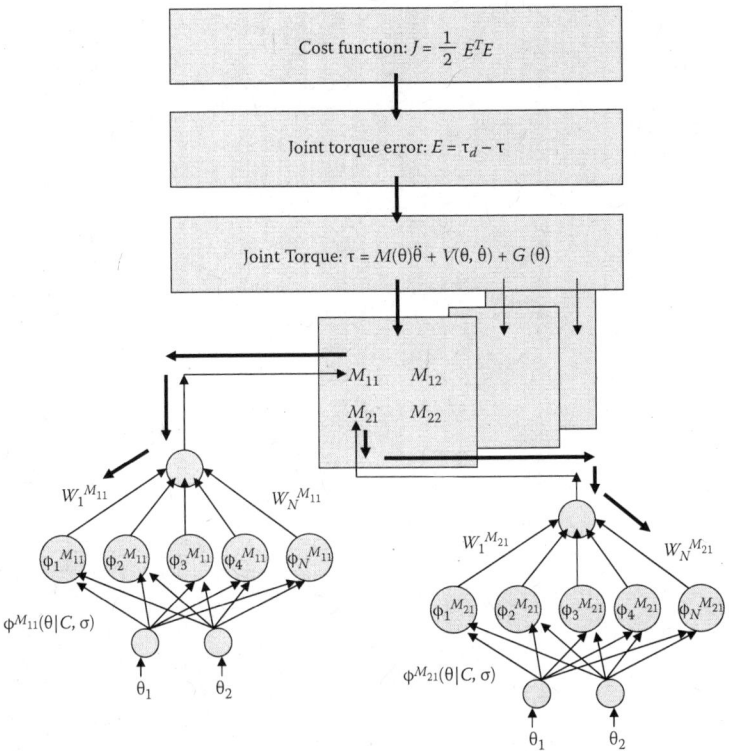

FIGURE 6.23
How artificial neural networks are organized to learn the dynamics of the two-link robot manipulator.

The selection of the primitive functions depends on our knowledge about the landscape of each element. Figure 6.23 shows how the parameters of a given NN can be updated using error feedback information.

6.2.1 An Approach to Estimate the Elements of the Mass, Centrifugal, Coriolis, and Gravity Matrices

We can write the dynamics of the two-link manipulator as follows:

$$\begin{pmatrix} \tau_1 \\ \tau_2 \end{pmatrix} = \begin{pmatrix} NN_{11}^M & NN_{12}^M \\ NN_{21}^M & NN_{22}^M \end{pmatrix} \begin{pmatrix} \ddot{\theta}_1 \\ \ddot{\theta}_2 \end{pmatrix} + \begin{pmatrix} NN_1^{CO} \\ NN_2^{CO} \end{pmatrix} \dot{\theta}_1 \dot{\theta}_2 + \begin{pmatrix} NN_{11}^{CE} & NN_{12}^{CE} \\ NN_{21}^{CE} & NN_{22}^{CE} \end{pmatrix} \begin{pmatrix} \dot{\theta}_1^2 \\ \dot{\theta}_2^2 \end{pmatrix} + \begin{pmatrix} NN_1^G \\ NN_2^G \end{pmatrix}$$

where NN_{ij}^M, NN_i^{CO}, NN_{ij}^{CE}, NN_i^G, $i = 1, 2$, $j = 1, 2$ are NNs approximating the elements of the inertial, Coriolis, centrifugal, and gravity matrices.

Let us look at how we can update the parameters of these NNs to evolve the required landscapes. We use a simple technique called the gradient-seeking method. The objective of the gradient-seeking method is to change the parameters of an NN to reduce the error of approximating a given landscape.

We start from a cost function to be minimized in the parameter space. In this case, the cost function is given by

$$J = \frac{1}{2} E^T E, \quad E = \begin{pmatrix} \tau_1^d - \hat{\tau}_1 \\ \tau_2^d - \hat{\tau}_2 \end{pmatrix}, \quad E = \begin{pmatrix} e_1 \\ e_2 \end{pmatrix}, \quad (6.4)$$

where $E, \tau_i^d, \hat{\tau}_i, i = 1, 2$ are the vector of errors, desired torque of joint i and the corresponding torque estimated by the approximate function shaped by the NN. Note that the cost given in Equation (6.1) is a function of the parameters of the NN. Therefore, the parameters should move from where they are to a location where the cost becomes minimum. How are we going to do this? What could be a mathematical model that will serve this purpose?

6.2.2 Optimum Parameter Estimation

We achieve this objective of moving the parameters to minimize the cost given in Equation (6.1) using a gradient-seeking algorithm. In the case of elements in the inertia matrix, the weight parameters of the NNs can be update as shown in Equations (6.5) through (6.7). Figure 6.23 shows how the error is backpropagated to the weights of the NN in a pictorial view.

$$W_k^{M_{ij}}(t+1) = W_k^{M_{ij}}(t) + \eta \frac{\partial J}{\partial W_k^{M_{ij}}} \quad (6.5)$$

$$\frac{\partial J}{\partial W_k^{M_{ij}}} = \frac{\partial J}{\partial e_i} \frac{\partial e_i}{\partial W_k^{M_{ij}}} = e_i \frac{\partial e_i}{\partial \hat{\tau}_i} \frac{\partial \hat{\tau}_i}{\partial W_k^{M_{ij}}} \quad (6.6)$$

$$\frac{\partial J}{\partial W_k^{M_{ij}}} = e_i(-1) \frac{\partial \hat{\tau}_i}{\partial M_{ij}} \frac{\partial M_{ij}}{\partial W_k^{M_{ij}}} = e_i(-1)\ddot{\theta}_i \varphi(\theta_1, \theta_2) = -e_i \ddot{\theta}_i \phi^{M_{21}}(\theta \mid C, \sigma) \quad (6.7)$$

From Equation (6.7), it is clear that the gradient in the cost landscape depends on that part of the model error attributable to the neuron concerned. In other words, the parameters of those neurons that were not active at the time concerned are not updated no matter how large the error is. However, if a neuron is active to a certain degree, the gradient for its parameter adaptation depends on the model error, activity of the respective neuron, and the corresponding joint acceleration at the time concerned.

Figure 6.24 shows how the network reduced the estimation error across trials. It is clear that the network dropped the error fast in the first few

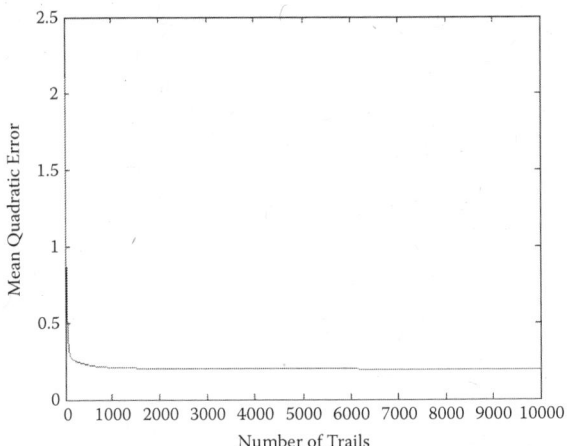

FIGURE 6.24
Learning history of the neural networks for Mexican hat neurons.

trials and then settled at a minimum error level. This is known as the local minima problem in any gradient-seeking learning algorithm. The shape of the cost landscape in the parameter space can have many local basins whose bottom is above that of the global minimum. If the parameters are initialized in the basin of attraction of a local minimum, the parameters will move along the gradients to reach the local minimum. Once reached, the gradient tends to zero. Therefore, according to Equation (6.2), the parameters cannot move any more. Therefore, the error also does not change. One way to jump out of the local minima is to introduce a disturbance to see if the algorithm finds another slope in the basin of attraction of the global minimum. Yet, this is a very time-consuming task. Therefore, gradient-seeking learning algorithms are good when we have a substantial amount of information about the model. However, our first simulation assumed minimum knowledge of the physical system. Wavelet neurons were selected because we know that the elements in the dynamic equation of the two-link manipulator have trigonometric functions that look like Mexican hat wavelets Figures 6.25 through 6.27 show the model error for three trajectories.

Figure 6.28 shows the convergence history of the neural network when the activation function was set to be a sinusodal function of joint angles.

Figures 6.29 through 6.31 show the model error of a trigonometric function-based neural network for the three test trajectories. Compare the errors with the corresponding figures for the mexican at neuron-based neural network in Figures 6.25 through 6.27.

Neural Network-Based Control

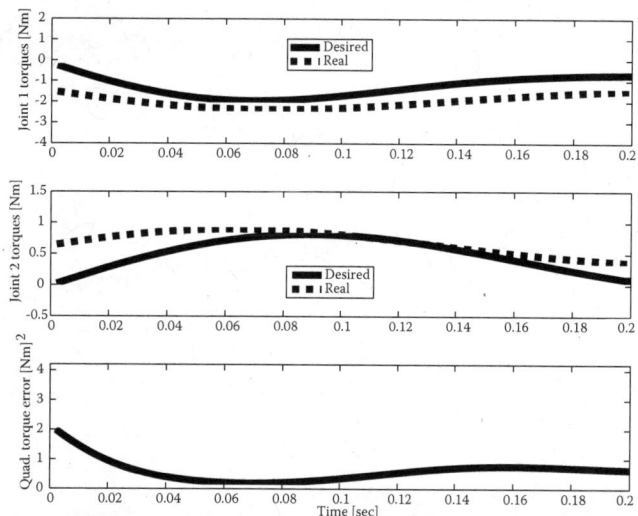

FIGURE 6.25
How the joint torques estimated by the learned model compare with the true torques of the physical system for the test trajectory—1.

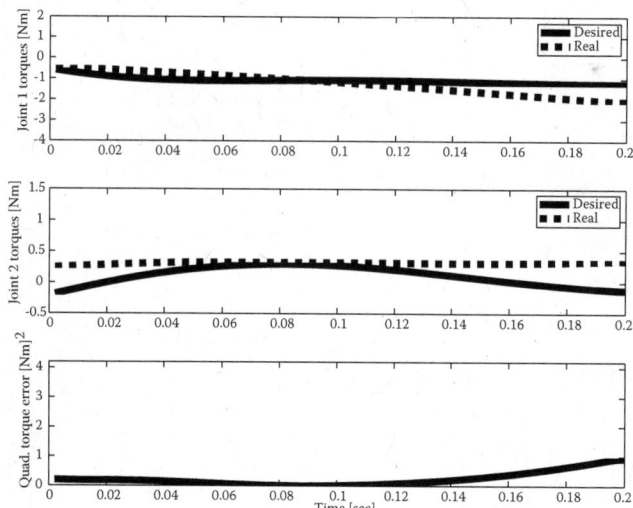

FIGURE 6.26
How the joint torques estimated by the learned model compare with the true torques of the physical system for the test trajectory—2.

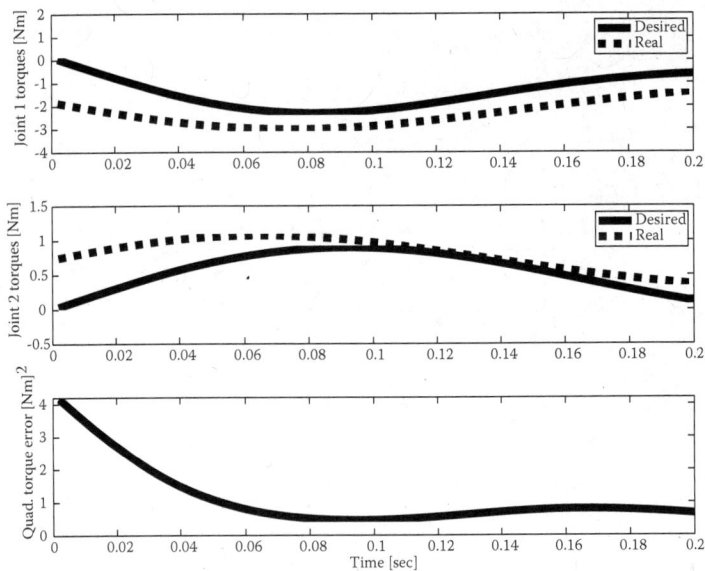

FIGURE 6.27
How the joint torques estimated by the learned model compare with the true torques of the physical system for the test trajectory—3.

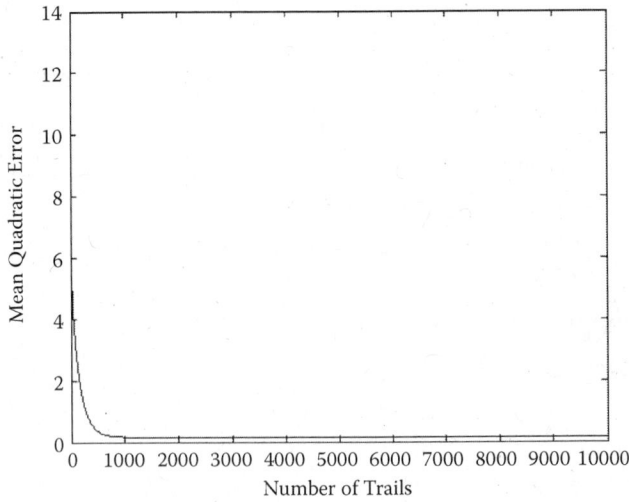

FIGURE 6.28
The learning history of the neural networks for trigonometric functions.

Neural Network-Based Control

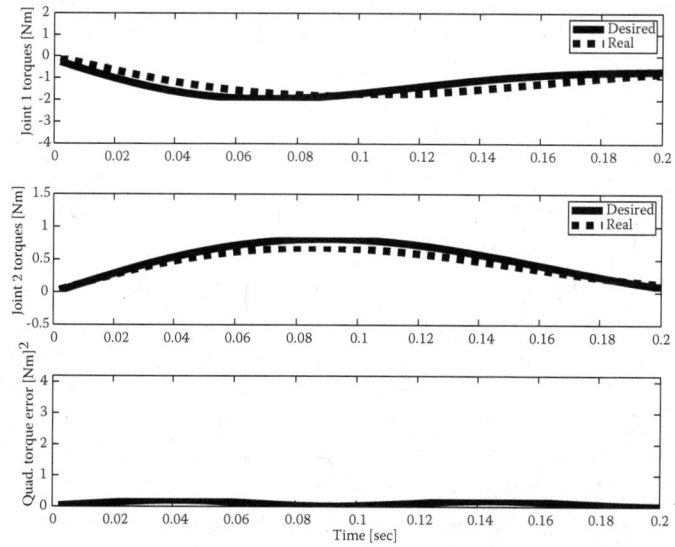

FIGURE 6.29
How the joint torques estimated by the learned model using the trigonometric functions compare with the true torques of the physical system for the test trajectory—1.

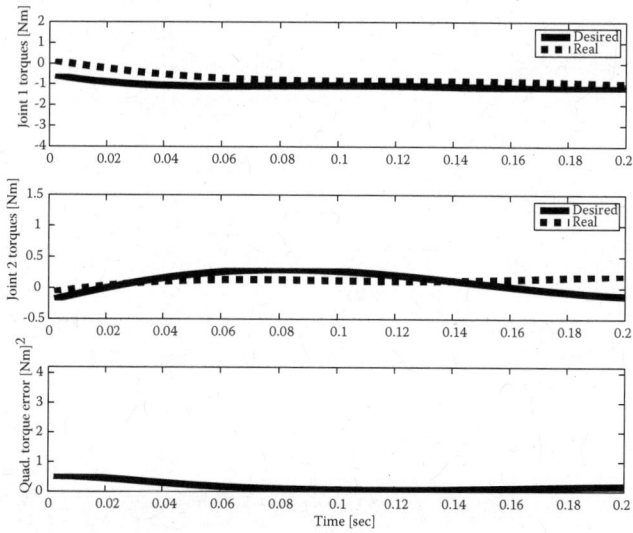

FIGURE 6.30
How the joint torques estimated by the learned model compare with the true torques of the physical system for the test trajectory—2.

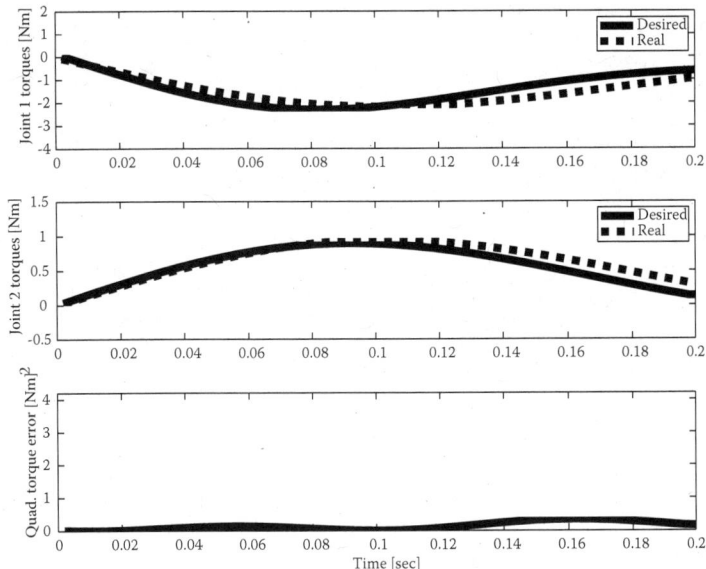

FIGURE 6.31
How the joint torques estimated by the learned model using trigonometric functions compare with the true torques of the physical system for the test trajectory—3.

6.3 Structure of NNs

An NN can cover a continuum from full knowledge to no knowledge of the structure of the process to be identified. However, it is very common to see standard NN structures being applied no matter to what degree the structure of the dynamics is known. This leads to suboptimal solutions because the learned parameters of the NN do not reflect the true structure of the parameters of the system.

Figure 6.32 shows the structure of the dynamics of the state-space equation of the inverted pendulum. It starts from the unknown parameters of the state-space equation and successively progresses toward more abstract elements of the dynamic equation. It is a good means of illustrating how parameters are related to higher-level elements in the dynamic equation. The state space equation can be used to define a cost function J that quantifies the estimation error given by

$$\dot{x} = Ax + bF$$

$$J = \frac{1}{2}\sum_{k=1}^{N}(\dot{x}_d(k) - \dot{x}(k))^2 = \frac{1}{2}\sum_{k=1}^{N}(e(k))^2$$

where $\dot{x}_d(k)$ and $\dot{x}(k)$ are the desired rate of change of the state vector and that estimated by the model, respectively, and N is the total span of sampling

Neural Network-Based Control

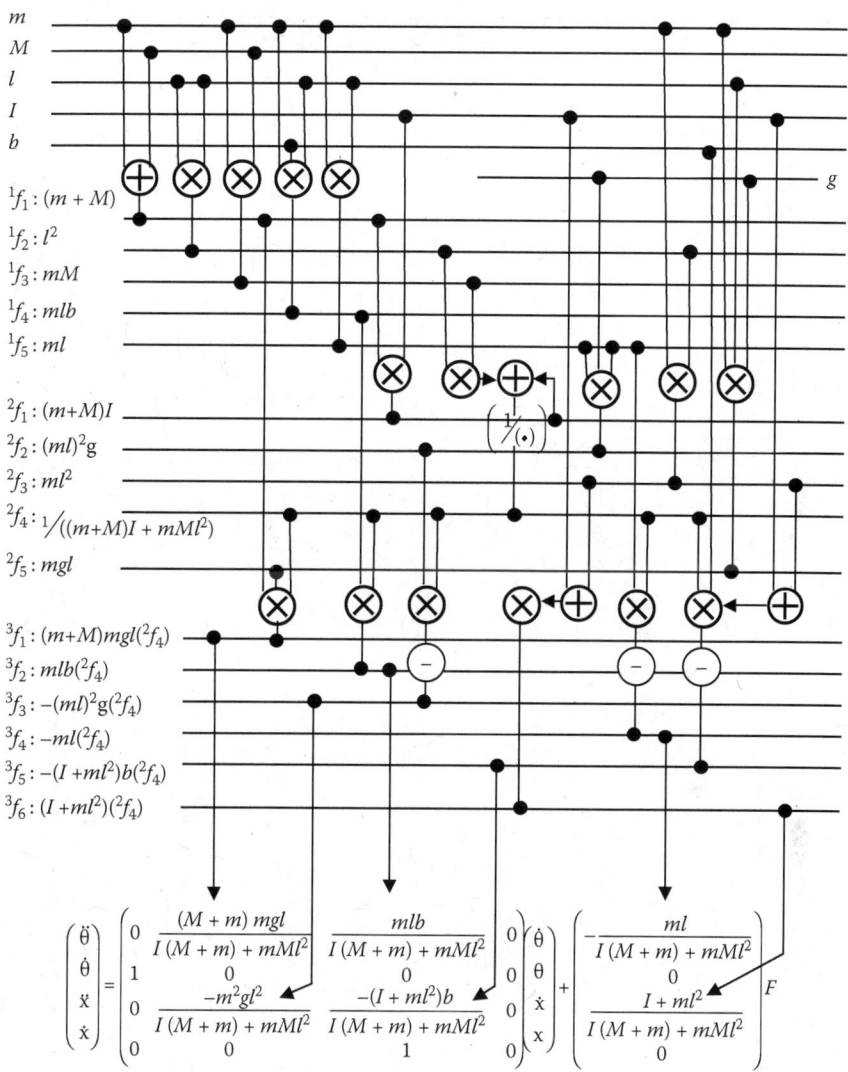

FIGURE 6.32
Structure of the dynamic model of the inverted pendulum.

steps being considered for estimation cost minimization. Furthermore, it helps to derive the error backpropagation equations shown in the following equation:

$$\frac{\partial J}{\partial \varphi} = \frac{\partial J}{\partial e} \frac{\partial e}{\partial \dot{x}} \frac{\partial \dot{x}}{\partial \varphi} = \sum_{k=1}^{N} e(k) \left(0 - \frac{\partial \dot{x}(k)}{\partial \varphi} \right)$$

where ϕ is the vector of parameters to be estimated. The rest of the gradient-based cost minimization in the parameter space is given by

$$\frac{\partial J}{\partial \varphi} = -\sum_{k=1}^{N} e(k)\left(\frac{\partial A}{\partial \varphi}x + \frac{\partial B}{\partial \varphi}F\right)$$

$$\frac{\partial J}{\partial \varphi} = -\sum_{k=1}^{N} e(k)\left(\begin{bmatrix} 0 & \frac{\partial^3 f_1}{\partial \varphi} & \frac{\partial^3 f_2}{\partial \varphi} & 0 \\ 0 & 0 & 0 & 0 \\ 0 & \frac{\partial^3 f_3}{\partial \varphi} & \frac{\partial^3 f_5}{\partial \varphi} & 0 \\ 0 & 0 & 0 & 0 \end{bmatrix} x + \begin{bmatrix} \frac{\partial^3 f_4}{\partial \varphi} \\ 0 \\ \frac{\partial^3 f_6}{\partial \varphi} \\ 0 \end{bmatrix} F\right)$$

$$\frac{\partial^3 f_1}{\partial \varphi} = \frac{\partial(^2 f_5\, ^2 f_4\, ^1 f_1)}{\partial \varphi}$$

$$\frac{\partial^3 f_1}{\partial \varphi} = [^2 f_4\, ^1 f_1 \quad ^2 f_5\, ^1 f_1 \quad ^2 f_5\, ^2 f_4] \begin{bmatrix} \frac{\partial(^2 f_5)}{\partial \varphi} \\ \frac{\partial(^2 f_4)}{\partial \varphi} \\ \frac{\partial(^1 f_1)}{\partial \varphi} \end{bmatrix}$$

A closer look at the nonlinear equations of the inverted pendulum reveals a very useful property of the dynamic equations that we can use to design an RBF NN. That is, the dynamics can be written in the linear form given by

$$F = \beta^T \lambda$$

where λ is a vector of scalar parameters, and β is a vector of variables.

Let us rewrite the nonlinear equations of the inverted pendulum given by

$$\begin{pmatrix} 0 \\ F \end{pmatrix} = \begin{pmatrix} I + ml^2 \cos^2\theta & ml\cos\theta \\ ml\cos\theta & M+m \end{pmatrix} \begin{pmatrix} \ddot\theta \\ \ddot x \end{pmatrix} + \begin{pmatrix} mgl\sin\theta \\ -ml\dot\theta^2 \sin\theta \end{pmatrix}$$

where M and m are the mass of the cart and the rod, respectively, and x, l, θ, and I are the cart position, the half length of the rod, the angle of the rod, and the moment of inertia of the rod around its center of gravity, respectively.

We can separate the variables and parameters given by

$$\begin{pmatrix} 0 \\ F \end{pmatrix} = \begin{pmatrix} \sin\theta & 0 & \ddot{\theta}\cos\theta & \ddot{\theta}\cos^2\theta & \ddot{\theta} \\ 0 & \ddot{x} & \ddot{x}\cos\theta & -\dot{\theta}^2\sin\theta & 0 \end{pmatrix} \begin{pmatrix} mgl \\ M+m \\ ml \\ ml^2 \\ I \end{pmatrix}$$

This format allows us to design various innovative NN structures to estimate the model parameters. Essentially, the vector

$$\begin{pmatrix} \sin\theta & 0 & \ddot{\theta}\cos\theta & \ddot{\theta}\cos^2\theta & \ddot{\theta} \\ 0 & \ddot{x} & \ddot{x}\cos\theta & -\dot{\theta}^2\sin\theta & 0 \end{pmatrix}$$

can be measured from the experiment. What we want to estimate are the elements of the vector

$$\begin{pmatrix} mgl \\ M+m \\ ml \\ ml^2 \\ I \end{pmatrix}$$

In fact, rigid body dynamics in the form $\tau = M(\theta)\ddot{\theta} + V(\theta,\dot{\theta}) + G(\theta)$ obey this linearity rule in the torque space. Therefore, one could simply design a layered NN for the pendulum dynamics as shown in Figure 6.33.

The weights of the above network can be tuned using the gradient-seeking equation given by

$$W_k(t+1) = W_k(t) + \eta \frac{\partial J}{\partial W_k}$$

where

$$J = \frac{1}{2} E^T E, \quad E = \begin{pmatrix} 0-\delta \\ F^d - \hat{F} \end{pmatrix}, \quad E = \begin{pmatrix} e_1 \\ e_2 \end{pmatrix}$$

6.4 Generating Training Data for an NN

Let us have a close look at the NN shown in Figure 6.33. Let us take the activation function $\ddot{\theta}\cos^2\theta$ for a detailed study. The landscape of this function is shown in Figure 6.34. Modalities of the landscape of these primitive functions should be taken into account before generating training data to make sure all modalities are covered. What we mean by modalities is the number of different types of geometrical features in the landscape of the primitive function. For instance, if the function has three peaks with unique shapes, we say that the particular function has three modalities. It can be seen from the landscape of $\ddot{\theta}\cos^2\theta$ that the angle is the most predominant variable deciding the modality of this function. Beyond a certain range, it repeats the same pattern. The variation along the angular acceleration axis is fairly linear. Therefore, we should be careful to take experimental data for the angle and the angular acceleration of the pendulum over a span to cover at least the most basic modalities. If not, the estimate of the parameter w_3 would not be very accurate.

Project. Use the linear dynamic equations of the inverted pendulum to generate training data for the above NN for two cases. Case 1: initial conditions given by $\theta = pi/20\,[rad], \dot{\theta} = 0\,[rad/sec], x = 0.02\,[m], \dot{x} = 0\,[m/sec]$. Case 2: initial conditions given by $\theta = pi/6\,[rad], \dot{\theta} = 0\,[rad/sec], x = 0.5\,[m], \dot{x} = 0\,[m/sec]$. Use the parameter values $M = 1\,[kg], m = 0.1\,[kg], l = 0.75\,[m]$. You may use the MATLAB® function "LQR" to design your linear-quadratic regulator in the form $F = -[K_1 \quad K_2 \quad K_3 \quad K_4][x \quad \dot{x} \quad \theta \quad \dot{\theta}]^T$.

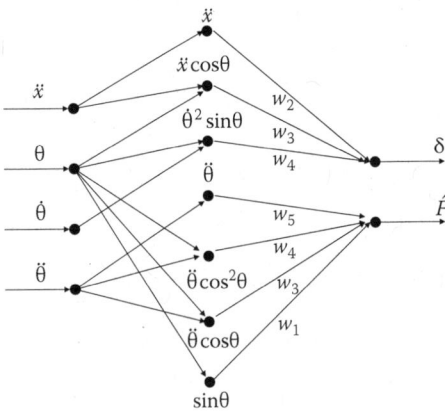

FIGURE 6.33
A neural network with trigonometric basis functions for a case where the structure of the dynamics is partially known.

Neural Network-Based Control

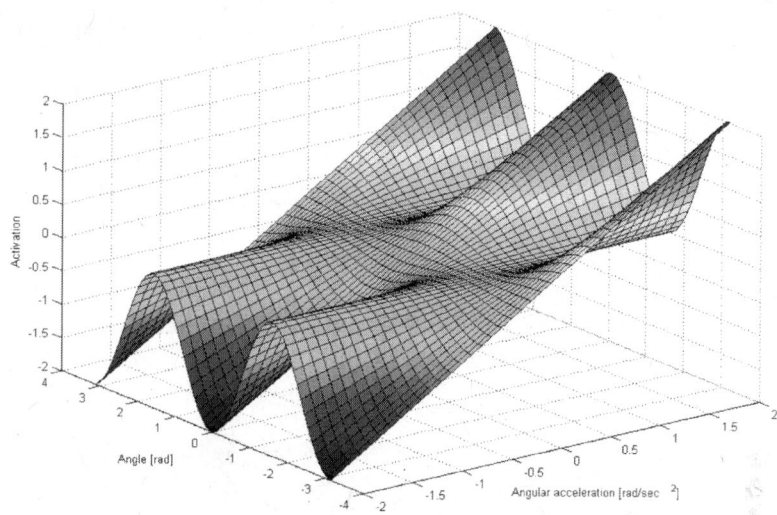

FIGURE 6.34
Landscape of the activation function $\ddot{\theta}\cos^2\theta$.

Use the above gradient-seeking parameter updating algorithm to estimate the weights of the NN. Compare the accuracy of the parameters for the above two cases.

6.5 Dynamic Neurons

You may have noted in Figure 6.33 that the NN requires higher-order information such as the acceleration of the cart and angular acceleration of the pendulum as input variables. However, in practice, it is very expensive to have sensors to measure such higher-order information. It would have been much better if we could use the same network with basic inputs such as the position of the cart and the angle of the pendulum as inputs because it is very easy to measure them. Likewise, in biological brains, the proprioceptive information sent by our peripheral sensory system is often low-order information such as position, velocity, tension, etc. Therefore, what would be a good mechanism to map relationships with higher-order information such as accelerations?

Recurrent networks that use internal temporal feedbacks have been a promising solution to this problem. To understand the idea better, let us look

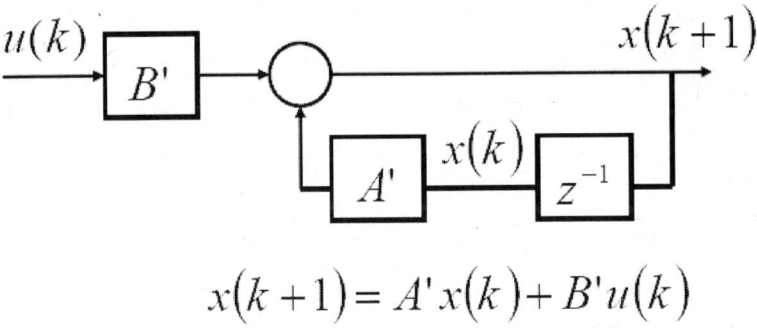

FIGURE 6.35
Discrete linear dynamical system, where $A' = (AT + I)$, $B' = BT$.

at a linear dynamical system given by $\dot{x} = Ax + Bu$. This system can be written as a discrete system given by $x(k+1) = (AT + I)x(k) + BTu(k)$, where T is the sampling time interval.

Figure 6.35 shows how a single input can drive the discrete dynamical system with internal state feedback. Likewise, NNs can define their own internal states that interact with other variables in the network to make it a dynamical system. This is the basic structural framework behind emergence. Emergence is a term used to explain the orchestration of complex phenomena out of an interaction among a simple set of structures and variables. Figure 6.36 shows a more general recurrent network that can develop more complex dynamical phenomena. It is believed that the most complex phenomena in biological brains are orchestrated in such recurrent networks evolved over a long period. A simple enactment of such a dynamic NN is the complex ground

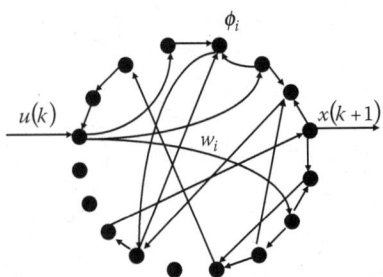

FIGURE 6.36
A general recurrent neural network.

scratching and pecking behavior of a 1-day-old chick. It is hard to believe that a few hours is enough to figure out the correct structure and parameters of a network that can make a few-hours-old chick scratch the ground and peck to eat food. It is more likely that the basic structure is genetically encoded so that the chick can tune the synapses through its own experience. However, there is much to be learned about how complex networks are formed in the human brain because it seems that the human brain is a more open plastic system where other phenomena such as aptitude may arise from genetically encoded hardwired network structures. The idea of recurrent neurons and recurrent NNs has been extended to model more complex phenomena such as chaotic systems.

6.6 Attractors, Strange Attractors, and Chaotic Neurons

6.6.1 Attractors and Recurrent Neurons

When we discussed about recurrent neurons and networks (Figure 6.36), you may have had questions about the stability of such complex dynamical systems. In fact, the main criterion for such a network to be useful is its stability under noisy inputs and external perturbations. The structure and parameters of such a network should ensure that internal states converge to attractors as shown in Figure 6.37. This is also important in the training of NNs where we experience this phenomenon in the way weights converge to a set of fixed values after which the training algorithm stops. If the initial set of weights is in the basin of attraction of the globally optimum set of weights, the gradient seeking algorithms will gradually converge to this attractor. Therefore, the size and properties of the basin of attraction of an attractor will decide most of the characteristics of the particular attractor.

Sometimes, we come across a class of attractors called *strange attractors*. In a strange attractor, although the rules of attraction are deterministic, the

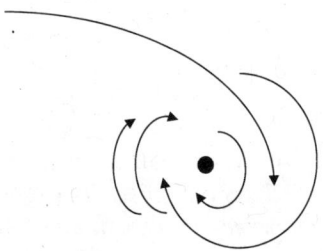

FIGURE 6.37
An attractor state.

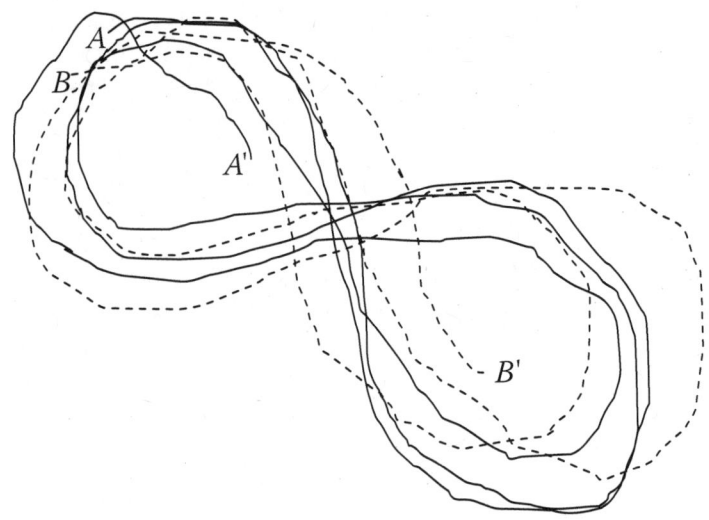

FIGURE 6.38
A strange attractor.

manner in which trajectories of states behave appears to be random. This phenomenon occurs because two state vectors that start very close to each other in the state space will travel in two completely different paths, and the difference will grow over time. Yet, the important point is that both of these trajectories do not go beyond a bounded region. This whole phenomenon is called *chaos*. You can see a strange attractor in Figure 6.38.

In Figure 6.38, you may notice that the two neighboring points A, B diverge to A', B' after a given span of time. By definition, if two neighboring states diverge over time, the original neighborhood is an unstable region. Yet, the fact that the trajectories of the diverging states remain inside a bounded region gives rise to the strange chaotic nature. In other words, in a strange attractor where chaotic properties are shown, we find repulsion between vectors of states, and at the same time, we find an overall attraction in the system as a whole that holds the state trajectories together. It can be viewed as a locally unstable but globally stable system. What you see in a waterfall can be an example (Figure 6.39). In a fall, if you trace two particular water particles originating almost at the same place, you will see that they travel in completely different trajectories because the ways that they bounce on the uneven rock are quite different from each other. Millions of particles like that form the beautiful chaotic pattern in a waterfall. Yet, on the other hand, gravity forces them not to bounce off completely out of order. Thus, gravity, as a force that tries to hold the trajectories together, and the collisions among water particles and with the rock, as another force that repels neighboring particles apart, work together to form this chaotic system that goes out of order within boundaries and rough shapes distinguish one fall from another.

Neural Network-Based Control

FIGURE 6.39
A waterfall in Sri Lanka.

6.6.2 A Chaotic Neuron

Chaos can arise by iterating a mathematical function. Consider the following equations characterizing the Navier-Stokes attractor:

$$\dot{x} = P(y - x)$$

$$\dot{y} = Rx - y - xz$$

$$\dot{z} = xy - By$$

$$P = 10, R = 28, B = 8/3$$

These three equations can be viewed as a recurrent network of three neurons that interact among each other as shown in Figure 6.40.

Figure 6.41 shows how the attractor behaves for a hundred different initial conditions randomly generated between 0 and 20. The most important thing to note here is that a network with three parameters given by P, R, and B can take the internal states $x(k)$, $y(k)$, and $z(k)$ defined at the kth sampling step along complex trajectories just by iterating the internal states in its structure.

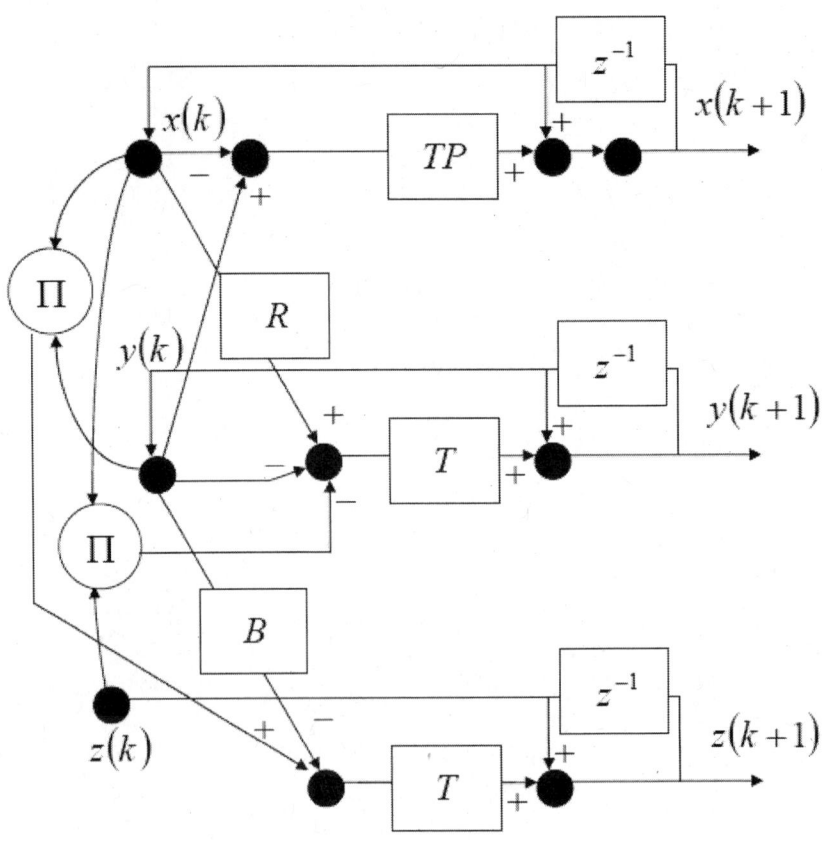

FIGURE 6.40
Neural network representation of Navier-Stokes attractor equations.

This is a good example of what can emerge from a recurrent NN given a structure and a set of parameters.

Project. Use the chap6/chaotic_neuron.m MATLAB® code to examine how the network develops state trajectories by changing the parameters P, R, and B around the given values. Observe the most sensitive parameters and guess why.

6.7 Cerebellar Networks and Exposition of Neural Organization to Adaptively Enact Behavior

Now that we have a basic understanding of what biological and artificial neurons are and what different types of NNs can do, let us have a look at a fairly well-known biological network in the cerebellum, which is thought to play

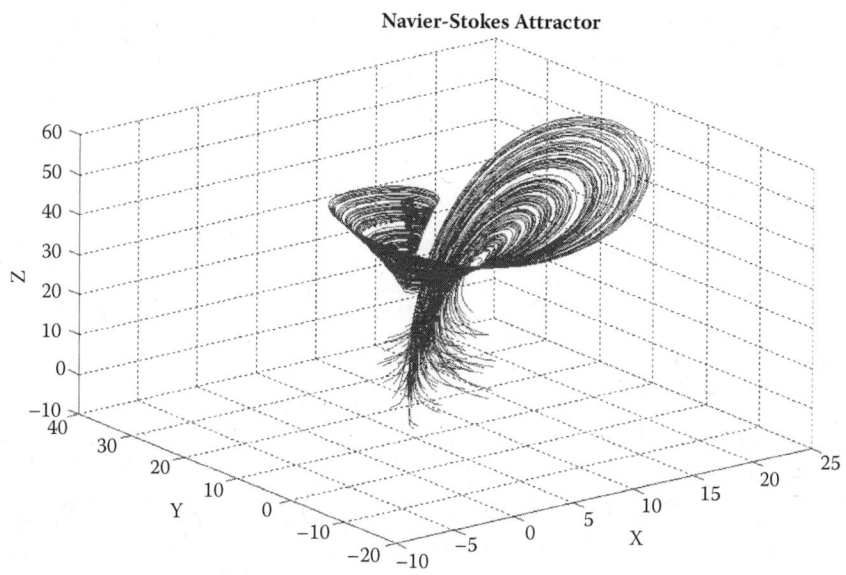

FIGURE 6.41
Navier-Stokes attractor for different initial states shown in the bottom part of the figure.

a major role in the learning dynamics of the hand and the task in order to perform accurate movements [4,5]. It is amazing to see how different neurons have organized to perform simple tasks so that, as a well-organized network, they perform a very complex task. First, let us briefly discuss the physical problem they try to solve. The first animals differed from their predecessor plants in the sense that the animals wanted to physically move from one place to another. This happened approximately 700 to 900 million years ago, when the atmosphere began to have an increased oxygen concentration so that moving beings could combust food more rapidly and release the required energy to actuate internal muscle-like structures. This also required building multicellular structures as opposed to their single-cell predecessors. It is thought that this process gave rise to more compartmentalized structures where groups of cells organized to perform specific duties to solve the mobility problem. Obviously, movement required the bodies to have actuators and controllers. We refer to the neural formation that organized to control the movement of the rest of the body as the motor system. This neural structure, known as the motor system, constantly faced the problem of adapting to control a changing body. The body changed in shape, degrees of freedom, size, and weight. Even today, we know that our hands change in length and mass every second, although very slowly. However, the effect of the change of the body can be significant in terms of the change in dynamic properties such as inertia, gravity effect, etc., if we consider two points of time separated by months.

Therefore, if we try to do some task that we have mastered well after the time of our last practice, we will first experience that we have lost some of our expertise because our body and the task itself may have changed in between. How does the brain as a controller cope with this? The only answer is to have an internal model in the brain that can retune its parameters to match the new dynamics of the hand and the task. Therefore, there should be neurons that can carry the information of the errors to the internal structures, there should be neurons that encode the desired behavior from previous memories, and there should be networks that map desired behaviors to descending motor commands to the muscles. On the other hand, there are reasons to believe that the brain requires an internal model of the task and the hand because the typical delays in the communication pathways between the proprioceptive sensors in the hand and the cortex (~120 ms) should introduce instability to the arm movements even if the motor command is perfectly accurate. In order for the brain to calculate predictive motor commands that will be relevant by the time it reaches the muscles, the brain should be able to use a model of the arm to assess the consequences of the just-computed motor command in order to correct the subsequent computations.

It is evident from the sketch in Figure 6.42 that the cerebellar networks have strong connections with the cortex, which makes us reinforce the belief that the cerebellar networks may provide important information about the dynamics of the hand and the task to the cortex to compute motor commands.

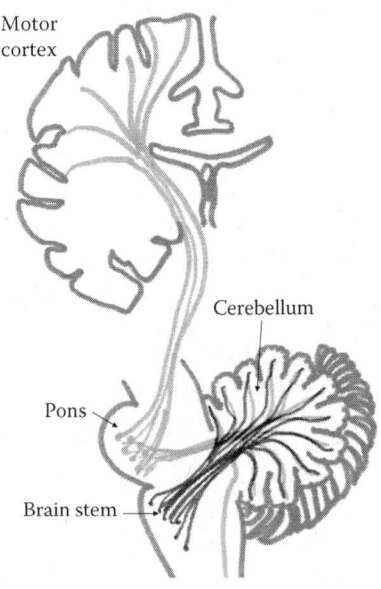

FIGURE 6.42
A sketch of how the cerebellum is connected to the motor cortex, pons, and the brain stem.

Neural Network-Based Control

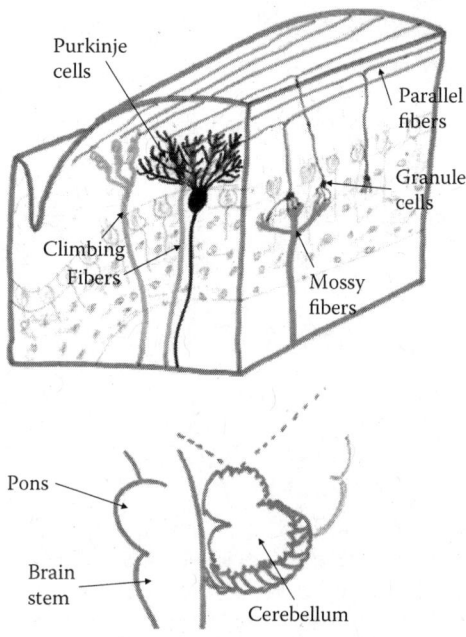

FIGURE 6.43
Neural structure in one segment of a cerebellar fol.

A close look at the details of the cerebellar networks shows how they are organized to model the dynamics of the hand and the task. Figure 6.43 shows the organization of the networks in the cerebellum. It is also shown in Figure 6.43 that there are two main sources carrying input information to the cerebellum. *Mossy fibers* are one source. They originate from the spinal cord and the brain stem, as shown in Figure 6.43. Therefore, they are thought to carry information about the goal and the desired force/position/velocity trajectory. Mossy fibers synapse with *granule cells* in the granule layer, also viewed as the hidden layer of the cerebellar network. Axons of the granule cells project toward the periphery of the cerebellum, where they branch off in a T junction to form *parallel fibers* that run virtually parallel to the outer surface of the cerebellar formation. These parallel fibers form a fabric that synapses with the Purkinje cells sending the final motor command out of the cerebellum. The other source of information is the movement error encoded in the *climbing fibers*. Climbing fibers start from a neural structure known as the inferior olive in the medulla. Climbing fibers wrap around Purkinje cells to change their synapses to correct the movement and muscle coordination accuracy as shown in Figure 6.43.

Figure 6.44 shows how the overall system is believed to be wired together to orchestrate a movement task. It should be noted here that our knowledge

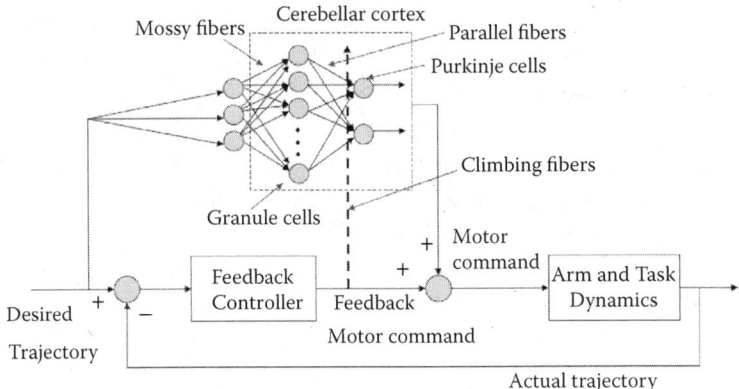

FIGURE 6.44
Organization of cerebellar networks to control the musculoskeletal system in the arm.

about the exact function and the exact neural structure is not yet complete. For instance, we are also aware of supportive mechanisms to the inverse dynamics-based control framework shown in Figure 6.44, such as forward models predicting the outcome of a descending motor command. For more detailed reading, we recommend the readers to refer to work done by Nikhil Bushman and Reza Shadmehr at http://www.bme.jhu.edu/~reza.

References

[1] Squire, L. R., Bloom, F. E., McConnell, S. K., Roberts, J. L., Spitzer, N. C., and Zigmond, M. J., *Fundamental Neuroscience*, Academic Press, San Diego, CA, 1999.

[2] Kandel, E. R., Schwartz, J. H., and Jessell, T. M., *Principles of Neural Science*, McGraw-Hill, Columbus, OH, 2000.

[3] Beatty, B., *The Human Brain—Essentials of Behavioral Neuroscience*, Sage Publications, Inc., Thousand Oaks, CA, 2001.

[4] Shadmehr, R., and Wise, S. P., *The Computational Neurobiology of Reaching and Pointing—A Foundation for Motor Learning*, MIT Press, Cambridge, MA, 2005.

[5] Barlow, J. S., *The Cerebellum and Adaptive Control*, Cambridge University Press, Cambridge, 2002.

7
System of Systems Simulation

In this chapter, a system of systems (SoS) simulation framework and related case studies are presented. First, we will introduce the SoS concepts and recent work on their simulation and modeling. Then, we will present our approach to SoS simulation framework based on discrete event systems specification (DEVS) simulation tools and Extensible Markup Language (XML). We present three case studies of the SoS simulation framework on robust threat detection and data aggregation using heterogeneous systems of rovers (systems). Finally, a real-time SoS simulation framework is introduced in agent-in-the-loop setting for testing and evaluating systems in an SoS.

7.1 Introduction

There has been a growing recognition that significant changes need to be made in government agencies and industries, especially in the aerospace and defense sectors [1–6]. Today, major aerospace and defense manufacturers such as Boeing, Lockheed-Martin, Northrop-Grumman, Raytheon, and BAE Systems include some version of "large-scale systems integration" as a key part of their business strategies [1–6]. In some cases, these companies have even established entire business units dedicated to systems integration activities [1].

In parallel, there is a growing interest in new SoS concepts and strategies. The performance and functioning of a group of heterogeneous systems have become the focus of various applications including military, security, aerospace, and disaster management systems [5–8]. There is an increasing interest in achieving synergy between these independent systems to achieve the desired overall system performance [5,6,9]. In the literature, researchers have addressed the issues of coordination and interoperability in an SoS [10–12]. To study SoS characteristics and parameters, one needs to have realistic simulation frameworks properly designed for SoS architecture. There are some attempts to develop simulation frameworks for multiagent systems (MAS) using discrete event simulation tools [13–21]. In these research efforts, the major focus is given to discrete event simulation architecture with Java. DEVS modeling is presented and discussed in detail by Zeigler et al. [13,14].

In the work of Mittal et al. [17,19], DEVS is combined with Service-Oriented Architecture (SOA) to create the DEVS/SOA architecture. The DEVS state machine approach is introduced Mittal [19]. Finally, DEVS Modeling Language is developed by using XML-based Java to simulate systems in a net-centric manner with relative ease [17].

Based on DEVS and Java, a discrete event XML-based SoS simulation framework was recently introduced by Sahin et al. [4,12,20]. In this chapter, we will extend this SoS simulation framework by designing an SoS problem with heterogeneous autonomous systems for threat detection and data aggregation. In addition, DEVS models are presented for real-time DEVS simulation that has real robots (agents) and virtual robots at the same time.

7.2 SoS in a Nutshell

The concept of SoS is essential to more effectively implement and analyze large, complex, independent, and heterogeneous systems working (or made to work) cooperatively [9,12,20,21]. The main thrust behind the desire to view the systems as an SoS is to obtain higher capabilities and performance than would be possible with a traditional systems approach. The SoS concept presents a high-level viewpoint and explains the interactions between each of the independent systems. However, the SoS concept is still at its developing stages [5,6,22,23].

Systems of systems are super systems composed of other elements that are independent complex operational systems themselves and interact among themselves to achieve a common goal [5,6,12,22,23]. Each element of an SoS achieves well-substantiated goals even if they are detached from the rest of the SoS. For example, a Boeing 747 airplane, as an element of an SoS, is not an SoS, but an airport is an SoS; or a rover on Mars is not an SoS, but a robotic colony (or a robotic swarm) exploring the red planet is an SoS [4,12]. Associated with SoS, there are numerous problems and open-ended issues requiring substantial fundamental advances in theory and verifications. In fact, there is not even a universal definition among system engineering community even though there are serious attempts to define SoS and create its standards.

Based on a survey of the SoS literature survey, there are several SoS definitions [5,6,24–30]. All definitions of SoS have their own merits, depending on their application and domain. However, we will list the more common definitions (with primary focus and application domain) as reported in the literature.

Definition 1: According to Sage and Cuppan [25], systems of systems exist when majority of the following five characteristics are present: operational independence, managerial independence, geographic distribution, emergent behavior, and evolutionary development. Primary focus: evolutionary acquisition of complex adaptive systems. Application: military.

Definition 2: Based on the work of Jamshidi [24] and Kotov [26], systems of systems are large-scale concurrent and distributed systems composed of complex systems. Primary focus: information systems. Application: private enterprise.

Definition 3: Carlock et al. [27] notes that enterprise systems of systems engineering is focused on coupling traditional systems engineering activities with enterprise activities of strategic planning and investment analysis. Primary focus: information intensive systems. Application: private enterprise.

Definition 4: Pei [28] describes SoS integration as a method of pursuing development, integration, interoperability, and optimization of systems to enhance performance in future battlefield scenarios. Primary focus: information intensive systems integration. Application: military.

Definition 5: According to Luskasik [29], SoSE involves the integration of systems into systems of systems that ultimately contribute to evolution of the social infrastructure. Primary focus: education of engineers to appreciate systems and interaction of systems. Application: education.

Definition 6: Manthorpe [30] notes that, in relation to joint war-fighting, SoS is concerned with interoperability and synergism of command, control, computers, communications, and information as well as intelligence, surveillance, and reconnaissance systems. Primary focus: information superiority. Application: military.

During the course of defining SoS, one may want to ask the following question: "How does an SoS differ from large-scale systems (LSS), MAS such as system of robotic swarms and unmanned air vehicle (UAV) swarm?" LSS is defined [31,32] as a system that can be decomposed into subsystems (leading to hierarchical control) or whose output information can be distributed (leading to decentralized control). Within these definitions, an LSS is not an SoS because the "systems" of an LSS as an SoS cannot operate independently like a robotic colony or an airport. In other words, an LSS does not create a capability beyond the sum of the individual capabilities of each system [33].

MAS, on the other hand, are special cases of SoS that have a further unique property aside from having an emergent behavior, homogeneous system members like similar robotic architecture, and UAV models. This observation is true even when agents in MAS do not communicate with each other. SoS and systems engineering are discussed further in several reports [5,6,21,30,34–45].

Our favorite definition is "Systems of systems are large-scale concurrent and distributed systems that are composed of complex systems." The primary focus of this definition is information systems, which emphasizes the interoperability and integration properties of an SoS [4,12].

Interoperability in complex systems (i.e., MAS) is very important because agents operate autonomously and interoperate with other agents (or nonagent entities) to take better actions [4,12,20]. Interoperability requires successful communication among the systems. Thus, the systems should carry out their tasks *autonomously* as well as communicate with other systems in the SoS to take better actions for the overall benefit of the SoS, not just for themselves. *Integration* implies that each system can communicate and interact (control) with the SoS regardless of their hardware and software characteristics. This means that they need to have the ability to communicate with the SoS or a part of the SoS without compatibility issues such as operating systems, communication hardware, and so on [4,12,20]. For this purpose, an SoS needs a common language its components can speak. Without a common language, the SoS components cannot be fully functional and the SoS cannot be adaptive in the sense that new components cannot be integrated to the SoS without a major effort. Integration also implies the control aspects of the SoS because systems need to understand each other in order to receive commands or signals from other SoS components [4,12,20].

One might argue that this explains why we cannot have a global state-space model of the SoS by combining the state-space models of the systems and subsystems in an SoS [4,12]. Designing a state-space mathematical model for complex LSS (e.g., MAS) is often difficult because of uncertainties in the operation and complex interactions among the systems (agents). In addition, the state-space representation of some systems may not be available. Such decentralized systems also require an efficient data handling (information processing) mechanism to capture the operational behavior of the system. The problem of characterizing the behavior and representation of such systems becomes even more difficult when each of these systems is heterogeneous and independently operational in nature. A naïve description of an SoS is multiple instances of such complex heterogeneous operational independent systems working in synergy to solve a given problem as described in Definition 2 [4,12].

In real-world systems, the problem is addressed in a higher level where the systems send and receive data from other systems in the SoS, and make a decision that leads the SoS to its global goals. Let us take the military surveillance example, where different units of the army collect data through their sensors trying to locate a threat or determine the identity of the target. In this type of situation, army command center receives data from these heterogeneous sensor systems such as airborne warning and control system ground radars, submarines, and so on [4,12]. These systems are parts of the SoS making the decision—say the command and control station. Still, they may be developed using different technologies. Thus, they will be using different hardware and/or software. This creates a huge barrier in data aggregation and data fusion when using the data received from these systems because they would not be able to interact successfully without hardware and/or software compatibility. In addition, data coming from these systems are not unified, which will add to the barrier in data aggregation [4].

One solution of the problem using SoS is to modify the communication medium among the SoS components. Two possible ways of accomplishing this task are explained below.

Create a software model of each system using the same software tool. In this approach, each component in the SoS talks to a software module embedded in itself. The software module collects data from the system and, through the software model, generates outputs and sends them to the other SoS components. If these software modules are written with a common architecture and a common language, then the SoS components can effectively communicate regardless of their internal hardware and/or software architectures [4].

Create a common language to describe data, where each system can express its data in this common language so that other SoS elements can parse the data successfully [4].

The overhead that needs to be generated to have software models of each system on an SoS is enormous and must be redone for new members of the SoS. In addition, this requires the complete knowledge of the state-space model of each SoS components, which is often not possible. Thus, data-driven approach would have better success in integrating new members to the SoS and also in applying the concept to other SoS application domains [4].

In this chapter, we present an SoS simulation architecture based on XML in order to wrap data coming from different sources in a common way. XML can be used to describe each component of the SoS and their data in a unifying manner. If XML-based data architecture is used in an SoS, the only requirement for the SoS components is to understand/parse XML file received from the components of the SoS. Most complex systems used by the military and government agencies have the processing and computational power to run an XML parser to process the data received from the components of the SoS.

In XML, data can be represented in addition to the properties of the data such as source name, data type, importance of the data, and so on. Thus, it does not only represent data but also gives useful information about the SoS to take better actions and to understand the situation better. XML has a hierarchical structure where an environment can be described with a standard and without a huge overhead. Each entity can be defined by the user in XML in terms of its visualization and functionality. For example, a hierarchical XML architecture shown in Listing 1 can be designed for an SoS so that it can be used in the components of the SoS and also be easily applied to other SoS domains. In Listing 1, the first line defines the name of the file that describes the functionality of the user-defined keywords used to define the SoS architecture. This file is mainly used for visualization purposes so that any of the SoS components can display the current data or the current status of the SoS to a human/expert to ensure that the proper decision is taken.

LISTING 1   AN XML-BASED SOS ARCHITECTURE [4,12]

```xml
<!--Created 11/8/2006 Author @ Ferat Sahin -->
<?xml-stylesheet type="text/css" href="genericxml.css"?>

<systemofsystem>
        <id> Id of the System of Systems </id>
        <name> The name of the System of System</name>
        <system>
                <id>Id of the first system</id>
                <name> The name of the first system </name>
                <description> The description of the first system
</description>
                <dataset>
                        <Output>
                                <id>Id of the first output</id>
                                <data>Data of the first output</data>
                        </Output>
                        <Output>
                                <id>Id of the second output</id>
                                <data>Data of the second output</data>
                        </Output>
                </dataset>
                <subsystem>
                        <id>Id of the subsystem of the first System</id>
<name>The name of the subsystem</name>
                        <description>This is a subsystem of the system
in a SoS</description>
                        <dataset>
                                <Output>
```

The first keyword of the XML architecture is "systemofsystem" representing an SoS. Everything described after this keyword belongs to this SoS based on the XML architecture. The following keywords, "id" and "name," are used to describe the SoS. Then, the keyword "system" is used to declare and describe the first system of the SoS. In addition to "id" and "name," two more keywords, "description" and "dataset," are used to describe the properties of the system and to represent the data coming out of this system. Data sources are denoted by "output" keyword and data are provided with the keyword "data." After representing data from two sources, a subsystem is described by the keyword "subsystem." The subsystem and its data are presented in a similar manner. Other subsystems can be described in this subsystem or in parallel to this subsystem as well as additional systems can be described in the SoS [4].

The next section describes the SoS simulation framework implementing an XML-based architecture with case studies and results.

7.3 An SoS Simulation Framework

In an SoS, systems and/or subsystems often interact with each other because of interoperability and overall integration of the SoS. These interactions can be achieved by efficient communication among the systems using either peer-to-peer communication or through a central coordinator in a given SoS. Since the systems within SoS are operationally independent, interactions between systems are generally asynchronous in nature. A simple yet robust solution to handle such asynchronous interactions (specifically, receiving messages) is to throw an event at the receiving end to capture the messages from single or multiple systems. Such system interactions can be effectively represented as discrete event models [4,12,20]. In discrete event modeling, events are generated in different time intervals as opposed to some predetermined time interval seen commonly in discrete-time systems. More specifically, the state change of a discrete event system happens only upon the arrival (or generation) of an event, not necessarily at equally spaced time intervals [4,12]. To this end, a discrete event model is a feasible approach in simulating the SoS framework and its interaction [4,12,20,46]. Several discrete event simulation engines [47–49] are available that can be used in simulating interaction among heterogeneous mixture of independent systems. We consider one such simulation framework—discrete event system specification (DEVS)—because of the available effective mathematical representation and its support to distributed simulation using Department of Defense's High-Level Architecture (HLA) [50].

7.3.1 DEVS Modeling and Simulation

Discrete event system specification [13] is a formalism that provides a means of specifying the components of a system in a discrete event simulation. In DEVS formalism, one must specify *basic models* and how these models are connected together. These basic models are called *atomic* models and the larger models obtained by connecting these atomic blocks in a meaningful fashion are called *coupled* models (Figure 7.1). Each of these atomic models has *inports* (to receive external events), *outports* (to send events), set of *state variables, internal transition, external transition,* and *time advance functions*. As stated formally by Zeigler et al. [13,51], using set theory notation, an atomic model in DEVS is a structure

$$M = X, S, Y, \delta_{int}, \delta_{ext}, \delta_{con}, \lambda, \text{ and } ta$$

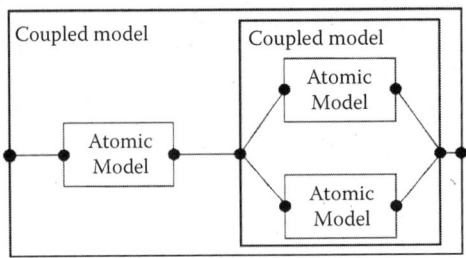

FIGURE 7.1
DEVS Model representing system and subsystems [4,12,20].

where
  X = set of input values
  S = set of states
  Y = set of output values
  $\delta_{int}: S \to S$ = internal transition function
  $\delta_{ext}: QX^b \to S$ = external transition function

$$\delta_{ext}: QX^b \to S$$

where
  $Q = \{(s,e) | s \in S, 0 \le e \le ta(s)\}$ = total state set and e is the time elapsed since last transition
  X^b = collection of bags over X (sets in which some elements may occur more than once)
  $\delta_{con}: QX^b \to S$ = confluent transition function
    $\lambda: S \to Y^b$ = output function
    $ta: S \to R^+_{0,\infty}$ = time advance function

The model's description (implementation) uses (or discards) the message in the event to do the computation and delivers an output message on the outport and makes a state transition. A Java-based implementation of DEVS formalism, DEVSJAVA [52], can be used to implement these atomic or coupled models. In addition, DEVS-HLA [52], based on HLA, will be helpful in distributed simulation for simulating multiple heterogeneous systems in the SoS framework [4].

As described in the SoS simulation framework, we use XML-based language to represent and transfer data among the systems in the SoS. To test the SoS architecture in the DEVS environment, the XML-based language needs to be embedded into the simulation environment. The following section explores how XML and DEVS environment can be combined in the simulation environment [4].

7.3.2 XML and DEVS

In DEVS, messages can be passed from one system (coupled or atomic model) to another using either predefined or user-defined message formats. Since

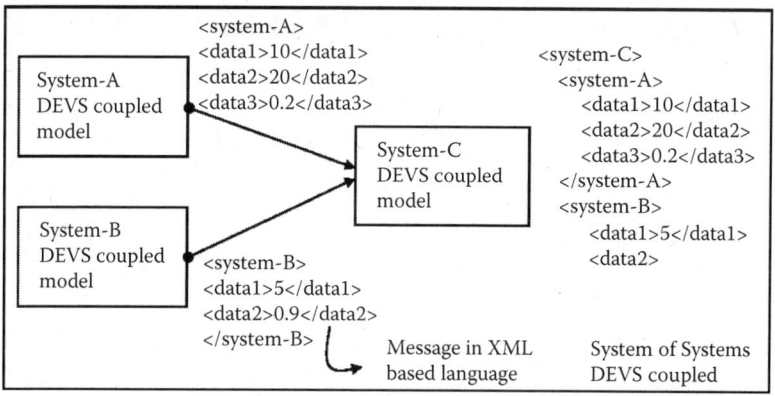

FIGURE 7.2
A three-system SoS simulation example and XML-like message passing [4,12,20].

the systems within SoS may be different in hardware and/or software, there is a need for a unifying language for message passing. Each system need not necessarily have the knowledge (operation, implementation, timing, data issues, and so on) of another system in an SoS. Therefore, one has to work at a high level (information or data level) to understand the present working condition of the system. One such good fit for representing different data in a universal manner is XML. Figure 7.2 conceptually describes an SoS simulation example to demonstrate the use of XML as a message passing paradigm using DEVS formalism [4].

In Figure 7.2, there are three systems in a hierarchy, where systems A and B send and receive data from system C. System C sends and receives data from a higher level as described in the message of system C. The data sent by system C have data from systems A and B. In addition, it has information about system A and B being system C's subsystems.

With the above-mentioned XML-based integration architecture and the DEVS environment, we can then offer solutions to SoS with heterogeneous complex systems such as mobile sensor platforms or microdevices [4]. Next, we present example scenarios and case studies that use the SoS simulation framework.

7.4 SoS Simulation Framework Examples

Based on the presented SoS architecture, we have developed and simulated three cases. The first case is a data aggregation scenario where there is a base robot, a swarm robot, two sensors, and a threat. The second case is a robust threat detection scenario where there is a base robot, two swarm robots, five sensors, and a threat. The last case is also a robust threat detection scenario with a colony of swarm robots and more sensors. In the last example, we also use realistic graphics and replace base robot with a command and control tower.

7.4.1 Case 1: Data Aggregation Simulation

In this scenario, the sensors are stationed such that they represent a border. The threat is moving in the area and it can be detected by the sensors. When the sensors detect the threat, they notify the base robot. Then, the base robot notifies the swarm robot about the location of the threat based on the information sent by the sensors [4].

7.4.1.1 DEVS-XML Format

As mentioned earlier, the hardware and/or software architectures of these systems will not be the same in reality. Thus, they may not able to talk to each other successfully even though they can operate independently. We have implemented an XML-based SoS message architecture in DEVSJAVA software [52]. In this XML-based message architecture, each system has an XML-like message consisting of their name and a data vector. The names are used in place of the XML keywords. The data vectors are used to hold the data of the systems. The length of the vectors in each system can be different based on the amount of data each system contains. For example, the XML message of the sensors has the sensor coordinates and the threat level. The threat level is set when a threat arrives in the coverage area of the sensors. On the other hand, the Base Robot's XML message has its coordinates, the threat level, and the coordinates of the sensor reporting a threat. Thus, the data vector length of Base Robot's XML message has five elements, whereas the data vector of an XML message of a sensor has three elements. Table 7.1 presents the names and length of the vector data of each system in the SoS.

The data vectors are made of "double" variables in order to keep track of the positions accurately. The "Threat" element in the data vector is a flag representing threat (1.0) or no threat (0.0). The elements X_t and Y_t are used for the destination coordinates in the XML messages of the Base Robot and the Swarm Robot. These represent the coordinates of the sensor that reported a threat.

TABLE 7.1

XML Message Components for Systems in SoS [4]

System	Name	Vector Data Length
Base Robot	"Base Robot"	5 (X, Y, Threat, X_t, Y_t)
Swarm Robot	"Swarm Robot"	5 (X, Y, Threat, X_t, Y_t)
Sensor	"Sensor 1"	3 (X, Y, Threat)
Sensor	"Sensor 2"	3 (X, Y, Threat)
Threat	"Fire"	2 (X, Y)

7.4.1.2 Programming Environment

In DEVSJAVA environment, a data structure called "XmlEntity" is created based on the "entity" data structure to create XML messages of each system as shown in Listing 2. This data structure is used to wrap/represent the data of each system.

LISTING 2 XMLENTITY DATA STRUCTURE IN DEVSJAVA [4,20]

```
public class XmlEntity extends entity{
Vector value;
String name;
/** Creates a new instance of XmlEntity */
public XmlEntity() {
```

The structures/behaviors of the systems in the SoS are created/simulated by DEVSJAVA atomic or coupled models. There are five atomic models in this scenario: BRobot (for Base Robot), SRobot (for Swarm Robots), Sensor (for stationary sensors), Threat (for threat), and PlotModule (this is for plotting the system components during the simulation). Figure 7.3 illustrates the DEVSJAVA development environment, all atomic models, and other components.

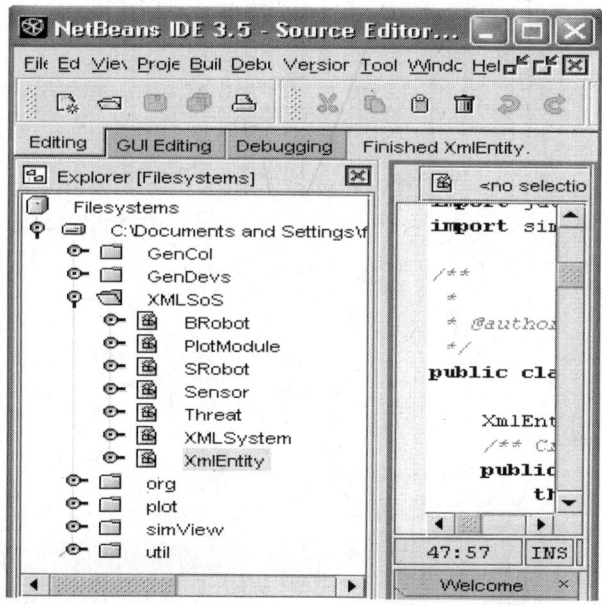

FIGURE 7.3
Atomic models and DEVSJAVA components [4].

The simulation framework is developed as a Java package, called XMLSoS. Under this package, we have atomic models and two more system components, XMLEntity and XMLSystem. XMLSystem is where the system components and their couplings are created or instantiated.

7.4.1.3 Simulation Results

The DEVS atomic and coupled models of the SoS in DEVSJAVA environment are shown in Figure 7.4. Figure 7.5 illustrates a DEVS simulation step with the XML messages sent among the systems in the SoS. Finally, Figure 7.6 exhibits the simulation environment created by the "Plot Module" atomic model in DEVS. In this environment, the two sensors are located near each other representing a border. The "Threat" (red dot) is moving in the area. When the threat is in one of the sensor's coverage area, the sensor signals the Base Robot. Next, Base Robot signals the Swarm Robot so that it can go and verify whether the threat is real. The behavior of the SoS components can also be seen in Figure 7.5 as the movements of the "Threat" and the "Swarm Robot" are captured. The green dot is representing the Swarm Robot. The Base Robot is not shown because it does not move. When the Threat enters into a sensor's area of responsibility, that sensor area is filled with red color by the PlotModule atomic model to show the threat level. Then, the Swarm Robot (green dot) moves into the sensor to check whether the threat is real.

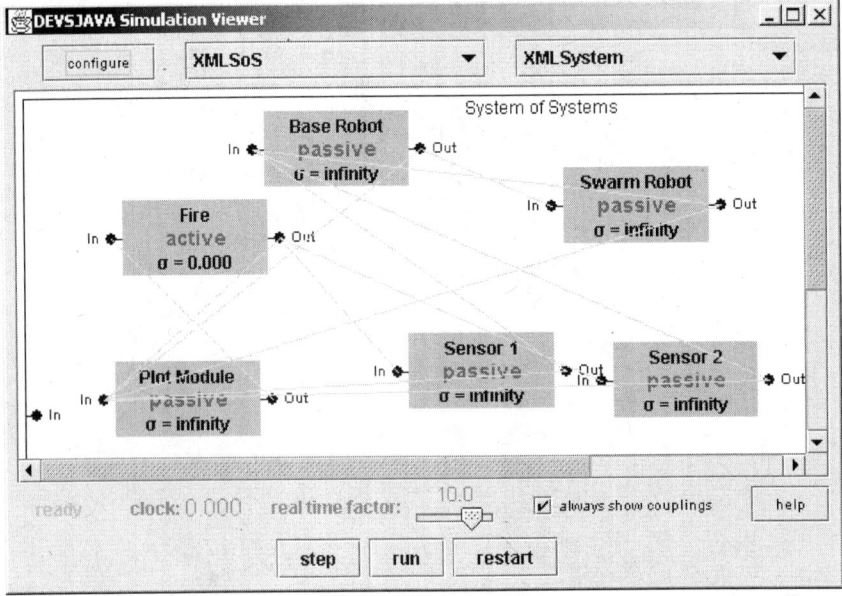

FIGURE 7.4
The DEVS atomic and coupled modules for XML-based SoS simulation [4,12].

System of Systems Simulation

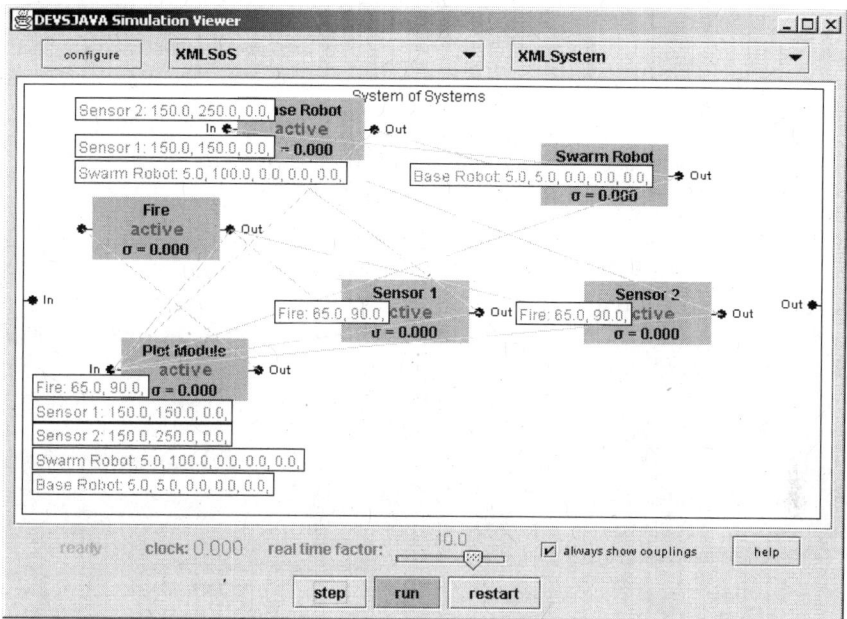

FIGURE 7.5
DEVS simulation with XML-based messages shown at the destination [4,12].

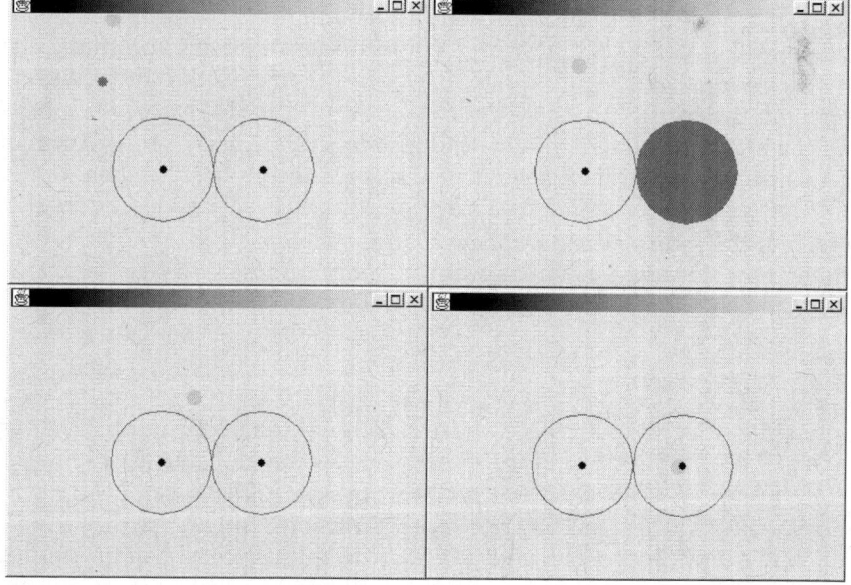

FIGURE 7.6
Progress of DEVS simulation on data aggregation [4,12].

When the Swarm Robot reaches the sensor, it reports the threat level to the Base Robot. If the threat is not real, the Swarm Robot moves away from the sensor's coverage area [4].

7.4.2 Case Study 2: A Robust Threat Detection System Simulation

In this case, we have extended the simulation framework such that multiple Swarm Robots and several Sensor nodes can be simulated for a robust threat detection scenario [4,12,20]. This robust threat detection (data aggregation scenario), there is one base robot, two swarm robots, five sensors, and one threat. The sensors are stationed such that they represent a sensor network covering an area. The threat is moving in the area and is detected by sensors when the threat is sufficiently close to a sensor node. All sensors communicate with a base robot, which does not actively seek threat. Instead, the base robot processes the information using sensor data and data sent by swarm robots. A command center can collect processed data (aggregated data) from the base robot without interacting with other components of the SoS such as sensors and swarm robots.

The robust threat detection works as follows. When a sensor detects the threat, they notify the base robot. Then, the base robot sends the swarm robots the location of the threat based on the information sent by the sensors. Swarm robots are assumed to be equipped with the same sensors as the stationary sensors so that they can verify the threat when they reach the sensor area. This is crucial in order to get rid of false detections as they can cause inefficient usage of system components [4]. For example, if a sensor sends false threat information to the base robot, the command center will automatically send systems designed to disable the threat without verifying the threat. This will cause spending valuable system resources and can make the system more vulnerable to real threats. Thus, verification of the threat by deploying several small-size swarm robots can help reduce false threat detections and save system resources. Thus, when a threat is detected, the swarm robots are notified by the base robot. Then, swarm robots move toward the threat area and communicate with each other so that they do not go to the same threat location [4].

7.4.2.1 DEVS-XML Format

As noted earlier, the hardware and/or software architectures of these systems will not be the same in reality. Thus, they may not able to talk to each other successfully even though they can operate independently. We have implemented XML-based SoS message architecture in DEVSJAVA software [52]. In this XML-based message architecture, each system has an XML-like message consisting of their name and a data vector. The name of each system represents an XML tag. The data vectors are used to hold the data of the systems. The length of the data vectors in each system can be different based on

TABLE 7.2

XML Components for Systems in SoS [4]

System	Name	Vector Data Length
Base Robot	"Base Robot"	5 (X, Y, Threat, X_t, Y_t)
Swarm Robot	"Swarm Robot 1"	6 (X, Y, Threat, X_t, Y_t, Threat V)
Swarm Robot	"Swarm Robot 2"	6 (X, Y, Threat, X_t, Y_t, Threat V)
Sensor	"Sensor 1"	3 (X, Y, Threat)
Sensor	"Sensor 2"	3 (X, Y, Threat)
Sensor	"Sensor 3"	3 (X, Y, Threat)
Sensor	"Sensor 4"	3 (X, Y, Threat)
Sensor	"Sensor 5"	3 (X, Y, Threat)
Threat	"Fire"	2 (X, Y)

the amount of data each system contains. For example, the XML message of the sensors has the sensor coordinates and the threat level. The threat level is set when a threat arrives in the coverage area of the sensors. On the other hand, the base robot's XML message has its coordinates, the threat level, and the coordinates of the sensor reporting a threat. Thus, the data vector of the Base Robot's XML message has five elements, whereas the data vector of an XML message of a sensor has three elements. Table 7.2 presents the names and the length of the vector data of each system in the SoS.

The data vectors are made of "double" variables in order to keep track of the positions accurately. The "Threat" element in the data vector is a flag representing threat (1.0) or no threat (0.0). Elements X_t and Y_t are used as the destination (target) coordinates in the XML messages of the Base Robot and Swarm Robot. These represent the coordinates of the sensor that reported a threat. The "Threat V" element in the data vector is a flag representing whether a swarm robot verified a threat [4].

7.4.2.2 Simulation Setup

Figure 7.7 shows a screenshot of the DEVS model of the SoS described above. Figure 7.8 shows a DEVS simulation step with the XML messages sent among the systems in the SoS. To evaluate the performance of the robust threat detection, we have plotted the locations of Base Robot, Swarm Robots, Sensors, and the Threat. The sensor coverage areas are represented by a circle to illustrate complete sensor network coverage area. The threat is represented as a red dot. The swarm robots are represented as green (Swarm Robot 1) and orange (Swarm Robot 2) dots. In addition, the sensor locations are presented as black dots. When a threat enters into a sensor area, the sensor location becomes red, meaning threat detection. The black dot at the upper left corner represents the Base Robot and it does not move during the simulation. Figure 7.9 shows the initial positions of all SoS elements.

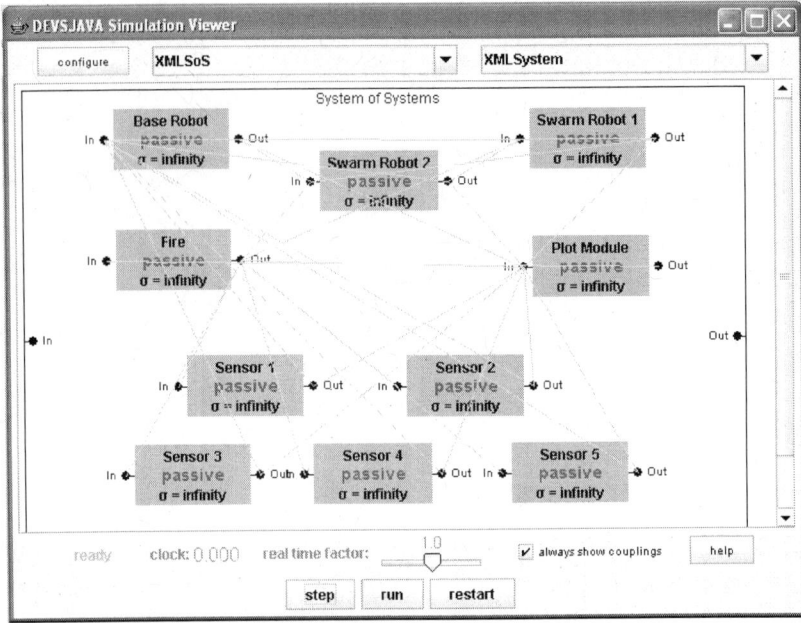

FIGURE 7.7
DEVS atomic and coupled modules for XML-based SoS simulation [4].

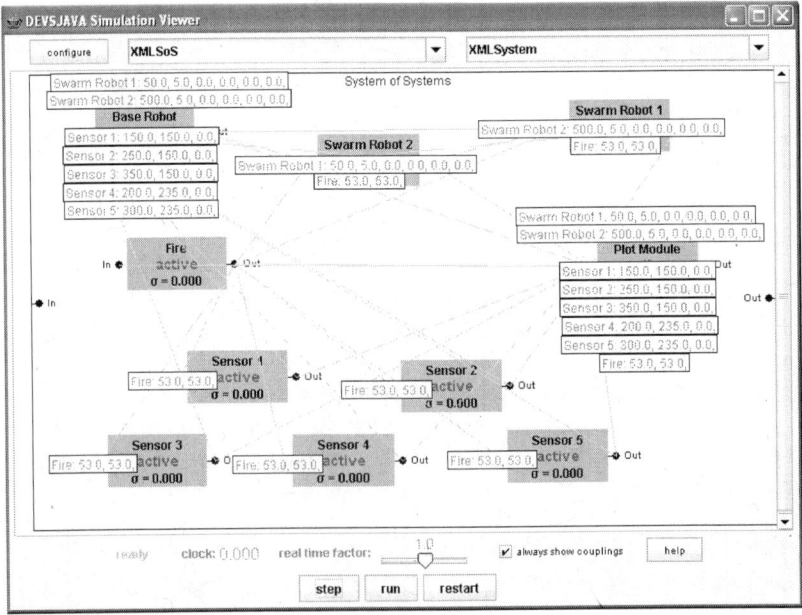

FIGURE 7.8
DEVSJAVA simulation with XML-based messages shown [4].

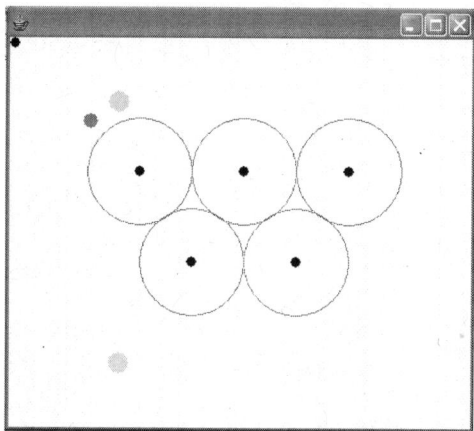

FIGURE 7.9
Initial conditions of XML system simulation [4].

Swarm robots are stationed on the top and bottom of the sensor field in order to have faster threat verification. A threat is generally verified by the closest swarm robot at the time of threat detection. If the swarm robots are far away from each other they both may go toward a threat location. However, when they are close enough, they can communicate and the second swarm robot generally goes back to its original location or tries to verify another threat. All five sensors can also be seen in the figure. Sensors 1 to 3 are on the top row from left to right, respectively. Sensor 4 and Sensor 5 are on the bottom row of the cluster, also from left to right.

7.4.2.3 Robust Threat Detection Simulation

In this section, we will present screenshots of the simulation to illustrate the behaviors of the robust threat detection SoS. We set the threat "Fire" moving randomly in the field in order to capture the behaviors of the swarm robots and sensors. Figure 7.10 shows the Threat moving within the range of Sensor 1. Sensor 1 shows its activation by changing its center from red to black. As soon as a threat is detected, a message is sent to the Base Robot, which has the X and Y coordinates of the sensor and the threat flag. This flag is normally "0" and it becomes "1" when a threat is detected. Once the Base Robot receives an XML message from Sensor 1, it checks the threat flag and sends the coordinates of Sensor 1 to Swarm Robot 1 and Swarm Robot 2 if the flag is set to "1." The swarm robots receive the XML message containing the threat location (X_t, Y_t) and the threat level. When a swarm robot receives an XML message from the Base Robot, it checks the threat level and moves toward the threat destination if the threat flag is on. In Figure 7.11, Sensor 1 shows up as red and swarm robots start to move toward Sensor 1 [4].

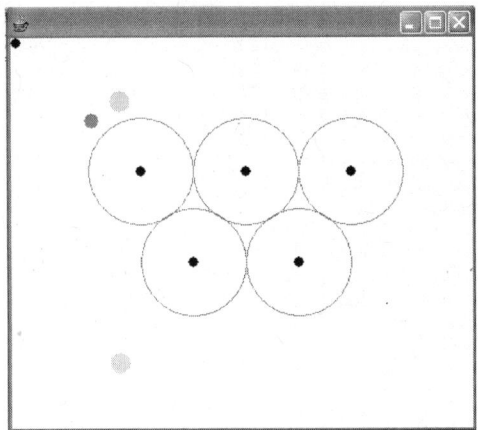

FIGURE 7.10
Sensor 1 detects a threat [4].

In Figure 7.11, we can still see that the Threat is within Sensor 1's range and that the Swarm Robots have moved to verify the threat's presence. Swarm Robot 1 has reached the sensor first and has verified the threat's presence using its own sensor. It will remain at Sensor 1 for as long as the Threat is within the range of Sensor 1. Swarm Robot 2 is still moving toward Sensor 1 because the two swarm robots are not yet within communication range of one another and it does not know that Sensor 1 has already reached the active sensor. In Figure 7.12, the Threat has moved out of every sensor's range and no threat is detected by Swarm Robot 1.

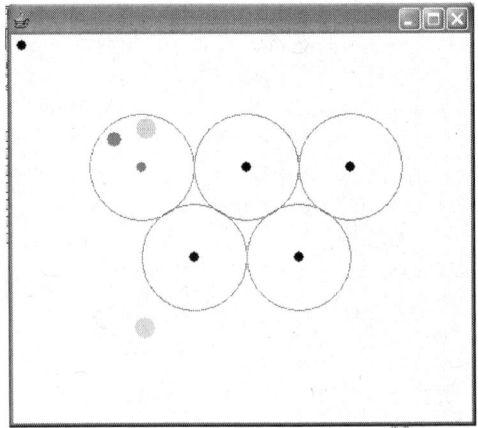

FIGURE 7.11
Swarm robots move to verify threat [4].

System of Systems Simulation

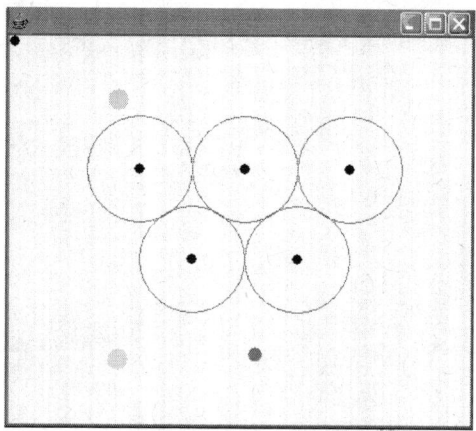

FIGURE 7.12
Swarm robots move to neutral location [4].

Once a Swarm Robot verifies that there is no threat, it sets "Threat V" flag to "0" and sends its XML data to Base Robot. This is the major reason for the robustness of the threat detection as false alarms are properly handled. Once the Threat is out of the sensor area, all five Sensors report to the Base Robot saying all is clear and the Swarm Robots are called off. Once they are called off (having their "Threat V" variables set to "0" and their X_t and Y_t values set to the neutral location), the Swarm Robots travel to their respective positions and await another threat [4].

In Figure 7.13, the Threat has moved into the range of Sensor 5 causing it to activate and signal the Base Robot. The Base Robot again signals the

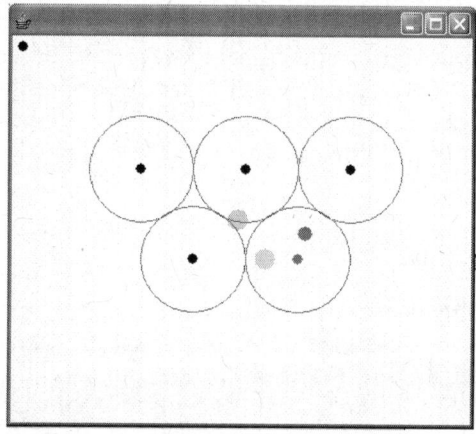

FIGURE 7.13
Swarm robots move toward Sensor 5 [4].

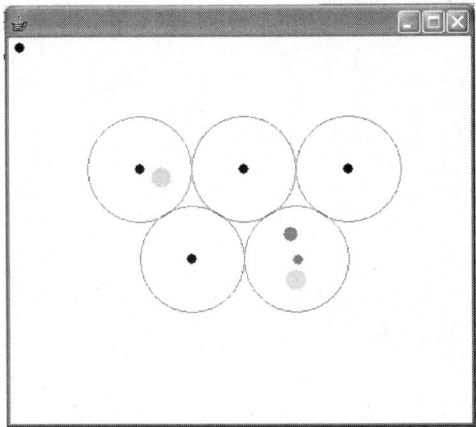

FIGURE 7.14
Swarm Robot 1 breaks off verification run [4].

Swarm Robots and they move in to verify the Threat's location much like in Figure 7.10.

Figure 7.13 differs from Figure 7.10 since in this scenario both swarm robots are within communication range of each other. Thus, the swarm robots will exchange their target locations (threat locations). Each swarm robot checks the other swarm robot's location and compares it to the threat target. If the swarm robot finds the other swarm robot's position closer to the threat destination, it decides to move away. It would go to its neutral position if there is no other threat in a different location. Thus, two swarm robots will never try to verify the same threat [4].

Figure 7.14 shows Swarm Robot 1, after communicating with Swarm Robot 2 breaking away from verifying the Threat at Sensor 5 because it determined that Swarm Robot 1 was already closer to the target. Swarm Robot 2 will now continue to verify the Threat at Sensor 5 and Swarm Robot 1 will travel to a neutral location until it is called again by the Base Robot [4].

7.4.3 Case 3: Threat Detection Scenario with Several Swarm Robots

This case study [53] is an extension of the previous case study where there are more swarm agents and the base robot is replaced by a command and control tower (called base station). In this scenario, there is a team of swarm robots that can detect a threat (shown as a fire) and are dispatched by a control tower (base station) to the location of a possible threat. The swarm robots are each stationed at a unique location in the map to provide a faster response time for detecting the threat. Stationary sensors have been deployed in the area to create a sensor network to warn of the presence of a threat. Although

TABLE 7.3
Summary of DEVS Models

System	DEVS Model
Swarm Robots	MobileSwarmAgent
Fire	MobileThreat
Stationary Sensors	Sensor
Base Station	BaseStation
SoS	XmlSystem v2
GUI	PlotModule

the stationary sensors are cost-effective for continuous operation, they cannot pinpoint the exact location of the threat and it is possible to have false positives. That is, a sensor reports the presence of a fire but the threat is not in the vicinity of the sensor. A base station processes the sensors' data and dispatches the swarm robots when a possible threat has been detected. The swarm robots then cooperate using basic swarm behavior and allow the swarm robot closest to the supposed fire location to take the lead investigating while the other robots in the swarm hold their positions a distance from the lead swarm robot and each other during this time. Each of the systems in this SoS example (the swarm robots, the fire, the sensors, and the base station) is presented with their DEVS models in Table 7.3.

The DEVS representation of the XML-based communication among these systems is presented. To provide a graphical view of the robust threat detection scenario during simulation, a DEVS model is also implemented that controls a window with color icons, which move as their corresponding DEVS model is updated in the simulation.

7.4.3.1 XML Messages

In the DEVSJAVA environment, the data structure `XmlStrEntity` extends the DEVS entity class to create XML messages for the component models. The XML message is stored as a `String` object containing both markup tags and embedded data as would be transferred between any two systems/subsystems in an SoS. This class is used to wrap data into the XML architecture and also provides methods to parse (extract) individual data fields out of the XML architecture. Thus, each model in the system is readily provided the ability to send and receive XML data from any source. In addition, `XmlStrEntity` can be written to an external file and parsed by any external program that understands the SoS XML hierarchy described in Listing 1. By creating the data structure `XmlStrEntity`, the messages among the components of the SoS are closer to the standard XML messages. In this manner, the designers do not need to create their own XML-like types such as a vector of doubles. Later in the chapter, we will present the XML message architecture in detail.

7.4.3.2 DEVS Components of the Scenario

7.4.3.2.1 Plot Module

The PlotModule does not represent an actual physical agent in the system, but provides a means to visually display the state of the SoS to the user during runtime. All models in the SoS send messages containing their state to the PlotModule model. The PlotModule parses the XML messages from the different systems and updates the graphical user interface (GUI) to reflect the current state of each of the components. The PlotModule is not expected to know the initial state of the SoS (this duplicates code management issues between the simulated system and this model). For each model, an XML message with the model's initial state facilitates a correct view of the system at time zero of the simulation.

7.4.3.2.2 Mobile Swarm Agent

Autonomous mobile agents are major entities in the SoS discussed in this scenario. MobileSwarmAgents receive target coordinates from the BaseStation model on the *baseStationIn* port to investigate and communicate with peer swarm agents on the *swarmIn*, and *swarmOut* ports as shown in Figure 7.15a. To perform actions such as threat verification (or possibly avoidance or tracking), the model receives XML messages from the MobileThreat component on the *threatIn* port. XML messages to other components, such as the PlotModule, are sent on the output port *Out*. The swarm agent's internal state is defined by the current x and y coordinates within the operational area of the SoS, the destination x and y coordinates, and the status of the sensors deployed on the agent (in this case, a sensor to "detect" the nearby presence of a threat). A home (initial) location is designated to each model upon its creation. When it has finished investigating a threat, each swarm agent automatically returns to this home location and waits for further dispatch commands from the central control tower. Figure 7.15b illustrates the event-driven logic of each MobileSwarmAgent. Each circle designates a DEVS *phase*; the double circle represents the initial phase of the model. An internal transition is shown with a broken line connecting two phases, whereas a solid line shows a phase transition due to an external event. Arrows pointing from a phase to circle to a box illustrate the XML message sent on an output port of the model as defined in the DEVS atomic model's *out()* function.

Like all models appearing in the graphical display of this simulation framework, the MobileSwarmAgent begins in the init phase and sends an XML message containing the initial state of the model to the PlotModule's *In* port. The swarm agent is in the idle phase whenever its current location is equal

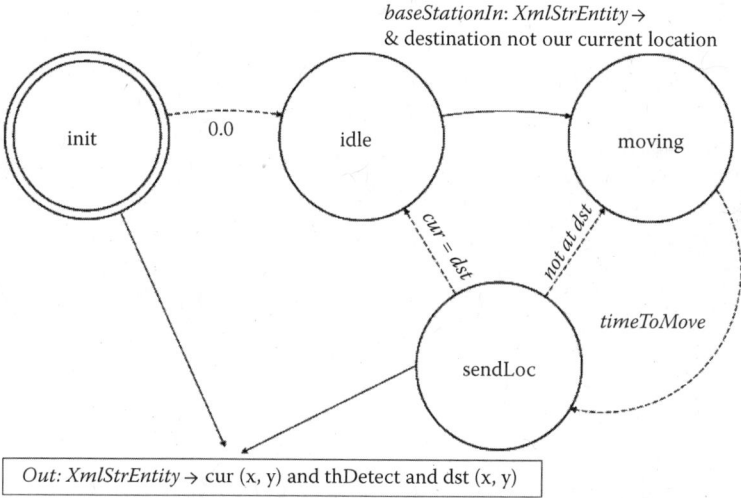

FIGURE 7.15
DEVS model of `MobileSwarmAgent` [53]. (a) `MobileSwarmAgent` model; (b) phases of the `MobileSwarmAgent` model.

to (or within a close threshold of) the destination coordinates. When an XML message arrives on the *baseStationIn* port, the agent's internal destination coordinates are set to be the coordinate data parsed from the incoming message. The moving phase is held for a time of *timeToMove* to simulate the time it would take a real-world robot to move in a single iteration. At the end of each movement (sigma = *timeToMove* has expired), the phase is changed (via the atomic model's *internal transition function*) to `sendLoc` and the new location of the agent is communicated in an XML message on the *Out* port. The message is also sent to other members on the same swarm team on the *swarmOut* port. If this last iteration of movement has relocated the model to its destination, it returns to the `idle` state. Otherwise, another cycle of `moving` and `sendLoc` is completed. The Boolean field `thDetect` of the `MobileSwarmAgent` is set to true when the distance between the threat and the agent is within a detection range. Each `MobileSwarmAgent` broadcasts its destination and current location to the swarm and yields to a fellow team member if the team member is closer to the destination than itself. As a result, the yielding members of the team hold their current position relative to the emerging lead swarm agent (who will go to the exact destination and investigate the threat) until the investigation is complete. Then, all `MobileSwarmAgents` return to their respective home locations.

7.4.3.2.3 Mobile Threat

The threat in this SoS is a fire and moves in a random fashion within the boundaries of the plotted area. The `MobileThreat` can receive initial starting coordinates on the *startIn* input port. From this location, the threat broadcasts its current location to the sensors and to the mobile swarm agents on its output port *Out*. Figure 7.16a shows the atomic model, and the phase transitions can be seen in Figure 7.16b. For simulation purposes, the mobile agents receive the location of the threat in order to verify its presence when investigating a location as directed by the control tower (base station). The `MobileThreat` then continues to move randomly and broadcasts its new location while in the `terrorizing` phase. The internal transition function is used to perpetuate the movement of the threat within the system; any additional XML message

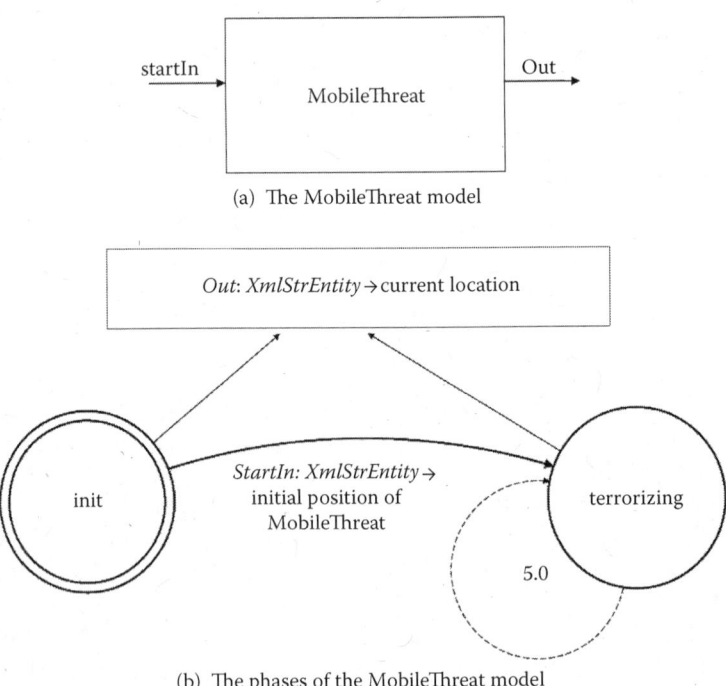

FIGURE 7.16
DEVS Model of `MobileThreat` [53]. (a) MobileThreat model; (b) phases of the `MobileThreat` model.

System of Systems Simulation

received while moving will be parsed for coordinate data to which the threat will instantly relocate to. Relocating the threat during a simulation provides a method for the user to test and debug specific case scenarios.

7.4.3.2.4 Sensor

To reduce communication costs, the sensor only transmits messages back to the BaseStation when a threat is detected and when a threat leaves the sensor's range. In other words, the Sensor remains idle until a threat is within range. Figure 7.17 shows a DEVS Sensor atomic model and its phase transitions. Input port *threatIn* receives messages from the MobileThreat model; these messages contain the current location of the threat. Output port *Out* transmits the sensor's location and threat detection status.

7.4.3.2.5 BaseStation

The BaseStation serves as a central coordination unit among the array of sensors and the swarm of mobile agents. The XML messages from the Sensor models are received on the input port *sensorsIn* and the XML messages from

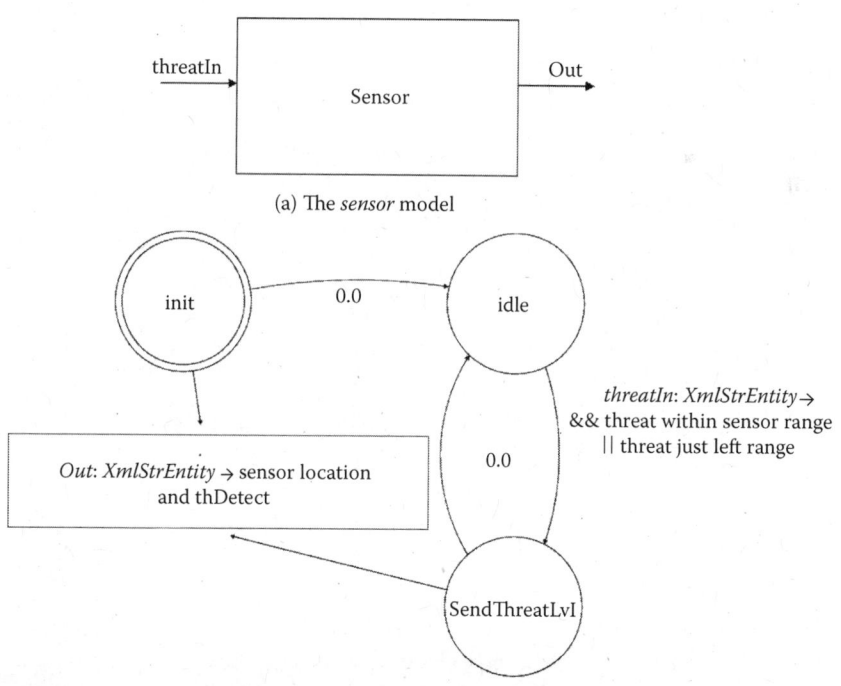

FIGURE 7.17
DEVS model of Sensor [53]. (a) *Sensor* model; (b) phases of the *Sensor* model.

the `MobileSwarmAgents` are received on the input port *mobileRobotsIn*. The output port *cmdOut* is the communication link to the agents in the swarm and the output port *Out* is used for couplings to other components (currently only a `PlotModule` model). The internal variables of the model hold the current position of a threat as detected by the sensors and a Boolean flag for the threat status (whether a threat is present or not). If an incoming XML message on port *sensorsIn* contains coordinate data for a present threat, the `BaseStation` "forwards" these coordinates to the mobile swarm robots via the *cmdOut* port. The `BaseStation` contains no other control logic for the robot team as each mobile robot is controlled by simplistic swarm behavior logic. Messages received from the robots on the *mobileRobotsIn* port do not affect the internal state of the base station in the chapter. Providing additional "leadership" abilities to the base station can be implemented according to the desired goals of the system. This could include establishing dynamic system, subsystem, or team formation as discussed by Hu et al. [54,55]. Similarly, this base station could add or remove agents as they join or leave the systems of systems as would be the case if supporting agents were called as reinforcements to contain a threat or to simulate system losses due to attacks from the threat.

7.4.3.2.6 System of Systems

At the topmost hierarchical level, the SoS is represented by a DEVS coupled model (`XmlSystem v2`) containing one or more instances of each of the models described previously. The coupled DEVS model provides the means for hierarchically grouping the entities and systems/subsystems in the SoS as well as specifying the communication links among them. This is accomplished with the `addCoupling(src,"portOut," dst,"portIn")` method. Each model's *Out* port is connected to `PlotModule`'s *In* port. For each `Sensor` in the system, the couplings are established as shown in Listing 3: (S stands for `Sensor`; B stands for `BaseStation`; PM stands for `PlotModule`; Th stands for `MobileThreat`).

LISTING 3   COUPLINGS OF SENSOR ATOMIC MODELS

```
addCoupling(S,"Out",B,"sensorsIn");
addCoupling(S,"Out",PM,"In");
addCoupling(Th,"Out",S,"threatIn");
```

Each sensor receives XML messages from the threat about its current location, sends the current threat detection status back to the base station, and also sends its state to the plotting module for the graphical representation. For each `MobileSwarmAgent` in the system, the couplings are established as shown in Listing 4 (MSA stands for `MobileSwarmAgent`):

LISTING 4 COUPLINGS OF `MOBILESWARMAGENT` **ATOMIC MODELS**

```
addCoupling(MSA,"Out",B,"mobileRobotsIn");
addCoupling(B,"cmdOut",MSA,"baseStationIn");
addCoupling(MSA,"Out",PM,"In");
addCoupling(Th,"Out",MSA,"threatIn");
addCoupling(MSA,"swarmOut",MSA2,"swarmIn");
```

Bidirectional couplings are initialized between every two mobile swarm agents, but connecting a model to itself is avoided by checking if the two models are equal. Figure 7.18 shows the atomic models and their couplings as seen in a SimView window while running the DEVSJAVA simulation. The `PlotModule` model is hidden from view in this case as it only serves to produce a visual of the SoS, and does not affect the state in any way.

7.4.3.3 DEVS-XML SoS Simulation

This section presents the modeling of a robust threat detection and interception example using the DEVS models presented in previous sections. This example extends the simulation in the second case study [20] so that the results will validate this XML-based SoS simulation framework. The coupled model `XmlSystem v2` represents an SoS containing a seven-sensor grid (atomic model `Sensor`) with a central command tower (atomic model `BaseStation`) to aggregate and process sensor data. The sensors detect the presence of a fire threat (a `MobileThreat` model). The central command tower notifies five swarm robots (`MobileSwarmAgent`), which cooperate to investigate and intercept the fire. Figure 7.19 depicts the graphics used in the simulation example for these systems in the SoS. (A circle surrounding a sensor represents the range at which it can detect the fire during a simulation.)

7.4.3.3.1 XML and DEVS

An XML hierarchical architecture is implemented in Listing 5 that conforms to the proposed architecture in Listing 1 and will also be compatible with the hardware deployed in real-time simulation. In real-time simulation (agent-in-the-loop simulation), GroundScout modular microrobots [56] are used as `MobileSwarmAgents`. The swarm robots only move in two dimensions for the simulation so the control logic can be deployed on a mobile robot with sensors meeting this requirement. Single character length XML tags are used in this framework due to memory constraints in the communication hardware of GroundScouts so that the XML architecture can be deployed as is for agent-in-the-loop simulations. The short-length XML tags are implemented at no loss of generality or functionality of the desired common communication in the SoS. An SoS, assigned the <q> XML tag, contains any number of systems defined by the <y> tag. Each independent system contains a unique identifier field (<i> tag)

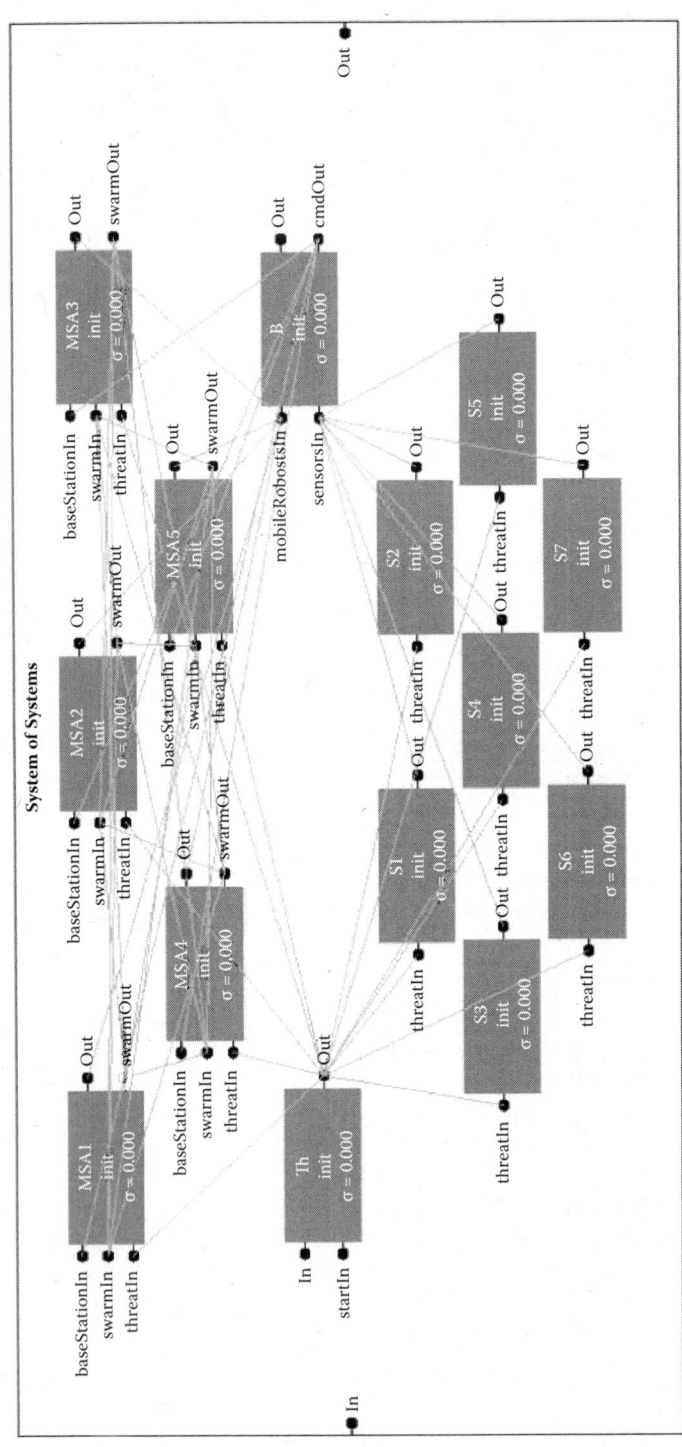

FIGURE 7.18
DEVS coupled model of an SoS [53].

System of Systems Simulation

(a) Radar   (b) GroundScout   (c) Fire   (d) Tower

FIGURE 7.19
Graphics used in simulations. (a) Radar; (b) GroundScout; (c) Fire; (d) Tower.

and possible sensor items, each containing an ID field and a data field. This architecture allows for any data to be provided and properly parsed by a system based on the expected data coming from a sensor of the type specified by the ID.

LISTING 5   XML-BASED SYSTEM ARCHITECTURE [53]

```
<!--Created 12/01/2008 Author @ Matt Hosking-->
<?xml-stylesheet type="text/css" href="genericxml.css"?>
<q>
    <y>
        <i>ID of the system </i>
        <s>
            <i>ID of the first sensor </i>
            <d>sensor data </d>
        </s >
        <s>
            <i>ID of the second sensor </i>
            <d>sensor data </d>
        </s >
    </y>
    <y>
        <i>ID of the 2nd system </i>
        <s>
            <i>ID of the sensor </i>
            <d>sensor data </d>
        </s >
    </y>
</q>
```

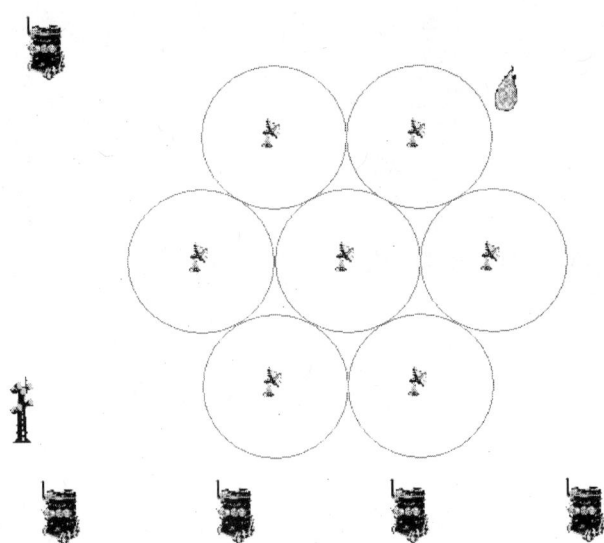

FIGURE 7.20
Initial positions of agents in SoS.

7.4.3.3.2 Simulation Results

Results of the threat detection and interception simulation are presented in this section to illustrate the behaviors of the components in SoS and the successful communication using XML data messages. Seven sensors form a sensor network area illustrated by the adjacent circles in Figure 7.20. The sensors are numbered left to right and from top to bottom so that station one S1 is on the top left, S4 is in the center of the network, and S7 is the lower right station. Figure 7.8 also shows four swarm robots (SR2–SR5) stationed south of the sensor network and a single swarm robot, SR1, stationed in the top left corner of the plot. The position of the central control tower (base station) in the lower left corner is not significant and the fire threat (F) is out of range of all sensors.

The simulation begins as shown in Figure 7.20; as time continues, the threat moves into the coverage area of S5. When the sensor detects the threat, the XML message is transmitted to the central tower, shown in Listing 6.

LISTING 6   XML MESSAGE SENT BY THE SENSORS TO THE BASE STATION

```
<y><i>S5</i><s>
<i>Coord</i><d>225,225</d>
<i>Th</i><d>true</d>
</s></y>
```

System of Systems Simulation 219

The central tower parses this message and finds that a threat has been detected and broadcasts the XML message in Listing 7 to the swarm robots.

LISTING 7 XML MESSAGE SENT BY THE BASE STATION TO THE SWARM ROBOTS

```
<y><i>B</i><s>
<i>Coord</i><d>225,225</d>
<i>Th</i><d>true</d>
</s></y>
```

The swarm robots receive this message and, because the *Th* data field is set to *true*, interpret the parsed coordinate data as the location of the threat to investigate. The swarm logic of the swarm robots allows each agent to move in to investigate until a member of the swarm emerges as the leader. The motion of all swarm robots to the threat area is illustrated in Figure 7.21. The swarm robots then communicate with one another using XML messages shown in Listing 8.

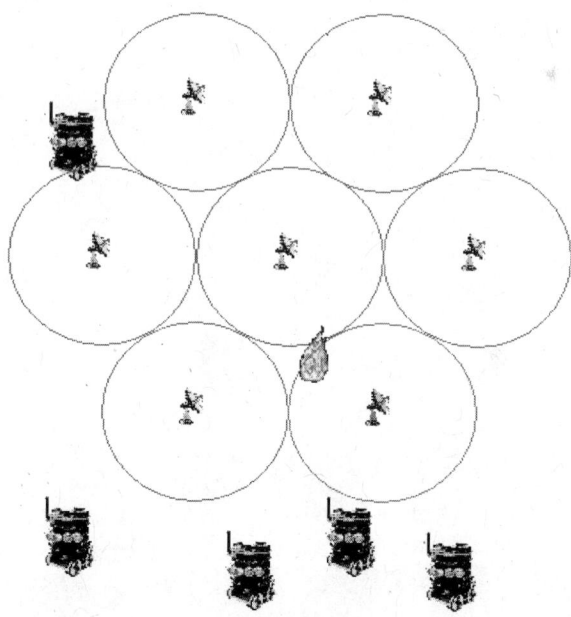

FIGURE 7.21
Swarm robots move toward reported threat.

LISTING 8 XML MESSAGE SENT BY THE SWARM ROBOTS

```
<y><i>H#</i><s>
<i>Loc</i><d>my X,my Y</d>
<i>Th</i><d>true/false</d>
<i>Dst</i><d>dst X,dst Y</d>
</s></y>
```

Each agent in the swarm knows the location of other agents to determine if they are near each other to begin cooperating. This simulation example is built around robust threat verification [20] so the critical data for cooperation are the current and destination coordinates of the other swarm agents. If the data tagged with ID *Th* are set to *true*, the swarm agents decide which of them is closest to the threat (and allow that agent to continue) while the rest hold their current pattern close to the leader agent.

Figure 7.22 shows a single swarm robot over S7 with the remaining swarm robots holding in close proximity. The sensor does not have the ability to pinpoint the threat location, so the fire could be anywhere in its coverage area and the swarm robot would still travel to the center of the station's detection area. The control tower's logic could be extended to triangulate a more precise location of the threat using data from multiple swarm agents and

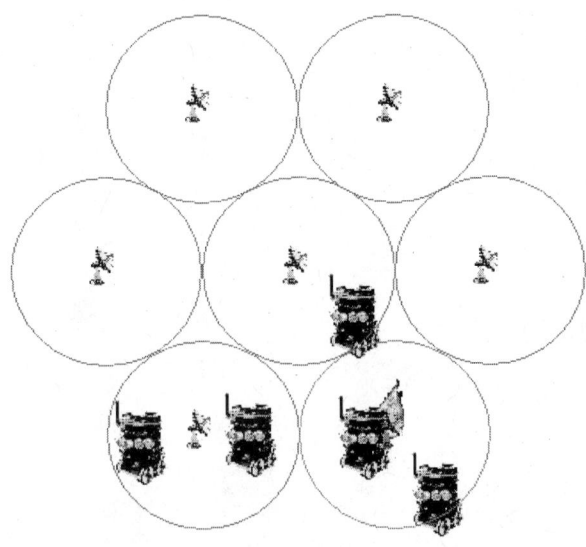

FIGURE 7.22
Swarm robots intercept threat.

sensors. This makes it useful to have the swarm agents hold in formation around the "lead" for this purpose—the threat is a moving target after all! A more precise location would be useful for a more efficient deployment of more expensive attack units as they enter the SoS.

Sometimes the sensors may falsely assert that a threat is present. It may also be the case that the threat crosses into the sensor's detectable range and then quickly retreats back out of contact. Swarm robots will, however, continue to the location to investigate only to return back to their initial locations after broadcasting an XML message with the *Th* field set to *false*. The behavior of the robotic swarm is the same as previously described: only a single swarm robot will travel to the exact sensor location and the others will maintain a formation at proximity. Figure 7.23 verifies this cooperative behavior in the event that the threat is no longer present.

As the fire threat moves in the coverage area of the seven sensors, the sensors send XML messages to the `BaseStation`, which in turn sends the location of the threat to the swarm robots. The swarm robots continue to investigate these locations with the closest swarm robot taking the lead. When the threat is no longer detected, and the lead swarm robot has investigated the last known location, all swarm robots return to their initial locations and wait for an XML message from the base station. All five swarm robots are returning to their initial locations in Figure 7.24.

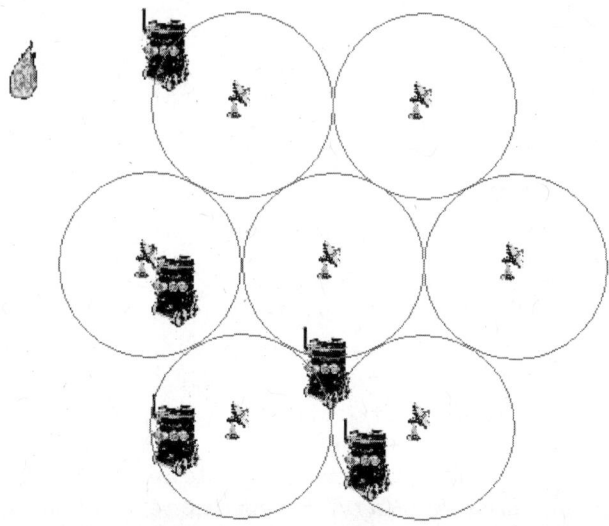

FIGURE 7.23
Swarm robots investigate false detection.

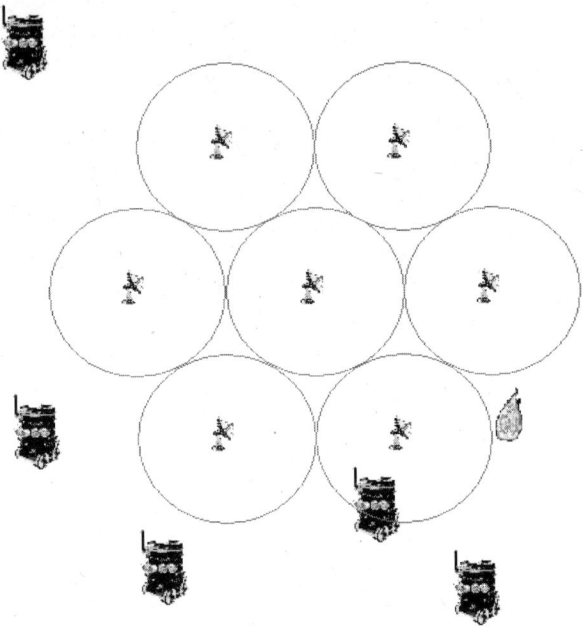

FIGURE 7.24
Swarm robots return to initial locations.

7.5 Agent-in-the-Loop Simulation of an SoS

Simulations play an important role in complex systems development. In an SoS consisting of heterogeneous independent systems (as is the case with cooperative robotics, for example), it is desirable for the systems to work together toward a common goal. Agent-in-the-loop simulations bridge the gap between computer simulations and system deployment testing by allowing a subset of the cooperative robotic team to interact with virtual computer models for performance testing in real time. In this section, a modular microrobot, called GroundScout [56], is deployed as an agent in the loop with simulated robots (virtual robots) to form a cooperative swarm. Details of GroundScouts' hardware and software are provided in Chapter 12. In this setup, the goal is to observe the behavior of GroundScout to verify the desired emergent swarm behavior as expected from previous computer simulations. To accomplish the agent-in-the-loop-simulation, an XML SoS real-time simulation framework is developed.

7.5.1 XML SoS Real-Time Simulation Framework

This section presents the modeling of an agent-in-the-loop robust threat detection and interception example using the DEVS models presented in Section 7.4. In several studies [54,55], DEVSJAVA has been used to support model continuity. This allows an easier migration of an SoS from centralized computer simulations to an SoS with some systems running on the computer simulation and other systems as real agents in the world interacting with their simulated counterparts. The time advance function discussed previously will be determined by actual performance metrics of the GroundScout robots. That is, the virtual systems will create and respond to events in the SoS, such as relocation commands, in the same general time as the real agents. As a benefit, manipulating the time constants for simulated agents will provide a means to test how systems interact with one another at faster or slower speeds.

7.5.1.1 Threat Detection Scenario

The robust threat detection scenario follows the third case study presented earlier. In this scenario, there is a team of GroundScout robots [56] that can detect fire and are dispatched by a control tower (base station) to the location of the possible emergency. The GroundScouts are each stationed at a unique location in the map to provide faster response time for detecting the fire. Stationary sensors have been deployed in the area to create a sensor network to warn of the presence of a fire. As noted earlier, the sensors cannot pinpoint the exact location of the fire and it is possible to have false positives. A control tower (base station) processes the sensors' data and dispatches the robot team when a possible fire has been reported. The GroundScouts cooperate by using basic swarming behavior to investigate the area. The robot closest to the reported threat location takes the lead while the other team members hold their positions at a distance from the lead robot. There is no specific formation that the other robots are required to adhere to. The graphics shown in Figure 7.19 will likewise be used in this agent-in-the-loop simulation.

7.5.1.2 Synchronization

The simulation framework must provide a means in which the virtual counterpart of any deployed agent can be updated as changes occur for effective and meaningful results when observing the interaction of a real agent with simulated systems. A DEVS simulation executed on a real-time simulator results in Java `threads` associated with each model and simulator. Of particular importance are DEVS `activities`. An activity serves to connect the DEVS simulated environment with real systems by providing a standard mechanism for inserting messages onto a DEVS model's input port. An activity runs as a thread and accesses sensors, sets actuators, and performs any other embedded driver task or threaded operation. The Java *Observer-Observable* pattern is used by these activity threads as well as threads to interact with operating system level functions for data transfer on a local serial port. The location of the control logic for a real robot could be implemented in DEVS, but is deployed on the real GroundScout itself to test a complete system. Synchronization requires messages to be passed seamlessly between virtual (simulated) and real robots, and relevant state data is kept up to date so that, for example, the graphical representation of the SoS simulation accurately reflects the relative locations of real and simulated robots.

7.5.2 DEVS Modeling

Interactions between the independent systems within an SoS are asynchronous in nature and can be effectively represented as discrete event models [21]. Discrete event system specification [51,54,55] is a formalism providing a means to specify the components in a discrete event simulation. Basic models, called *atomic* models, can be connected together to form larger, more complex models called *coupled* models. Coupled models are used to model the concurrent and independent tasks that complex systems, such as the GroundScout robot, often exhibit. Each of the systems in this SoS scenario is presented as a DEVS model here. The example scenario contains a seven-sensor grid (atomic model Sensor) with a central command tower (atomic model BaseStation) to aggregate and process sensor data. The sensors detect the presence of a fire (MobileThreat model). The central command tower notifies five members of the robot team (MobileSwarmAgent), which cooperate to investigate and intercept the threatening fire.

7.5.2.1 DEVS Components

Because we are allowing DEVS real-time simulation environment to have virtual and real swarm robots, MobileSwarmAgent and BaseStation atomic models are updated according to the agent-in-the-loop-conditions. Thus, PlotModule, MobileThreat, and Sensor atomic/coupled models are kept similarly as described in Section 7.4.

7.5.2.2 Mobile Swarm Agent

Autonomous mobile agents are major entities in an SoS containing a cooperative robot team as in this example. A mobile agent, in our robust threat detection scenario, is able to receive target coordinates from a base station as well as communicate with fellow swarm agents. GroundScout robots are constructed in a modular fashion, so that each function of the system is contained in its own layer. These modules include a control layer, multiple sensor layers, a locomotion layer, and a communications layer [56]. Detailed description of the GroundScouts is presented in Chapter 12. In our cooperative robotic scenario, and indeed any SoS, communication among agents is key for interoperability. GroundScouts use a secondary PIC microprocessor to handle the sending and receiving of data transmission over radio frequency wireless channels [56]. (The primary processor is an 8051-based microcontroller.) The communication layer thus operates independently of the main control unit to connect, synchronize, transmit, and receive data packets. To effectively model this behavior in DEVSJAVA, the virtual GroundScout is a coupled model. This coupled model consists of an atomic communications layer model and an atomic main control unit model. The communications model contains input and output ports to receive and send data to external systems as well as to receive and send data to the main control unit model. Figure 7.25 depicts a block view of this model.

The control sub-microdevices model of the GroundScout robot implements the swarm behavior logic necessary to cooperate with other team members in the threat detection and verification scenario. Figure 7.26 illustrates the event-driven logic of each GroundScout's control layer. Each circle designates a DEVS phase; the double circle represents the initial phase of the model. An internal transition is shown with a broken line connecting two phases, whereas a solid line shows a phase transition due to an external event. Arrows pointing from a phase to circle to a box illustrate the message sent on an output port of the model as defined in the DEVS atomic model's *out()* function. This message is then wrapped with the appropriate XML tags in the communication layer and subsequently transmitted to the destination system(s). If the GroundScout model is serving as the virtual counterpart of a deployed GroundScout robot during an agent-in-the-loop simulation scenario, the communication layer model must also support sending and receiving data to the real world. A DEVS activity is used to facilitate this functionality. The communication

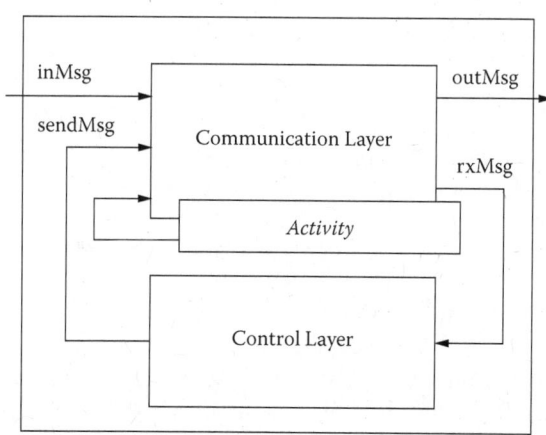

FIGURE 7.25
DEVS model of MobileSwarmAgent communication layer.

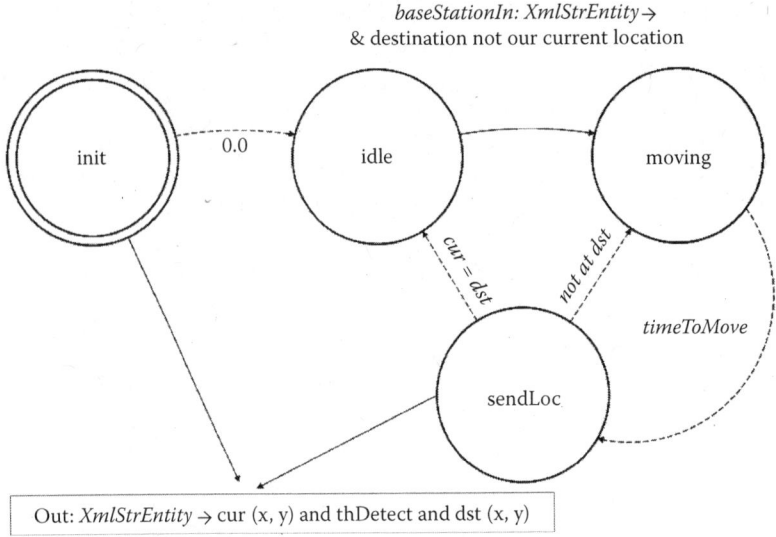

FIGURE 7.26
DEVS model of `MobileSwarmAgent` control layer phases.

model's activity class implements the `java.util.Observer` interface and listens for updates from the `serialPortThread` referenced in the system's base station. This activity class also contains references to the transmit and receive buffers created in the base station; this model checks if the new packets added to the receive buffer are intended for the GroundScout using the unique identification assigned to each GroundScout (real and simulated). The information is transferred into the DEVS simulation loop by posting a message on the `inputFromActivity` port to invoke the external transition function of the communication layer's model.

7.5.2.3 Base Station

The base station is responsible for aggregating the sensor network data and informing the robot team of new threat locations to investigate. It also serves as the virtual counterpart to the GroundScout base station used to communicate with the robot team. This real base station, wired serial port connection, and virtual counterpart serve as the gateway between the DEVSJAVA simulation environment and the real-world robots. This is accomplished by using a transmit buffer and a receive buffer to store outgoing and incoming data packets, respectively. The base station Java class contains accessor methods to return a pointer to these packet buffers so other objects may read and write to the storage structures. The base station also creates a thread to handle the serial port interface utilities. `serialPortThread` is an object that implements the `javax.comm.serialPortEventListener` interface

System of Systems Simulation

and extends the `java.util.Observable` class. A reference to the `serialPortThread` object is also available via an accessor method in the base station class and is monitored by the other agents in the system using the *Java Observer pattern*. The serial port thread aggregates the raw data bytes from the serial port into packetized objects used in our simulation framework. Most malformed packets and partial packets due to data loss are filtered out at this initial level of programming and decrease the amount of invalid data entering into the simulation loop. When a valid sequence of data forms a packet, the packet object is added to the receive buffer and the event listeners (GroundScout communication layer models) are notified.

7.5.3 Agent-in-the-Loop Simulation

As stated previously, there will be a real GroundScout in the simulation and four simulated (virtual) GroundScouts. The real GroundScout moves on the floor while receiving information about other virtual GroundScouts via the communication unit attached to the PC where DEVS real-time simulation is running. The virtual GroundScouts are informed by the real one as it sends its location and sensor. All the communications in the simulation and on the real GroundScout use the same XML-based SoS message architecture described previously (see Listing 5). When the threat is detected by a sensor, the sensor informs the base station. Then, the base station informs the real and virtual

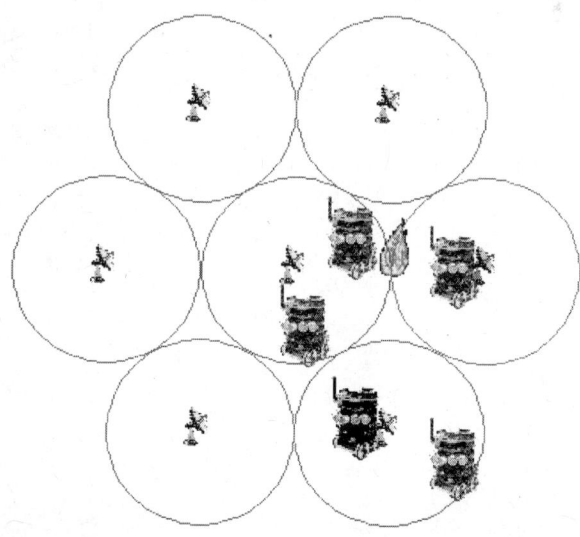

FIGURE 7.27
Real-time agent-in-the-loop simulation with five GroundScouts.

GroundScouts. In Figure 7.27, four virtual and one real GroundScout are shown with the threat (fire) and sensors. In this particular case, one of the virtual robots reaches the threat location. The rest of the robots are keeping a specific distance to the lead robot as part of their swarming behavior. The real GroundScout is at the center of Sensor 7. It is also shown with a different background color.

As shown in Figure 7.27, the simulation works seamlessly with real and virtual robots because the realistic models of GroundScouts are implemented in DEVSJAVA and the communication interface is handled via DEVS `Activity` functionality.

7.6 Conclusion

In this chapter, we have presented extensions to an XML-based SoS simulation framework and discussed three simulation case studies of robust threat detection and data aggregation with multiple autonomous systems working collaboratively as an SoS. Through multiple swarm robots in the field, the simulation results showed that false alarms can be avoided by verifying the threats with swarm robots. This would enhance the robustness of data aggregation and lead to the efficient usage of the system resources. DEVS formalism helps to represent the structure of an SoS, whereas XML provides a way to represent the data generated by each system. Together, DEVS formalism and XML form a powerful tool for simulating any given SoS architecture.

To make the simulation results more realistic and reliable, we have implemented real-time DEVS simulation with virtual and real robots (agents). This agent-in-the-loop simulation setup allows users to evaluate the performance of their swarm robots (or any system in an SoS) without physically implementing or buying numerous robots. This also opens opportunities for testing and evaluation of the swarm characteristics of a robot realistically without completely depending on simulation environment and assumptions.

We are currently working on extending the XML data representation and making it more generic and dynamic so that when a new system is added to an SoS simulation, it will generate its XML message automatically and send it to other components of the SoS.

Acknowledgment

The author would like to express his appreciation to Matthew R. Hosking, a graduate of Department of Computer Engineering and a research assistant in the Multiagent Bio-Robotics laboratory at Rochester Institute of Technology.

References

[1] Crossley, W. A., "System of systems: An introduction of Purdue University Schools of Engineering's signature area," *Engineering Systems Symposium*, March 29–31 2004, Tang Center-Wong Auditorium, MIT.

[2] Lopez, D., "Lessons learned from the front lines of the aerospace," in *Proceedings of IEEE International Conference on System of Systems Engineering*, Los Angeles, April 2006.

[3] MIL-STD-499B, *Military Standard, Systems Engineering*, Draft, Department of Defense, 1992.

[4] Sahin, F., and Jamshidi, M., "A System of systems simulation framework and its applications," *System of Systems Engineering—Principles and Applications* (M. Jamshidi, Ed.). CRC Press, Boca Raton, FL, 2008.

[5] Jamshidi, M. (Ed.), *System of Systems—Innovations for the 21st Center*, Wiley & Sons, New York, 2008.

[6] Jamshidi, M. (Ed.), *System of Systems Engineering*, CRC Press, Boca Raton, FL, 2008.

[7] Lopez, D., "Lessons learned from the front lines of the aerospace," *Proceedings of IEEE International Conference on System of Systems Engineering*, Los Angeles, April 2006.

[8] Wojcik, L. A., and Hoffman, K. C., "Systems of Systems Engineering in the Enterprise Context: A Unifying Framework for Dynamics," in *Proceedings of IEEE International Conference on System of Systems Engineering*, Los Angeles, April 2006.

[9] Azarnoush, H., Horan, B., Sridhar, P., Madni, A. M., and Jamshidi, M., "Towards optimization of a real-world robotic-sensor system of systems," in *Proceedings of World Automation Congress (WAC) 2006*, July 24–26, Budapest, Hungary.

[10] Abel, A., and Sukkarieh, S., "The coordination of multiple autonomous systems using information theoretic political science voting models," *Proceedings of IEEE International Conference on System of Systems Engineering*, Los Angeles, April 2006.

[11] DiMario, M. J., "System of systems interoperability types and characteristics in joint command and control," *Proceedings of IEEE International Conference on System of Systems Engineering*, Los Angeles, April 2006.

[12] Sahin, F., Jamshidi, M., and Sridhar, P., "A discrete event xml based simulation framework for system of systems architectures," in *Proceedings of the IEEE International Conference on System of Systems*, April 2007.

[13] Zeigler, B. P., Kim, T. G., and Praehofer, H., *Theory of Modeling and Simulation*, Academic Press, New York, NY, 2000.

[14] Zeigler, B. P., Fulton, D., Hammonds, P., and Nutaro, J., "Framework for M&S-based system development and testing in a net-centric environment," *ITEA Journal of Test and Evaluation* 26(3), 21–34, 2005.

[15] Mittal, S. "Extending DoDAF to allow DEVS-based modeling and simulation," Special issue on DoDAF, *Journal of Defense Modeling and Simulation JDMS*, 3(2), 2006.

[16] DUNIP: A Prototype Demonstration, http://www.acims.arizona.edu/dunip/dunip.avi.

[17] Mittal, S., Risco-Martin, J. L., and Zeigler, B. P., "DEVSML: Automating DEVS simulation over SOA using transparent simulators," *DEVS Symposium*, 2007.

[18] Mittal, S., Risco-Martin, J. L., and Zeigler, B. P., "DEVS-based web services for net-centric T&E," *Summer Computer Simulation Conference*, 2007.

[19] Mittal, S., "DEVS unified process for integrated development and testing of service oriented architectures," PhD dissertation, University of Arizona, 2007.

[20] Parisi, C., Sahin, F., and Jamshidi, M., "A discrete event XML based system of systems simulation for robust threat detection and integration," in *IEEE International Conference on System of Systems*, Monterey, CA, June 2008.

[21] Sahin, F., Sridhar, P., Horan, B, Raghavan, V., and Jamshidi, M., "System of systems approach to threat detection and integration of heterogeneous independently operable systems," in *Proceedings of IEEE Systems, Man, and Cybernetics Conference* (SMC 2007), Montreal, October 2007.

[22] Meilich, A., "System of systems (SoS) engineering and architecture challenges in a net centric environment," in *Proceedings of IEEE International Conference on System of Systems Engineering*, Los Angeles, April 2006.

[23] Abbott, R., "Open at the top; open at the bottom; and continually (but slowly) evolving," in *Proceedings of IEEE International Conference on System of Systems Engineering*, Los Angeles, April 2006.

[24] Jamshidi, M., "Theme of the IEEE SMC 2005, Waikoloa, Hawaii, USA," http://ieeesmc2005.unm.edu/.

[25] Sage, A. P., and Cuppan, C. D., "On the systems engineering and management of systems of systems and federations of systems," *Information, Knowledge, Systems Management* 2(4), 325–334, 2001.

[26] Kotov, V., "Systems of systems as communicating structures," Hewlett Packard Computer Systems Laboratory Paper HPL-97-124, pp. 1–15, 1997.

[27] Carlock, P. G., and Fenton, R. E., "System of systems (SoS) enterprise systems for information-intensive organizations," *Systems Engineering* 4(4), 242–261, 2001.

[28] Pei, R. S., "Systems of systems integration (SoSI)—a smart way of acquiring Army C4I2WS Systems," in *Proceedings of the Summer Computer Simulation Conference*, pp. 134–139, 2000.

[29] Luskasik, S. J., "Systems, systems of systems, and the education of engineers," *Artificial Intelligence for Engineering Design, Analysis, and Manufacturing* 12(1), 11–60, 1998.

[30] Manthorpe, W. H., "The Emerging Joint System of Systems: A Systems Engineering Challenge and Opportunity for APL," *John Hopkins APL Technical Digest* 17(3), 305–310, 1996.

[31] Jamshidi, M., *Large-Scale Systems—Modeling and Control*, North-Holland Publishing Company, New York, NY, 1983 (also 2nd edition, Prentice-Hall, 1997).

[32] Jamshidi, M., *Large-Scale Systems—Modeling, Control and Fuzzy Logic*, Prentice-Hall, Saddle River, NJ, 1997.

[33] Jamshidi, M., Class notes on System of Systems Engineering Course, University of Texas, San Antonio, TX, Spring 2006.

[34] Blanchard, B., and Fabrycky, W., *Systems Engineering and Analysis*, 3rd ed., Prentice-Hall, Upper Saddle River, NJ, 1998.

[35] Checkland, P., *Systems Thinking, Systems Practice*, 1st ed., Wiley, Chichester, 1981.

[36] Checkland, P., *Systems Thinking, Systems Practice*, 2nd ed., Wiley, New York, 1999.
[37] ECSS-E-10-01, System Engineering Process, European Cooperation for Space Standardization (1996).
[38] EIA/IS 632, Systems Engineering, EIA (1994). EIA/IS 731.1, Systems Engineering Capability Model, EIA, 1998.
[39] Grady, J. O., *Systems Engineering Deployment*, CRC Press, Boca Raton, FL, 2000.
[40] IEEE P1220, Standard for Application and Management of the Systems Engineering Process, IEEE, 1994.
[41] Martin, J. N., *Systems Engineering Guidebook*, CRC Press, Boca Raton, FL, 1997.
[42] MIL-STD-499B, Military Standard, Systems Engineering, Draft, Department of Defense, 1992.
[43] Rechtin, E., and Maier, M., *The Art of Systems Architecting*, 2nd ed., CRC Press, Boca Raton, FL, 2000.
[44] Sage, A. P., *Systems Engineering*, Wiley, New York, 1992.
[45] Keating, C., Rogers, R., Unal, R., Dryer, D., Sousa-Poza, A., Safford, R., Peterson, W., and Rabadi, G., "System of systems engineering," *Engineering Management Journal* 15(3), 36–45, 2003.
[46] Mittal, S., Zeiglar, B. P., Sahin, F., and Jamshidi, M., "Modeling and simulation for systems of systems engineering," *System of Systems—Innovations for the 21st Century* (M. Jamshidi, Ed.). Wiley & Sons, New York, 2008.
[47] MATLAB® Simulink, http://www.mathworks.com/products/simulink/.
[48] OMNET++, http://www.omnetpp.org/.
[49] NS-2, http://www.isi.edu/nsnam/ns/.
[50] HLA, https://www.dmso.mil/HLAComplianceTesting.html.
[51] Zeigler, B., "Devs today: Recent advances in discrete event based information technology," in *Proceedings 11th IEEE/ACM International Symposium on Modeling, Analysis and Simulation of Computer Telecommunications Systems MASCOTS 2003*, pp. 148–161, 2003.
[52] Zeiglar, B. P., and Sarjoughian, H., "Introduction to DEVS Modeling and Simulation with JAVA A Simplified Approach to HLA-Compliant Distributed Simulations," ACIMS, http://www.acims.arizona.edu.
[53] Hosking, M. R., and Sahin, F., "An XML based System of Systems Discrete Event Simulation Communications Framework," submitted to *Proceedings of the 2009 DEVS Integrative M&S Symposium, Spring Simulation Multiconference*, March 2009.
[54] Hu, X., and Edwards, D. H., "Behaviorsim: A simulation environment to study animal behavioral choice mechanisms," in *Proceedings of the 2005 DEVS Integrative M&S Symposium, Spring Simulation Multiconference*, April 2005.
[55] Hu, X., and Zeigler, B., "Model continuity in the design of dynamic distributed real-time systems," *IEEE Transactions on Systems, Man, and Cybernetics, Part A* 35(6), 867–878, 2005.
[56] Sahin, F., "GroundScouts: Architecture for a modular micro robotic platform for swarm intelligence and cooperative robotics," in *Proceedings IEEE International Conference on Systems, Man and Cybernetics* 1, 929–934, 2004.

8
Control of System of Systems

8.1 Introduction

As discussed in Chapter 3, almost all aspects of system engineering need to be revisited when it comes to systems of systems (SoS). Two aspects are sensing and control. From the control problem point of view, the difficulty arises that each system's control strategy cannot solely depend on its own onboard sensory information, but must also depend on communication links among all neighboring systems or between sensors, controllers, and actuators.

The main challenge in the design of a controller for SoS is the difficulty or impossibility of developing a comprehensive SoS model, either analytically or through simulation. By and large, SoS control remains an open problem and is, of course, different for each application domain. Should a mathematical model be available, some control paradigms are available, which will be the focus of this chapter. Moreover, real-time control—which is required in almost all application domains—of interdependent systems poses an especially difficult problem. Nevertheless, several potential control paradigms are briefly considered in this chapter. These control paradigms are hierarchical, decentralized, consensus-based, cooperative, and networked.

8.2 Hierarchical Control of SoS

First, as illustrated in Figure 8.1, hierarchical control of an SoS assumes that it can be characterized by a finite set of subsystems, say N, each of which could be separately optimized by a classical optimal control approach using, for example, an optimal control paradigm such as a linear quadratic regulator based on either continuous or discrete time theory. Through an iterative process of modeling the interactions between the coordinator system and the n subsystems, a convergent optimal solution could be obtained. In a real-time implementation of hierarchical control, a number of additional issues need to be resolved. For example, data transmission among the systems that constitute an SoS could be achieved through the use of Extensible Markup Language to code or decode data exchanges between them [1].

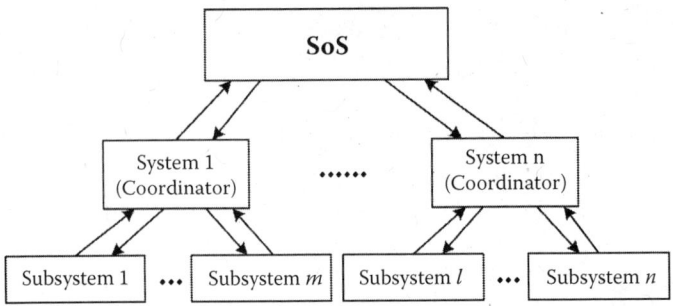

FIGURE 8.1
A hierarchical structure for control of an SoS.

Assume that an N-system SoS at time t is described by a linear time-invariant model

$$dx(t)/dt = A.x(t) + B.u(t) \tag{8.1}$$

$$x(t_0) = x_0 \tag{8.2}$$

where A and B are $n \times n$ and $n \times m$ state and input matrices, respectively, and Equation (8.2) represents the initial condition or history of the state of the SoS. The traditional optimal control paradigm for Equation (8.1) will be to find a control function $u(t)$ that satisfies Equations (8.1) and (8.2) while minimizing a linear quadratic cost function,

$$J = \tfrac{1}{2} x'(t_f) Q_f x(t_f) + \tfrac{1}{2} \int (x'(\tau) Q x(\tau) + u'(\tau) R u(\tau)) d\tau \tag{8.3}$$

where t_f is the final time and Q_f, Q, and R matrices are the weighting matrices for penalizing extreme variations of state and control vectors. Assume that system S_i of this SoS is described by

$$dx_i/dt = A_i.x_i + B_i.u_i + z_i \tag{8.4}$$

where x_i and u_i are the state and control vectors of system S_i, respectively, and z_i is the total communication (wireless or otherwise) inputs from all neighboring systems of S_i, S_j, $j = 1, 2, \ldots, i-1, i+1, \ldots, n_i$. The interaction or data from neighboring systems z_i is assumed to be given by

$$z_i = \sum_{j=1}^{N} (G_{ij} x_j), \quad j = 1, 2, \ldots, i-1, i+1, \ldots, N \tag{8.5}$$

where G_{ij} represents the data (communication packets) being received from the jth system to the ith system in SoS. The optimal control problem of system

Control of System of Systems

S_j is to find a control function $u_j(t)$ such that constraints (8.2)–(8.3) are satisfied, while minimizing a quadratic cost function,

$$J_i = \int \{x_i'(\tau)Q_i x_i(\tau) + u_i'(\tau)R_i u_i(\tau)\}d\tau = \int \{L_i(x_i, u_i)\}, \ \tau \text{ from } o \text{ to } t_f \quad (8.6)$$

This problem, via interaction prediction [2,3], begins by introducing a Lagrange multiplier α_i associated with constraint (8.3), defining a Hamiltonian function H_i:

$$H_i = x_i'Q_i x_i + u_i'R_i u_i + \alpha_i' z_i - \sum (\alpha_j' G_{ji} x_i) + p_i'(A_i x_i + B_i u_i + z_i) \quad (8.7)$$

for system S_i, using the theory of minimum principle of optimality [3], the following necessary conditions are obtained:

$$dx_i/dt = \partial H_i/p_i = A_i.x_i + B_i.u_i + z_i \quad (8.8)$$

$$p_i = -\partial H_i/\partial x_i = -Q_i x_i + A_i' p_i + \Sigma_j (G_{ji}' \alpha_j), \ j = 1, \ldots, N(j=/i) \quad (8.9)$$

with

$$x_i(t_o) = x_{io} \text{ and } p_i(t_f) = 0 \quad (8.10)$$

and finally,

$$0 = \partial H_i/\partial u_i = R_i.u_i + B_i'.p_i \quad (8.11)$$

In Equations (8.7–8.11), p_i is the ith system S_i co-state vector and α_i is a Lagrange multiplier of communication constants between all the systems S_j, $j \neq i$ with S_i.

Solving u_i from Equation (8.11), assuming that matrix R_i is nonsingular, and substituting it into Equations (8.8) and (8.9), we obtain the following two-point boundary value (TPBV) problem for system i.

$$dx_i(t)/dt = H_i/\delta p_i = A_i.x_i - S_i.p_i + z_i \quad (8.12)$$

$$p_i(t) = -\delta H_i/\delta x_i = -Q_i x_i + A_i' p_i + \sum_j (G_{ji}' \alpha_j)$$

$$j = 1, \ldots, N(j=/i), \ p_i(t_f) = 0 \quad (8.13)$$

where matrix $S_i = B_i R_i^{-1} B_i'$. This TPBV problem can be easily solved by using a Riccati formulation of the linear state regulator problem of optimal control theory [3], by the transformation

$$p_i = K_i(t) + g_i(t)$$

where $K_i(t)$ is the $n \times n$ symmetric positive definite Riccati matrix and $g_i(t)$ is an adjoint vector for ith system of the SoS:

$$dK_i(t)/dt = -K_i(t)A_i - A_i'K_i(t) + K_i(t)S_iK_i(t) - Q_i \quad (8.14)$$

$$dg_i(t)/dt = -(A_i. - S_i.K_i)g_i(t) - K_i(t)z_i + \sum_j (G_{ji}'\alpha_j) \quad (8.15)$$

$$j = 1, \ldots, N(j \neq i)$$

where

$$K_i(t_f) = 0 \text{ as is } g_i(t_f) = 0 \quad (8.16)$$

The hierarchical control algorithm works as follows [2,3].
Define a Lagrangian function,

$$L_i = \int_0^{t_f} \{x_i'(t)Q_ix_i(t) + x_i'(t)Q_ix_i(t) + u_i'(t)R_iu_i(t) + \alpha_i'(t)z_i(t)$$

$$- \sum^N \alpha_i'(t)G_{ji}x_i(t) + p_i'(t)[-dx_i/dt + A_ix_i(t) + B_iu_i(t) + z_i(t)]\}dt \quad (8.17)$$

The coordinating (communication data) among systems of the SoS are obtained from Equation (8.17) and necessary conditions of Lagrangian,

$$0 = \partial L_i/\partial z_i(t) = \alpha_i'(t) + C_i'p_i(t) \quad (8.18)$$

$$0 = \partial L_i/\partial \alpha_i(t) = z_i(t) - \sum_{j=1}^{N} \alpha_i'(t)G_{ji}x_i(t) \quad (8.19)$$

which provide $\alpha_i'(t) = -C_i'p_i(t)$ and $z_i(t) = \sum_{j=1}^{N}\alpha_i'(t)G_{ji}x_i(t)$. The coordinator's (e.g. ground station's) paradigm to enhance a *feasible* and optimal solution will be

$$[\alpha_i'(t)]^{m+1} = [-C_i'p_i(t)]^m \quad (8.20a)$$

$$[z_i(t)]^{m+1} = \left[\sum_{j=1}^{N}\alpha_i'(t)G_{ji}x_i(t)\right]^m \quad (8.20b)$$

The above development leads to the following algorithm.

ALGORITHM 8.1
Step 1: Solve N independent differential matrix Riccati Equation (8.14) with final condition (8.16) and store $K_i(t)$, $i = 1, 2, \ldots, N$ and $0 \leq t \leq t_f$.

Step 2: For assumed initial values $\alpha'_i(t)$, $z_j(t)$, solve "adjoint" Equation (8.15) with final condition (8.16). Evaluate and store $g_i(t)$, $i = 1, 2, \ldots, N$,.

Step 3: Solve the state equation

$$dx_i(t)/dt = (A_i - S_i K_i)x_i - S_i g_i(t) + z_i(t), \quad x_i(0) = x_{io} \qquad (8.21)$$

and store $x_i(t)$, $i = 1, \ldots, N$ and $0 \le t \le t_f$.

Step 4: At the coordinator (ground station in a system of unmanned aerial vehicles, as an example) level, use the results of steps 2 and 3 and Equation (8.20) to update the coordination (communication) vector $[\alpha'_i(t) \mid z'_i(t)]$.

Step 5: Check for the convergence at the coordinator level by evaluating the overall interaction (communication) error

$$\text{Error} = \sum_i^N \int_0^{t_f} \left\{ \left[z_i(t) - \sum_{j=1}^N \alpha'_i(t) G_{ji} x_i(t) \right]' \left[z_i(t) - \sum_{j=1}^N \alpha'_i(t) G_{ji} x_i(t) \right] dt \right\} \bigg/ \Delta t \qquad (8.22)$$

where Δt is the step size of the integration. If Error is sufficiently small, then stop. Otherwise, set $m = m + 1$ and go to step 2.

Example 8.1 Hierarchical Control of Four-System SoS

Consider the four-system SoS control structure given in Figure 8.2. The solid lines represent sensor feedbacks of the systems, whereas the dashed lines represent the wireless sensor and/or data transmission from each neighboring system to any given system. The dynamic state equation of

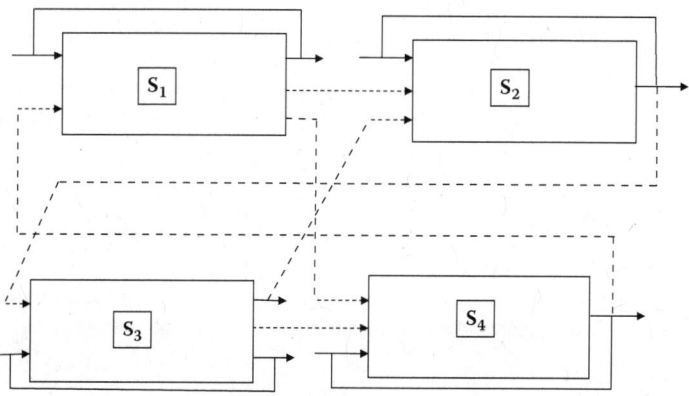

FIGURE 8.2
A four-system SoS.

this system is given by Jamshidi [2,3] as

$$\dot{x} = \begin{bmatrix} 0 & 1 & 0 & & & & & & & 0 & 0 & 0 \\ 0 & 0 & 1 & & 0 & & & 0 & & 0 & 0 & 0 \\ -3 & -2 & -1 & & & & & & & 1 & 0 & \\ \hline 0 & 0 & 0 & 0 & 1 & 0 & 0 & 0 & 0 & & & \\ 0 & 0 & 0 & 0 & 0 & 1 & 0 & 0 & 0 & & 0 & \\ 1 & 0 & 0 & -1 & -3 & -2 & 0 & 1 & 0 & & & \\ & & & 0 & 0 & 0 & 0 & 1 & 0 & & & \\ & 0 & & 0 & 0 & 0 & 0 & 0 & 1 & & 0 & \\ & & & 0 & 1 & 0 & -1 & -2 & -3 & & & \\ \hline 0 & 0 & 0 & & & & 0 & 0 & 0 & 0 & 1 & 0 \\ 0 & 0 & 0 & & 0 & & 0 & 0 & 0 & 0 & 0 & 1 \\ 0 & 1 & 0 & & & & 1 & 0 & 0 & -3 & -2 & 1 \end{bmatrix} x$$

$$+ \begin{bmatrix} 0 & 0 \\ 0 & 0 \\ 1 & 0 \\ \hline 0 & 0 \\ 0 & 0 \\ 0 & 0 \\ \hline 0 & 0 \\ 0 & 0 \\ 0 & 1 \\ \hline 0 & 0 \\ 0 & 0 \\ 0 & 0 \end{bmatrix} u \quad y = cx = \begin{bmatrix} 1 & 0 & 0 & 0 & 0 & 0 \\ 0 & 1 & 0 & & & \\ \hline & & 1 & 0 & 0 & 0 & 0 \\ & 0 & 0 & 1 & 0 & & \\ \hline & & & & 1 & 0 & 0 & 0 \\ & 0 & & 0 & & 0 & 1 & 0 \\ \hline & & & & & & 1 & 0 & 0 \\ & 0 & & 0 & & 0 & 0 & 1 & 0 \end{bmatrix} x$$

The cost function's parameters and matrices were chosen as $Q_i(t) = [0\ 1\ 1]'$, $R_i(t) = 1, i = 1, \ldots, 4, t_0 = 0$, and $t_f = 20$ s.

When this SoS is put in the form of a hierarchical control system, it is put in the format depicted in Figure 8.3 [3]. In this figure, there are six interactions (wireless communication links), which are initially disconnected, and gradually, through the *interaction prediction principle* of large-scale systems theory, they will be strengthened and reconnected. This disconnection in communication causes an error, which will be accounted for shortly. The parameters $\alpha_i^k, i = 1, 2, \ldots, 4$, are the Lagrange multipliers accounting for feasibility of the communication constraints.

The interaction error vector (due to breakage of communication links or cutoff of the links) among the four systems of the SoS is given by

$$e = [e_1, e_2, e_3, e_4, e_5, e_6]' = [(z_1 - x_{11}), (z_2 - x_1), (z_3 - x_7), (z_4 - x_5), (z_5 - x_8), (z_6 - x_2)]$$

At the system level of this SoS control problem, four differential matrix Riccati equations such as Equation (8.8) are performed from 0

Control of System of Systems

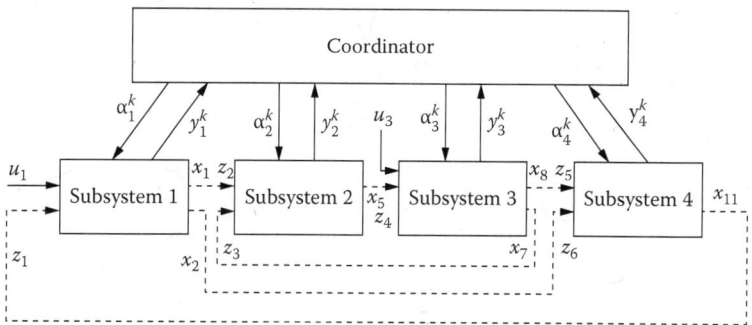

FIGURE 8.3
A hierarchical control reformulation of SoS control problem of Example 8.1.

to 20 s. A sample of the solutions of the Riccati matrix for system 1 is given in Figure 8.4. The adjoint vector components $g_i(t)$ for system 1 (see Algorithm 8.1) are shown in Figure 8.5.

It is noted that the Riccati matrix and adjoint vector elements need to be integrated in negative time, as theory calls for it. The 12 optimal states of this SoS are shown in Figure 8.6. The two curves per subfigure here represent the optimal centralized control solution and the one obtained from the interaction prediction approach to this SoS control problem. Note that

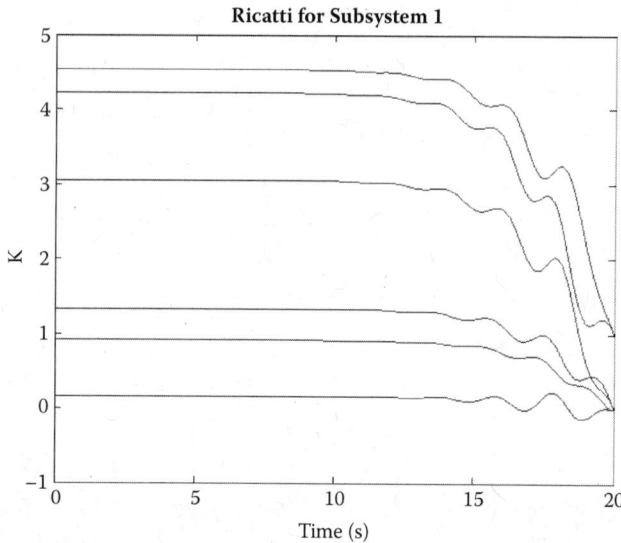

FIGURE 8.4
Riccati matrix elements for the optimal hierarchical control of system 1 in Example 8.1.

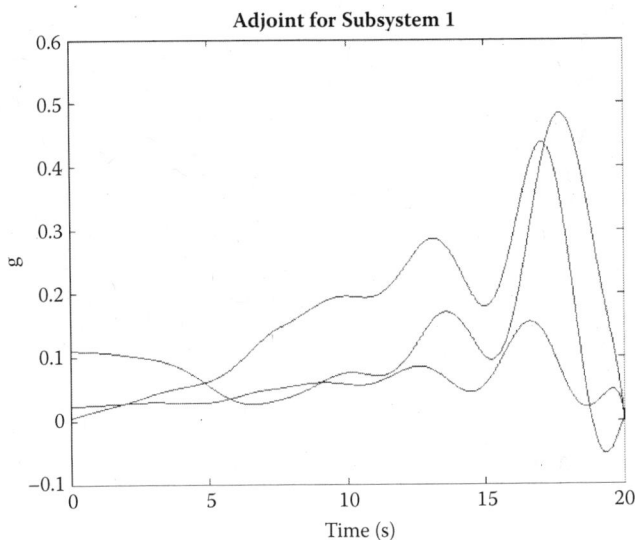

FIGURE 8.5
System 1's adjoint vector elements for the optimal hierarchical control of system 1 of Example 8.1.

the centralized control solution is normally impossible to obtain, but in this simple 12th-order system, it was possible to obtain them.

The interaction (communication balance) error measure of this hierarchical control of the SoS problem reduced to zero in 15 iterations of Algorithm 8.1, as shown in Figure 8.7. At this point, where all error components e_i, $i = 1, \ldots, 6$, are reduced to zero, the interaction among systems of SoS is said to be balanced and optimal control has been achieved. The control and output signals of the problem are shown in Figure 8.8.

8.3 Decentralized Control of SoS*

The second paradigm, decentralized control of an SoS, as illustrated in Figure 8.9, assumes that it can be characterized by a great multiplicity of input and output variables, with each subset of variables or systems exercising local control. For example, an electric power grid (i.e., an SoS) has numerous substations (i.e., systems), with each substation being responsible for the operation of a portion of the grid. The designer of a decentralized SoS needs to determine a control structure that assigns system inputs to a given set of local controllers, each of which observes only local system outputs. In

---

* This section is partially based on the paper by Kumar Ray et al. [4] and Jamshidi [2,3].

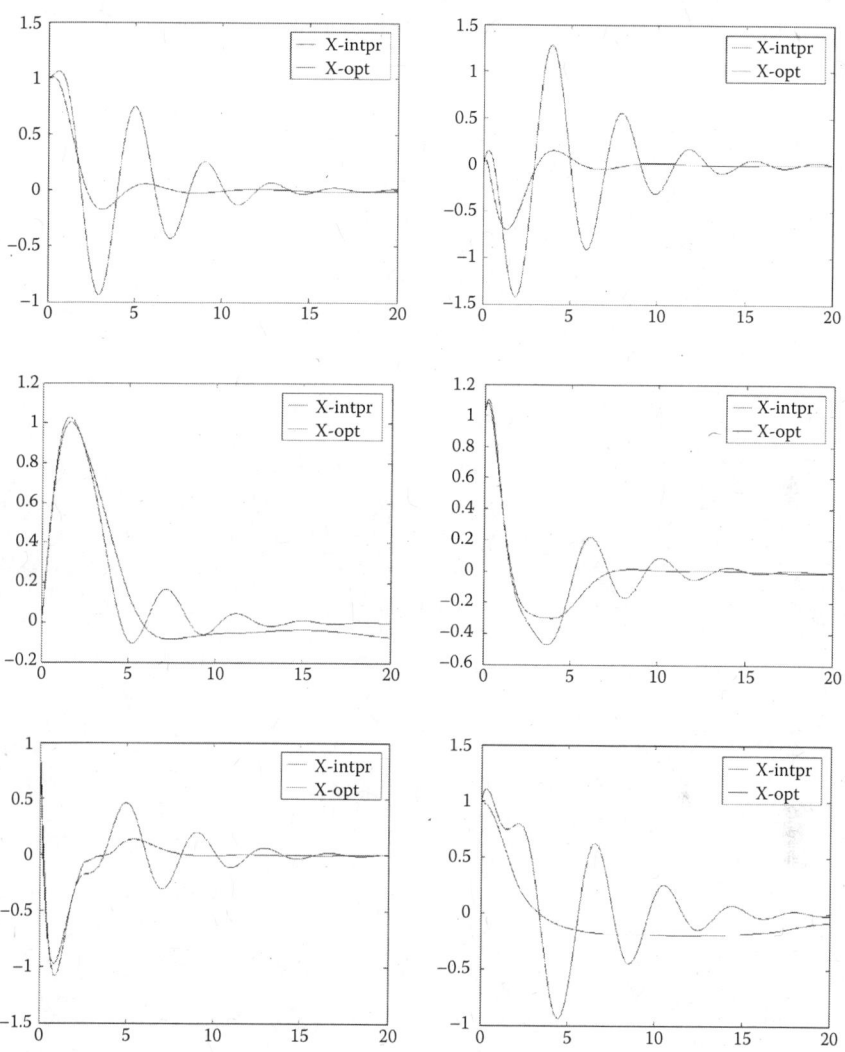

FIGURE 8.6
SoS control problems 12 state variables shown with the centralized optimal regulator solutions in Example 8.1.

essence, decentralized control attempts to avoid difficulties in data gathering, data storage, and system debugging. As in the case of hierarchical control, although the literature on the subject is filled with classical, steady-state approaches to decentralized control [3], it is lacking in real-time considerations. In this section, two decentralized schemes will be discussed following Kumar Ray et al. [4,5]—one is the decentralized control of a group of

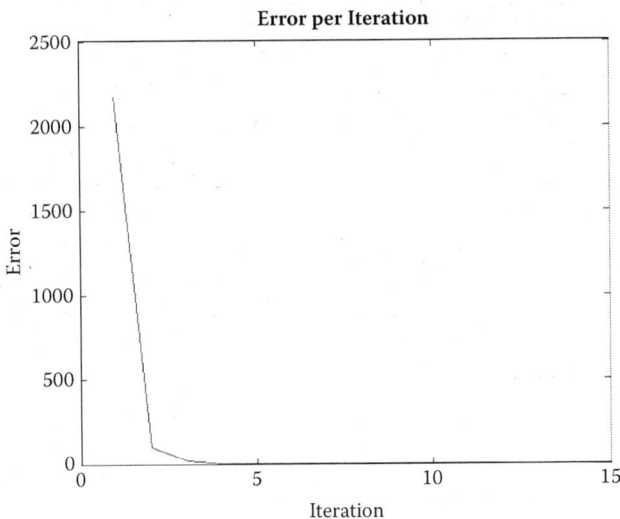

FIGURE 8.7
Interaction (communication) error measure convergent behavior for Example 8.1.

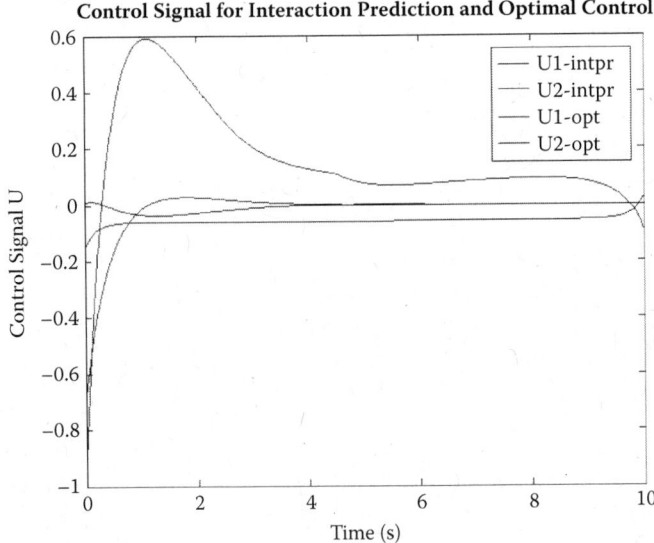

FIGURE 8.8
SoS control problems control and output variables shown with the centralized optimal regulator solutions in Example 8.1.

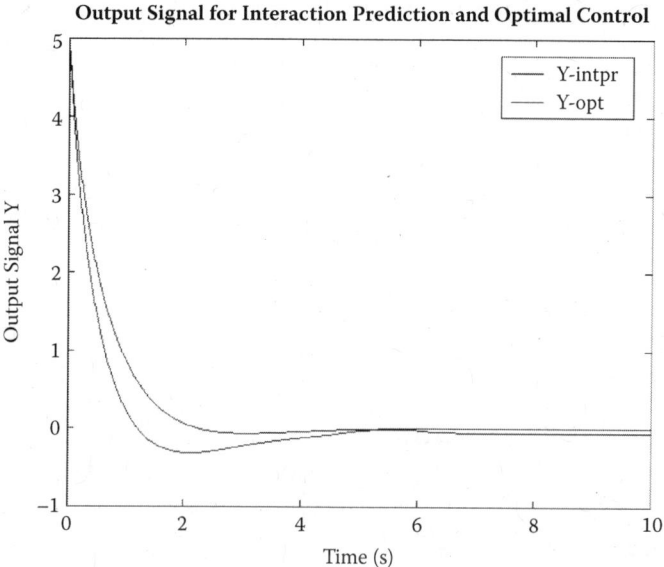

FIGURE 8.8
(*Continued*)

rovers, whereas the second one is the decentralized formation control of a system of rovers.

8.3.1 Decentralized Navigation Control

Consider N nonholonomic rovers (agents) with acceleration u_i and rotational velocity ω_i chosen as control inputs for each agent, and find a control law that moves each rover from any feasible initial configuration (position and orientation) to its goal configuration, while avoiding any obstacles of other rovers [6].

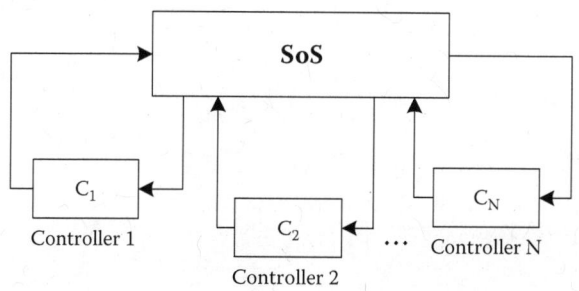

FIGURE 8.9
Decentralized control paradigm for an SoS.

The dynamic model of the N rovers is given by the following set of differential equations:

$$\dot{x}_i = v_i \cos\theta_i$$
$$\dot{y}_i = v_i \sin\theta_i$$
$$\dot{\theta}_i = \omega_i \quad , \quad i \in \{1, \ldots, N\}$$
$$\dot{v}_i = u_i$$

where ω_i and v_i are translational and angular velocities of the ith rover (agent), the x_i, y_i pair represents the position of the center of gravity of the ith rover, and θ_i is the orientation of the ith rover. The desired objective function (to be minimized or maximized) is given by [6]:

$$\varphi_i = \frac{\gamma_{di} + f_i}{((\gamma_{di} + f_i)^k + H_{nh_i} \cdot G_i \cdot \beta_{o_i})^{1/k}}$$

where λ_{di} is the desired ith rover objective, which is convergence to a desired destination; f_i is an encoding function guaranteeing cooperation between the ith and other $N - 1$ agents; G_i is the collision scheme in which the ith rover could get involved; β_{o_i} is the workspace bounding obstacle; H_{nhi} has a pseudo-obstacle form; and ϕ_i is the objective function of the ith rover [7].

8.3.2 The Decentralized Control Law

Dimarogonas and Kyriakopoulos [7] have proposed a decentralized control paradigm,

$$u_i = -v_i \{|\nabla_i \varphi_i \cdot \eta_i| + M_i\} - g_i v_i - \frac{v_i}{\tanh(|v_i|)} K_{v_i} K_{z_i}$$

$$\omega_i = -K_{\theta_i}(\theta_i - \theta_{d_i} - \theta_{nh_i}) + \dot{\theta}_{nh_i}$$

where

$$\lambda_{d_i} = \|q_i - q_{d_i}\|^2 \quad f_i(G_i) = \begin{cases} a_0 + \sum_{j=1}^{3} a_j G_i^j, & G_i \leq X \\ 0, & G_i > X \end{cases}$$

$$H_{nh_i} = \varepsilon_{nh} + \eta_{nh_i}$$

$$\beta_{o_i} = r_{world}^2 - \|q_i - q_{d_i}\|^2$$

$$q_i, q_{d_i} \in R^2$$

$$\eta_{nh_i} = \|(q_i - q_{d_i}) \cdot \eta_{d_i}\|^2$$

$$\eta_{d_i} = [\cos(\theta_{d_i}) \ \sin(\theta_{d_i})]^T$$

$$\phi_{nh_i} = \arg\left(\frac{\partial \varphi_i}{\partial x_i} \cdot s_i + i \frac{\partial \varphi_i}{\partial y_i} \cdot s_i\right)$$

$$s_i = \text{sgn}((q_i - q_{di}) \cdot \eta_{di})$$

$$\eta_i = [\cos \phi_i \ \sin \phi_i]^T$$

$$\eta_{di} = [\cos \phi_{di} \ \sin \phi_{di}]^T$$

$$K_{zi} = \|\nabla_i \varphi_i\|^2 + \|q_i - q_{di}\|^2$$

$$M_i > \left|\sum_{j \neq i} \nabla_i \varphi_j \cdot \eta_i\right|_{\max}$$

$$\nabla_i \varphi_j = \left[\frac{\partial \varphi_j}{x_i} \ \frac{\partial \varphi_j}{y_i}\right]$$

$K_{v_i}, K_{\phi_i}, g_i > 0$ are positive gains

The authors used the Lyapunov stability to prove that this control law would asymptotically stabilize the multiple Nonholonomic rovers.

A Simulated Example

Kumar Ray et al. [5] have simulated a four-rover system using the above development. Figures 8.10 and 8.11 show two sets of simulated trajectories and angular and translational velocities of the four rovers.

Motion coordination for a formation of rovers. Kumar Ray et al. [4] have proposed a coordinated formation of rovers (system of rovers) for navigation and movement. The approach will concentrate on an autonomous, collision-free, minimum-communication protocol among systems (or agents), while the formation would navigate through an unstructured environment. Each system will have a unique identification (ID) through which any agent, acting as the "master," can coordinate its motion. Should the master system (agent) fail for any reason, it will be replaced by another agent in the formation.

8.3.2.1 Motion Coordination

Figure 8.12 shows the rovers' formation. As noted, in order to coordinate the formation, one of the agents is assigned as the *master*, which

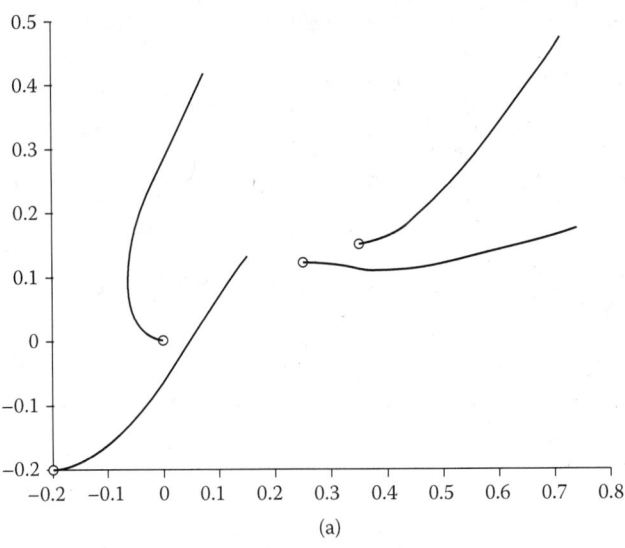

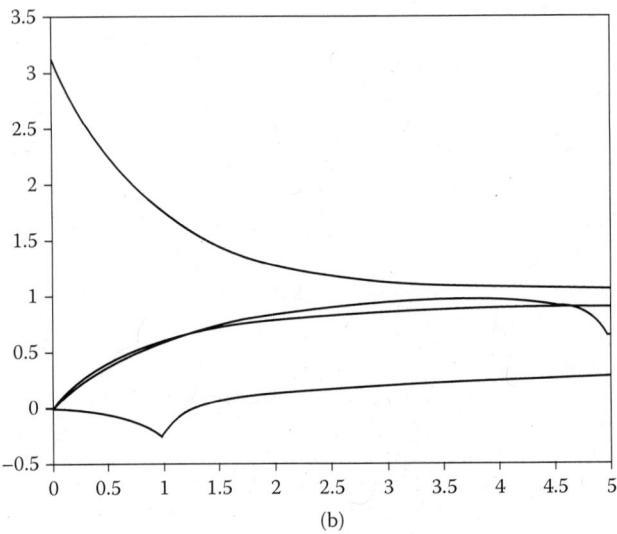

FIGURE 8.10
Multiagent movement with decentralized navigation control: case study I. (a) Planar trajectories, (b) angular velocity plots, (c) translational velocity plots.

Control of System of Systems

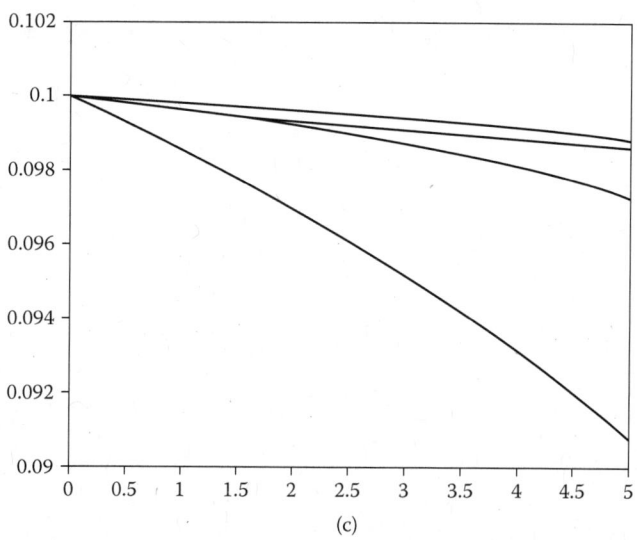

FIGURE 8.10
(*Continued*)

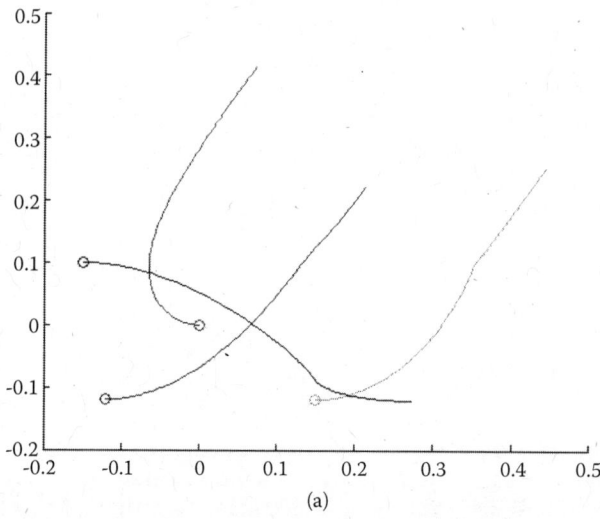

FIGURE 8.11
Multiagent movement with decentralized navigation control: case study II. (a) Planar trajectories, (b) angular velocity plots, (c) translational velocity plots.

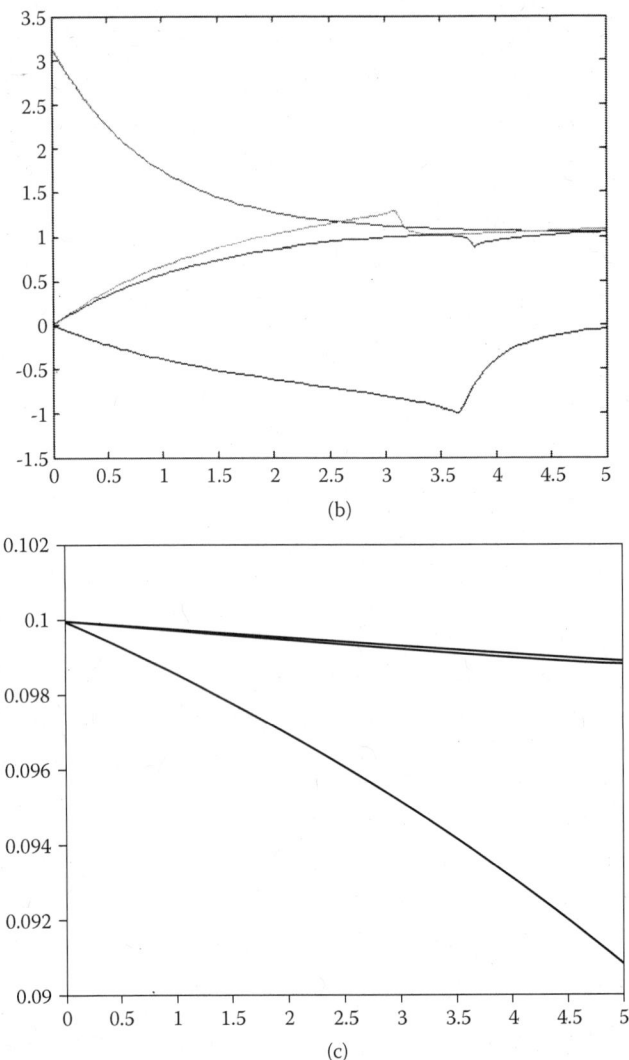

FIGURE 8.11
(Continued)

communicates with a remote host or a human-operated control center, and the other agents receive most of the information from the master agent and adjust their behavior to complete the coordinated task [5]. The main issue here is twofold: agent communication and collision-free navigation of the rovers. The approach emphasizes collision-free navigation of the formation satisfying various constraints while making the formation suitable for real-time operation in an unknown environment.

Control of System of Systems 249

FIGURE 8.12
A rover formation—a joint work between UTSA and IIT-Kanpur (courtesy of ACE Center, University of Texas, San Antonio, USA).

The reaction with the environment is not considered in this work [4]; rather, it is assumed that the main agent has the information of motion initiative depending on the environmental situation. Figure 8.13 shows how the formation of rovers is making a right turn. Figure 8.14 shows the front envelope of the formation of rovers having the motion initiative command from the master agent. The agents are assigned a row-column ID, where the first row refers to the front envelope of the formation

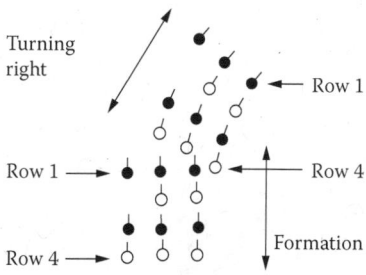

FIGURE 8.13
Navigation of the formation of multiagent rovers [4,5].

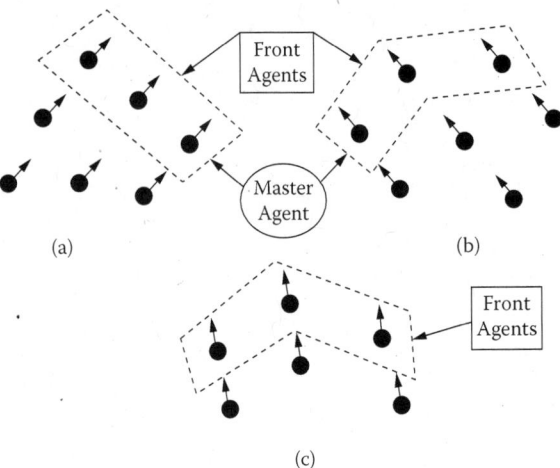

FIGURE 8.14
Front envelope of the formation of the system of rovers [4].

(see Figure 8.13). One of the first-row agents is assigned as the master of the formation and the other first-row agents act as their corresponding column's agent guides. Now, the general front agents may be at the left side of the master agent (Figure 8.14a), at the right side of it (Figure 8.14b), or at either side of it (Figure 8.14c). Depending on the positions of those agents, the requirements of navigation of the formation change, which is modeled by Kumar Ray et al. [4]. The numbers of agents in any two or more columns (or rows) do not have to be equal. Their codes can be in any sequence and not in any particular order. This ID assignment scheme will allow great flexibility in the formation. For a new configuration of the formation, not all of the IDs need to be reassigned, thereby saving a lot of computational time and adding robustness.

Figure 8.15 indicates that a continuous turning with the same linear and angular velocities may lead to new front agents in the formation,

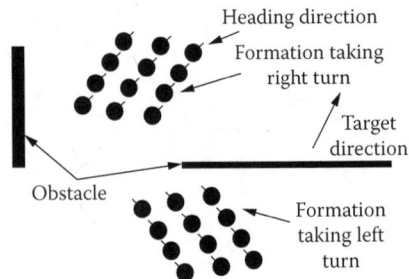

FIGURE 8.15
Environmental situation where the formation requires varying ± angular velocity [4].

whereas the actual front agents will become followers of columns. In this situation, although the formation satisfies the collision-free movement constraints [4], it falls short of keeping the collision-free movement, as well as the form of the formation. Under such conditions, assigning and reassigning the IDs of the agents will create chaos in the formation and would increase the communication burden for a coordinated task.

The authors [4] have started, from each agent (rover), local and global coordinate systems and have provided physical conditions and scenarios for various movements of the formation, such as turning and heading in the directions of the front agents and turning toward the left and turning toward the right for both cases of when the front agents are to the right and left of the "master" agent. The authors have tested two formation movement approaches with four different cases of navigational situations via simulation. Here, numerous angular and translational velocity responses for various formation movement scenarios have been presented. Figure 8.15 shows a situation where the formation requires varying positive/negative angular velocity and another situation where the formation requires varying angular velocity over a certain period, and after that, zero angular velocity.

Kumar Ray et al. [4] have proposed a novel approach for the formation of rovers' movements via ID assignments of agents that would guarantee collision-free situation during various maneuvers in the formation. Under their regime, agents adjust themselves during navigation, with more spaces between them during turns. In the proposed scheme, each agent can analyze the behavior of the formation on board, getting the data from the *master* agent. Moreover, in real-time implementation, should there be any mismatch due to odometery, it can be easily corrected by the agents with the existing techniques. Figure 8.16 provides rover formations while confronting wall-type obstacles.

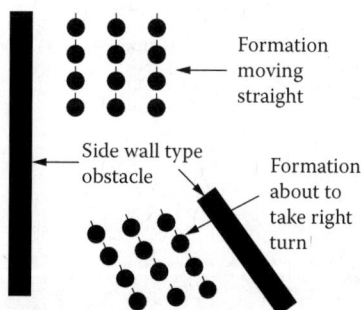

FIGURE 8.16
Environmental situation where the formation requires varying angular velocity over a certain period, and after that, 0 angular velocity [4].

8.4 Other Control Approaches

8.4.1 Consensus-Based Control

Ren and Beard [8] have proposed a cooperative control paradigm based on "consensus" among systems in an SoS. The primary motivation for this control paradigm, like all other paradigms, is extracting greater benefit from constituencies of an SoS. In fact, the entire motivation of any SoS is to increase benefits, increase robustness, lower cost, etc. In the case of a system of rovers (land, sea, or air), it is conceived that all would agree on and follow a common goal. Subgoals of individual rovers (or systems) may be different, but through shared communication, the family of rovers would reach a "consensus" eventually and follow certain pattern in navigating, say, an unstructured environment such as the surface of Mars.

Communication among systems has limited bandwidth and connectivity, and questions arise such as what, when, and to whom a system (rover) should communicate with. One notes that, as usual, computational resources are limited. Modeling of vehicles (systems) interactions are based on graph theory. Figure 8.17 shows two scenarios for communication among three rovers. In the first scenario, there is limited communication between rover 1 and the other two rovers.

In the case of implementing consensus-based control paradigm for a set of underwater and simulated robots, one needs to know where the rovers are and which way they are facing. The underwater rover used in this research was the VideoRay, shown in Figure 8.18. As seen in this figure, these rovers are tethered, and because of this, they are easy to control with a PC and easy communication is provided between VideoRays.

A hardware-in-the-loop simulation of the VideoRay following a path described by four waypoints set in a square 2 × 2 m (all are at different depths) is shown in Figure 8.19. Each underwater rover needs to know at any time where all the

Communications

- How well can the robots talk to each other?

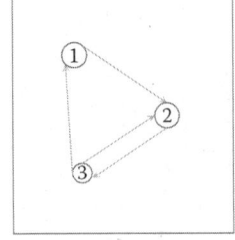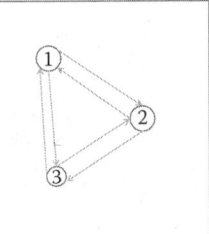

FIGURE 8.17
Graph of theoretic representation of communication protocol among three rovers.

Control of System of Systems

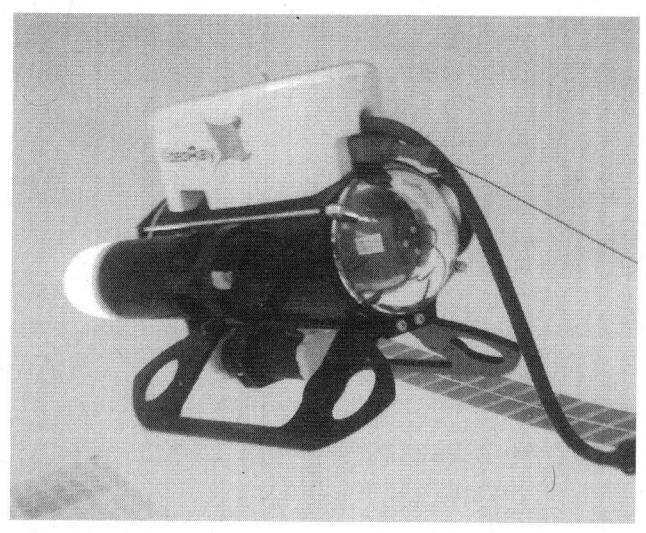

FIGURE 8.18
A commercial VideoRay underwater rover (submarine) use in consensus-based control.

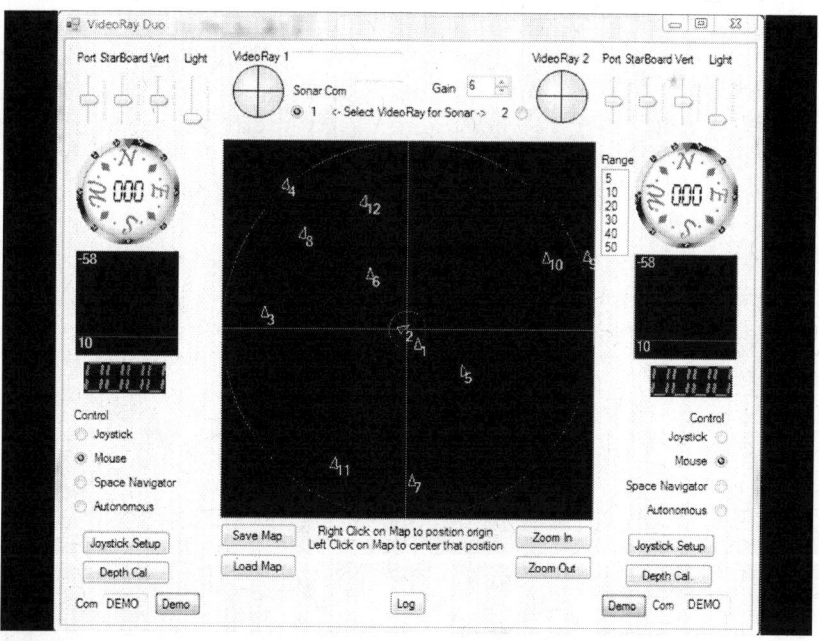

FIGURE 8.19
Hardware in the loop simulation of VideoRay rovers.

other rovers are, know what its own goal is, communicate its position to others, and decide on its own action independently or vote on an overall consensus. The following step summarizes a consensus among the rovers:

Each rover must agree to have equal space from its neighbors along a path; they must share information about themselves and identify their coordinates and ID. The objective is given by

$$J = p/n - \left(\sum_{i=2}^{n} VR_i - VR_{i-1} \right) / n$$

where p/n is the distance between rovers and the second term represents the average distance between consecutive rovers. Consensus is reached when $J = 0$. The strategy is to maintain distance between rovers in a line and be able to make decisions when the communication among rovers is poor.

Rovers need to know what data to share, need to look at the constraint equation (J), and need to know the positions (X and Y) and ID of other rovers. Communication among them will use a token system, that is, the rover with the lowest ID starts with the token; when it transmits, it loses the token; the rover with the next ID takes the token; and if no transmission occurs for some time, the rover with the lowest ID takes the token. There will be one transmission every tenth of a second.

Centralized strategy—This comes from the constraint equation for J; the distance between rovers must be the patrol path length divided by the number of rovers; the speeds of the rovers change based on the distance between rovers.

Consensus building—Breaking the centralized strategy down, each rover knows the total number of rovers by counting unique IDs it hears via transmissions and it knows which rover is in front of it by looking at transmitted positions. To control speed, modify the constraint equation:

$$J = p/n - (FR - Me)$$

where FR is front robot. At this point, each rover controls its speed to be higher or lower based on J.

Joordens and Jamshidi [9] have performed a simulation for a VideoRay rover and ten simulated ones (as hardware in the loop). Figure 8.20 shows a scene of this simulation. In this simulation, each rover transmits its location and all other rovers use that to keep their distance.

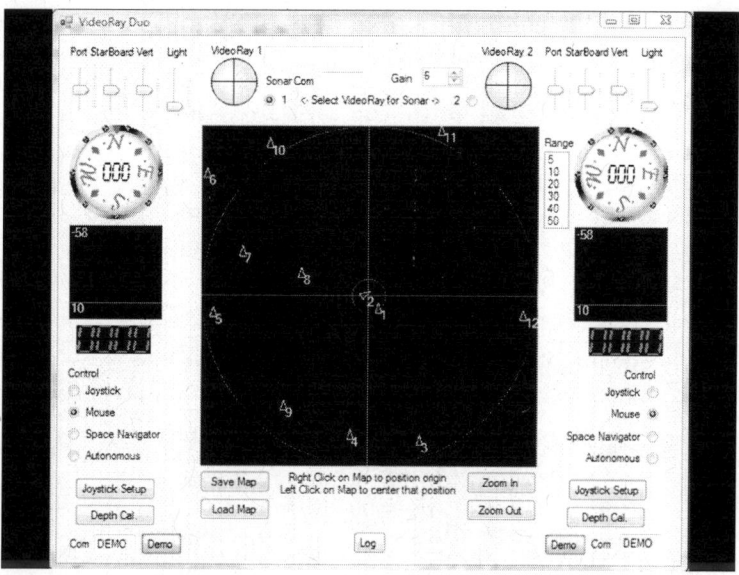

FIGURE 8.20
Another scene of the hardware in the loop simulation of VideoRay rovers.

8.4.2 Cooperative Control

Cooperative control of an SoS assumes that it can be characterized by a set of interconnected systems or agents with a common goal. Classical techniques of control design, optimization, and estimation could be used to create parallel architectures for, as an example, coordinating underwater gliders [1]. However, many issues dealing with real-time cooperative control have not been addressed, even in non-SoS structures. A critical issue concerns controlling an SoS in the presence of communication delays to and among the SoS systems. Ren and Cao [10] have presented some categorization for cooperative control. These include approaches such as leader-follower, behavioral, virtual structure/leader, etc. Application areas of cooperative control, just as in any viable control paradigm of SoS, are autonomous or semiautonomous vehicles, satellites, spacecrafts, automated highways, earth observation, air traffic, border/port security, environment (oil spills, rural areas, forest fires, wildlife), etc.

8.4.3 Networked Control

A control system where a real-time communication network is in the feedback path is called a *network control system* (NCS) [11]. The *ad hoc* network can be implemented through a number of alternatives such as Ethernet, FireWire, etc. However, one often has a time-varying channel dependent on a fixed capacity for the total amount of information that can be communicated at any one time instance to the collection of autonomous unmanned vehicle controllers.

One of the main challenges in NCS is the loss of or delays in transmission and receipt of data from sensors to controllers τ_{sc} and from controllers to actuators τ_{ca}. The challenge in SoS networked control is to develop an SoS distributed control system that can tolerate lost packets, partially decoded packets, delays, and fairness issues—that is, add robustness to the control paradigm. Here, by fairness, it is meant that certain systems are more capable of maximizing their total wireless system capacity if we transmit to their controllers more often. Values of parameters such as sensor-controller and controller-actuator delays will be keys to how much the fairness issue must be practiced. These communication infractions can be compensated by [12]:

Adjusting control power and controlling distances between systems (power control)

Trading off modulation, coding, and antenna diversity versus throughput (adaptive modulation coding)

The (nonwireless) intrafeedback (onboard hardware) loop of the autonomous control within S_i is lower latency than the interwireless distributed control loop between S_i and S_j or the interwireless SoS controller and the S_i controller

As an alternative approach to checking on the communication infractions and guarantee a high level of quality of service (QoS) and quality of control performance (QoP), the design of a wireless networked control system needs to fully take all aspects of the attributes of the *ad hoc* network into account. When the sensor measurements reach the controller (see Figure 8.21), the following tasks need to be completed: (1) compute the control action $u(kT)$ and (2) adjust the control action depending on QoS parameters, where T is the sampling period [13]. Here, at each sampling period, the distributed control will generate two components, that is, $u(kT) = u_l(kT) + u_c(kT)$, where the first component $u_l(kT)$ is the local controller using classical or modern control techniques, such as proportional integral derivative or linear quadratic Gaussian (LQG), etc., and $u_c(kT)$ is the correction component of the controller determined to compensate for *ad hoc* network QoS parameters. The latter correction control component will depend on the sampling period $T(k + 1)$ and the sampling policy of the wireless networked control SoS (WNCSOS). One way to ensure that the correction component $u_c(kT)$ is determined to devise a performance index in order to minimize the effects of *ad hoc* net parameters on QoS is to use an LQG problem and, at the same time, use the sampling policy to determine the next period $T(k+1) = T(k) + \Delta T(k)$. The combination of a local controller, a correction component, and an adaptive sampling period will enhance both the stability and robustness of the WNCSOS. Figure 8.22 shows this modification with an adaptive sampler [13].

Gupta [14] has also presented comprehensive work in his doctoral dissertation on distributed estimation and control of networked systems. He noted that advances in wireless communication, sensor technology, and

Control of System of Systems

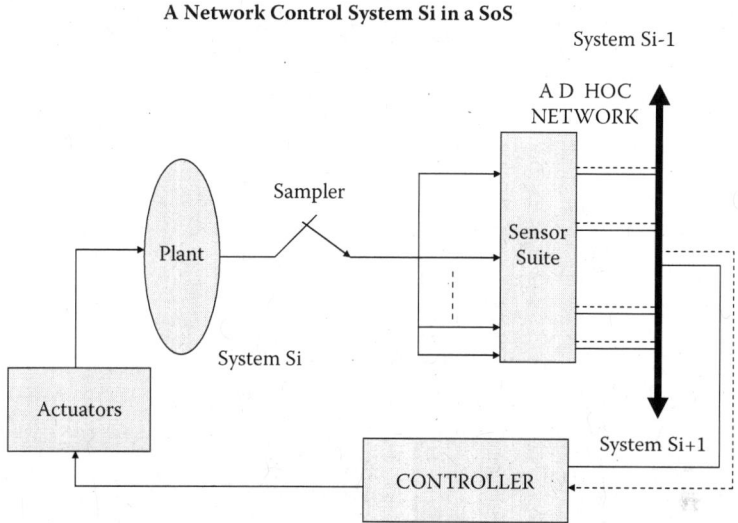

FIGURE 8.21
A network or distributed control system for system S_i of an SoS (lines in the figure are communication lines).

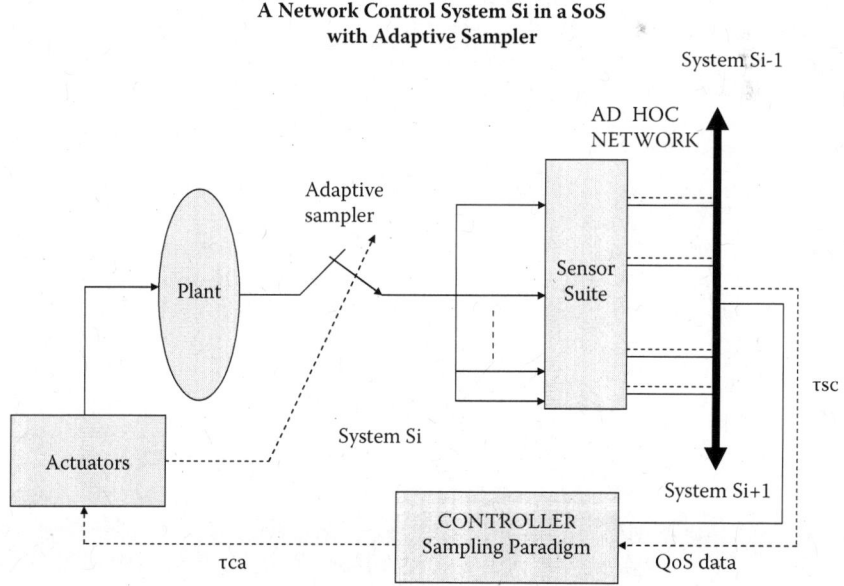

FIGURE 8.22
Adaptive sampling interval for wireless NCS.

information processing make estimation and control of a network of systems (or subsystems) possible. The objective is what it is for an SoS—increase robustness and reliability, reduce cost, and enhance potential. The challenge here, as it has been all along, is the extension of many single-system theories to multiple systems or an SoS. The nature of these challenges as far as control is concerned is the imperfect information coming from remote systems (sensors and/or actuators), as well as a lack of a central depositor of information; hence, each system has to go with partial information.

Gupta's [14] main contribution here is the simultaneous design of information flow and the control law. Unlike traditional control design tools, which concentrate on calculating the optimal control input by assuming a particular information flow between the components, Gupta's [14] approach seeks to synthesize the optimal information flow along with the optimal control law that satisfies the constraints of the information flow. Thus, the questions of "what should an agent do?," "whom should an agent talk to?," "what should an agent communicate?," "when should an agent communicate?," and so on also have to be answered. This idea is exactly what the decentralized formation control of Kumar Ray et al. [4] is all about, which was briefly discussed earlier.

Gupta [14] considered a networked (distributed) system of the form shown in Figures 8.21 and 8.22. Here, again, similar to the work of Kumar Ray et al. [4], communication among agents in a formation is the key idea in cooperation, while optimizing some measure of performance or cost. Gupta [14] has provided synthesis tools to optimal control involving linear equations and constraints. A topology is designed through which the agents can communicate, as in Section 8.3. Gupta [14] has further studied the effects of communication channels on control performance.

8.5 Conclusions

This chapter sets forth the stage for one of the most important issues in SoS—the control problem. The challenge for any control engineer and designer is to stabilize the SoS based not only on its own onboard sensors, but, just as important, also on data transmissions from other sensors and actuators. For that reason, the control design needs to be flexible to account for lack or shortfall of communication packets from other systems or on usual needs, such as unmodeled dynamics, disturbances, noise, etc. Therefore, we propose that two controllers be designed simultaneously—one for the integrity and stability of each system (of an SoS) and another to account for communication and information processing among partners or systems in the SoS. We call the former component a local controller and the latter a global controller. This idea is not unique and has been around for many years [2]. In recent books by one of the authors of this text [15,16], a control problem was missing due to the challenges in this problem.

References

[1] Hipel, K. W., Jamshidi, M., Tien, J. M., and White, C. C., "The future of systems, man and cybernetics: Application domains and research methods," *IEEE Transactions on SMC, Part C* 37(5), 726–743, 2007.

[2] Jamshidi, M., *Large-Scale Systems—Modeling and Control*, Elsevier North-Holland, New York, NY, 1983.

[3] Jamshidi, M., *Large-Scale Systems: Modeling, Control, and Fuzzy Logic*, Prentice-Hall, Englewood Cliffs, NJ, 1997.

[4] Kumar Ray, A., Behera, L., and Jamshidi, M., "Motion coordination for a formation of rovers," to appear in *IEEE Systems Journal*, Vol. 3, 2009.

[5] Kumar Ray, A., Behera, L., and Jamshidi, M., "Autonomous and decentralized navigation and control of multiple rovers," to appear, 2010.

[6] Dimarogonas, D. V., Zavlanos, M. M., Loizou, S. G., and Kyriakopoulos, K. J., "Decentralized motion control of multiple holonomic agents under input constraints," in *Proceedings of IEEE International Symposium on Intelligent Control*, Maui, HI, December 2003.

[7] Dimarogonas, D. V., and Kyriakopoulos, K. J., "A feedback stabilization and collision avoidance scheme for multiple independent nonholonomic non-point agents," in *Proceedings of IEEE International Symposium on Intelligent Control*, Limassol, June 27–29, 2005.

[8] Ren, W., and Beard, R. W., *Distributed Consensus in Multi-vehicle Cooperative Control, Communications and Control Engineering Series*, Springer-Verlag, London, 2008.

[9] Joordens, M. A., and Jamshidi, M., "Consensus control for a system of underwater swarm robots," submitted for publication, 2009.

[10] Ren, W., and Cao, Y.-C., "Simulation and experimental study of consensus algorithms for multiple mobile robots with information feedback," *Intelligent Automation and Soft Computing (AutoSoft) Journal* 14(1), 73–87, 2008.

[11] Zhang, W., Branicky, M. S., and Phillips, S. M., "Stability of networked control systems," *IEEE Control Systems Magazine* 21(1), 84–99, 2001.

[12] Kelley, B., Jamshidi, M., and Akopian, D., "MRI development of a test-bed swarm of autonomous unmanned vehicles ad-hoc networked control via secure 4G wireless," NSF MRI Proposal. University of Texas, San Antonio, January 2009.

[13] Colandairaj, J., Irwin, G. W., and Scanlon, W., "Wireless network control systems with QoS-based sampling," *IET Control Theory and Applications* 1(1), 430–437, 2007.

[14] Gupta, V., "Distributed estimation and control in networked systems," PhD dissertation, California Institute of Technology, Pasadena, CA, 2006.

[15] Jamshidi, M. (Ed.), *System of Systems Engineering—Principles and Applications*, Taylor Francis CRC Publishers, Boca Raton, FL, 2008.

[16] Jamshidi, M. (Ed.), *System of Systems Engineering—Innovations for the 21st Century*, Wiley & Sons, Inc., New York, 2009.

9

Reward-Based Behavior Adaptation

9.1 Introduction

In previous chapters we learned how a neural network can learn to improve its performance by comparing its outputs with a desired value. This is known as supervised learning because there is a desired value for the output of the network that can act as a supervisor of the value being computed by the network. To avoid any confusion, let us take two scenarios to clarify this issue.

Figure 9.1 shows how a neural network learns to approximate the dynamics of a system by working parallel to it. Since there is a measurable control input and an output state from the dynamical system, we can feed the neural network with a copy of the control input and compare the computed output state with the actual output state from the dynamical system. If the neural network has accurately approximated the dynamics of the system, the computed output state should come close to the actual output state. Otherwise, the error of estimation can be used to update the internal parameters of the neural network. We call this a supervised learning scheme because there is a quantifiable desired value against which the output of the neural network can be compared.

In Figure 9.2, the neural network is trying to learn the inverse dynamics of the system so that, given a desired state, it can compute a control command to the dynamical system expecting it to follow the desired profile of states. However, there is no way of knowing if the control input that the neural network computes is accurate because there is no "teaching" controller as in the example shown in Figure 9.1. The simple trick we can design in this case is to feed the resulting state of the dynamical system back to a copy of the same neural network to compute a control command corresponding to the resulting state. Then, we argue that the resulting state should be close to the desired state if the controller was computing the correct control input to the system. Therefore, if there is an error between the desired state and the resulting state, it should be reflected in the two control inputs computed by the same neural network for the desired state and for the resulting state. Hence, even in this case, there is a quantifiable reference that can be used to compute the error signal to update the parameters of the neural network.

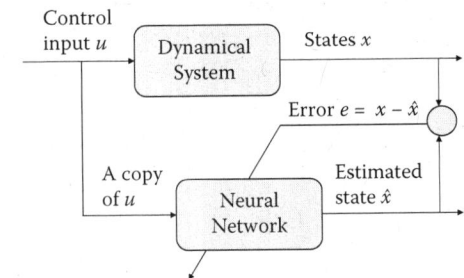

FIGURE 9.1
Supervised learning to identify the dynamics of an unknown system.

Now, let us compare the above two cases with a scenario where there is no such quantifiable reference signal to compute an error signal. Imagine a neural network is supposed to take proximity readings from a sonar sensor mounted on a robot as input, and compute two torque commands that will be given to the two motors to enact an obstacle avoidance behavior as shown in Figure 9.3. In this case, we see three different behaviors for three different parameter sets of the neural network. Can we compute an error signal to tune the parameters of the neural network in this case? It is hard to do that because there is no reference path against which we can compare. All that we can say is that one behavior is "better" than another. We may be guided by such notions as "good" behavior is one that neither overreacts nor goes too close to the obstacle. Therefore, such evaluations are more of qualitative judgments than quantifiable error signals.

In a natural learning scenario, we meet more of such qualitative feedback-based learning than supervised learning. We call this type of learning

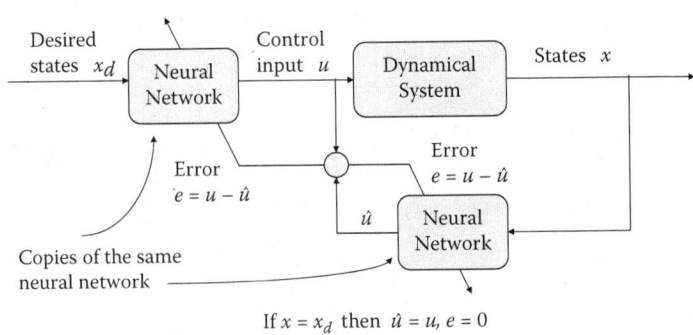

FIGURE 9.2
Supervised learning to control a dynamical system with unknown inverse dynamics.

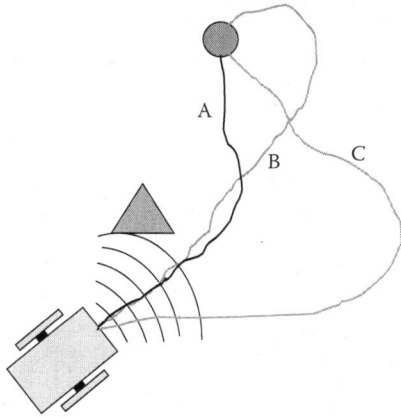

FIGURE 9.3
How a robot responds to an obstacle under three different behavior controllers.

"reinforcement-based" learning or reward-based learning, where one takes an action in response to an environmental situation, and the interaction between the action and the environment leads one taking the action to change its state relative to the environment [1–4]. This relative state change may be evaluated by some criteria to give a reward or a punishment. This enables the one taking actions to learn to maximize rewards. Let us define some formal terms to characterize this learning-by-interaction paradigm for robots [5].

9.1.1 Embodiment

A robot is a machine with a body. The physical shape of the body and its dynamics such as inertia around different axes, mass, etc., have a physical meaning in the way the control commands are translated to robot behaviors. Furthermore, the physical space occupied by the robot's body decides how the environment interacts with the robot. For instance, if it is a swimming robot, the hydrodynamic forces acting upon the robot depend on the shape of the surface of the robot that moves relative to the liquid. Therefore, we define a term called an *agent* to denote only the controller that maps sensory inputs to a control command. We define *environment* to be the robot's body and the rest of the environment as a whole. This can be generalized to any machine with a physical body.

9.1.2 Situatedness

A robot or a machine with a physical body and a physical shape located in a real environment has to deal with real physical phenomena, such as forces exerted by the environment on the robot, posture decided by the shape of the terrain, disturbances, etc.

9.1.3 Internal Models

The robot or the machine learning to interact with the environment can try to model the world in terms of memory primitives. A memory primitive is a very simple association between a cause and an effect in a locality. It may construct more complex representations of the environment by combining such primitives.

9.1.4 Policy

A *policy* maps a situation to an action. The internal mechanism of this mapping may consist of internal models of the world. For instance, if somebody is allowed to lift a cup filled with a liquid of a particular color, he/she will learn to exert the correct force to lift it up over time. After learning, if we show that person an empty cup painted in the same color so that the person will be under the illusion that the cup is filled with the same liquid, he/she will exert more force than is required to lift the empty cup. This happens because the individual learned to associate the color cue to the required force over time and built an internal model of the dynamics of the cup that he/she used as a resource to construct a policy to lift the cup. When the cup was empty, his/her internal models were no longer valid. Yet, the individual went ahead and lifted the cup with the wrong force because he/she was guided by the internal models. However, in nature, it is very rare that we get encounter illusions. Perhaps that is why nature has encouraged the brain not to be reactive all the time because it is computationally very expensive and may lead to unstable behavior in some cases. Behavior based on internal models is very robust due to its ability to reject disturbances. We will study these concepts in more detail later.

9.1.5 Reward

A reward is a value assigned to assess the performance of the policy at a given time. Assessment is done by a reward function that takes the states of the agent and the environment as inputs to calculate a scalar reward value [6]. Later, we will discuss how vector reward functions are used in practical cases to evaluate concurrent behaviors.

9.1.6 Emergence

Intelligence emerges when a learning machine interacts with the environment with a view to achieve a particular set of goals. As we discussed above,

Reward-Based Behavior Adaptation

the policy it learns or the internal models it constructs directly depend on how the agent feels about the environment and how it associates its actions with reactions from the environment.

9.2 Markov Decision Process

9.2.1 A Markov State

A Markov state is a state that gives all required information of a process to make an informed decision about the future. In dynamical systems, we tend to think that we should know how the states evolved in the past to decide on an action to force the states to a desired value. However, if we can come up with one single vector of variables that summarizes the past at any given time, it is enough to make a decision for the next time step. We call such a vector a Markov state.

A learning process that considers a Markov state to take actions to transit to the next state is called a Markov decision process. This phenomenon is shown in Figure 9.4, where we identify the whole learning process as a sequence of state transitions. At each state $S(t)$, the agent will compute an action $a(t)$ that will make a state transition to state $S(t+1)$ while obtaining a reward $r(t)$. The probability of state transition can be written as $P\{S(t+1) = S', r(t+1) = r \mid S(t), a(t)\}$. The physical meaning of this notation is that the probability of the next state becoming S' and the next reward becoming r depends only on the current state $S(t)$ and the current action $a(t)$. It is very important that there is some mechanism to compute a scalar reward value based on the quality of the state transition. Sometimes, it can be a physical observer who decides the reward value or it can be just a mathematical function that takes the states and actions into account to compute the reward. We call it a *reward function* [6–8].

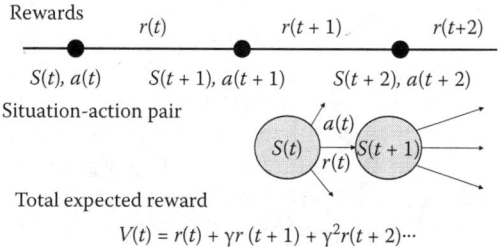

FIGURE 9.4
Elements of a reward-based learning paradigm.

9.2.2 Value Function

In certain cases, it can be a delayed reward where the reward is obtained once the whole behavior is completed. Still, in such cases, the above notation holds because the reward at any given state transition can hold a zero value. The most important feature in a reward-based learning paradigm is the estimate of the total expected future rewards at any given state. The function that estimates this is known as the *value function*. A reinforcement-based learning system spends a fair portion of its learning time to learn this value function. Once a value function is known, the agent can decide the action at any given state that will accrue the most future rewards.

The value function shown in Figure 9.4 can be written in two forms. One is known as the state value function, where we assign a total expected future reward to a particular state assuming that we know a policy π. The argument is that, if we know a policy that maps a particular Markov state to an action and we know how actions make state transitions, we can start from a particular state and continue to iterate toward an end goal. Therefore, the knowledge of a state is enough to estimate total future rewards. Therefore, we write such a state-value function as

$$V^\pi(s) = E_\pi \left\{ \sum_{k=0}^{\infty} \gamma^k r_{t+k} \mid S(t) = s \right\}$$

Similarly, we can define a value function based on an action given a state. In a state-value function, we assume that at each state we choose the action that will maximize the total future reward because we know the policy π. Here, we consider the expected total future rewards for each action given a state. However, this can be applied only for those cases with a discrete set of actions. The action-value function can be written as

$$Q^\pi(s, a) = E_\pi \left\{ \sum_{k=0}^{\infty} \gamma^k r_{t+k} \mid S(t) = s, a(t) = a \right\}$$

Let us consider the simple grid world navigation problem shown in Figure 9.5. The black blocks are obstacles. Each grid location is a state in this state space. The location information of a particular grid point is a Markov state because that information is the only relevant information for the agent to make a decision as to which grid location it should move to next so that it will not run against an obstacle while moving toward the goal. On the right side of the grid world, we show the relationship between a state-value function and an action-value function. At each state $S(t)$, the policy $\pi(s, a)$ determines the probability of taking an action to move to the next grid location. In this case, the discrete action space is limited to eight alternatives. This probability can be updated later depending on the actual rewards obtained for each alternative. We will discuss how this can be done in policy iteration.

Reward-Based Behavior Adaptation

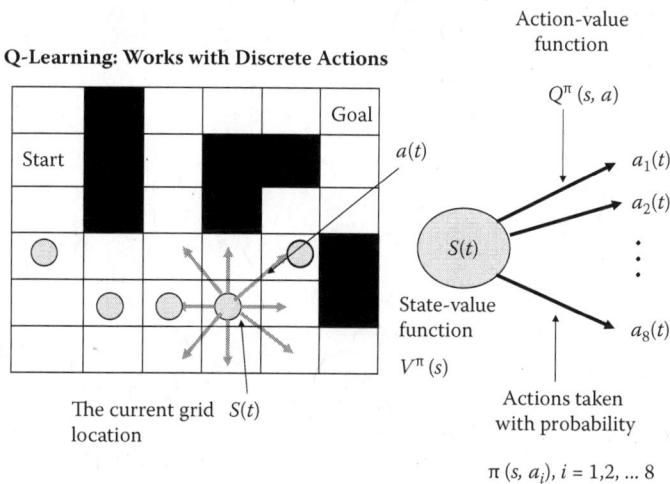

FIGURE 9.5
The difference between state-value function and action-value function in a discrete action-based learning paradigm.

9.3 Temporal Difference-Based Learning

Now it should be clear that the accuracy of action selection depends on the accuracy of the value functions, because at each state, the agent takes the action with the best total expected future rewards. Let us study one popular prediction method known as *temporal difference-based learning*. Figure 9.6 shows how it works. Here, at each time step, we use some function (it can be a neural network) to predict the total expected future rewards at that particular state given a policy. We denote this prediction as $\hat{V}(t)$ given by

$$\hat{V}(t) = \sum_{k=0}^{\infty} \gamma^k r(t+k)$$

which can be expanded to give

$$\hat{V}(t) = r(t) + \sum_{k=1}^{\infty} \gamma^k(t+k)$$

$$= r(t) + \gamma \sum_{k=0}^{\infty} \gamma^k r(t+k)$$

$$= r(t) + \gamma \hat{V}(t+1)$$

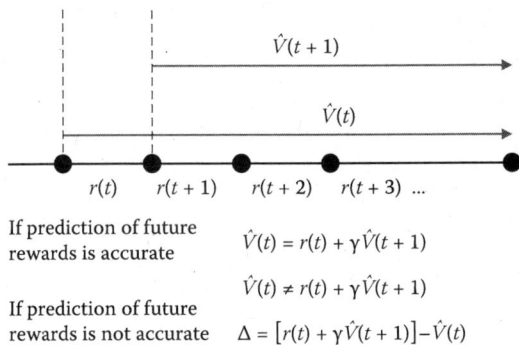

FIGURE 9.6
Temporal difference-based learning to predict.

This means that when we move to the next state, the agent gets a real reward $r(t)$, and with that updated knowledge, we can make another prediction at the next state. The two predictions and the reward can be related by $\hat{V}(t) = r(t) + \gamma \hat{V}(t+1)$ only if the prediction function is accurate [9]. If not, the prediction $\hat{V}(t)$ would have either overestimated or underestimated the real reward $r(t)$. In that case, $\hat{V}(t) \neq r(t) + \gamma \hat{V}(t+1)$. We call this prediction error a temporal difference of prediction because the computation considered the prediction accuracy between two consecutive time steps. It is commonly denoted as TD error given by $\Delta = (r(t) + \gamma \hat{V}(t+1)) - \hat{V}(t)$.

An optimum value function can be learned by updating the internal parameters of the function to minimize the temporal difference error Δ. Since Δ is a known quantity, learning the value functions is strictly a supervised learning mechanism. However, it learns how a qualitative reward function behaves with states and actions. It should be noted that the ability to predict is an essential ingredient of intelligence because it allows us to explore for better behavioral policies without waiting for the final outcome. When a human child learns to do various things such as walking, jumping, cycling, etc., the parents often use their experience to predict the outcomes of their actions. This helps the child to learn till he/she builds up his/her own internal models of the world to predict the outcomes of actions. This internal or external predictor is commonly known as a *critic*. The job of the critic is to estimate the value function $\hat{V}(t)$. Now, let us discuss how a critic can help an agent to learn an optimum policy.

Assume that a critic function has learned to predict total expected future rewards at any given state using a temporal difference-based learning scheme. Figure 9.7 shows how a critic can be used to evolve an optimum controller or a policy. In this case, the action space is a continuous one. We start with a particular situation. In our terms, a situation can be simplified to a Markov state. Given this state, the trained critic predicts the total expected

Reward-Based Behavior Adaptation

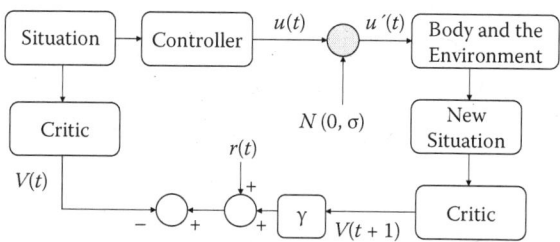

FIGURE 9.7
Actor-critic-based learning.

future rewards known as the value of the state $V(t)$. On the other hand, the policy can take an action $u(t)$. For the time being, we do not know if this action is the best one to take given the current state. Therefore, we add a normally distributed exploratory disturbance $N(0,\sigma)$ to the action, where σ is the width of the distribution. Since this distribution is centered at the origin, the disturbance can increase or decrease the value of the action. If it is a robot, the action is the torque/force commands to the actuators. Once this modified action is fed to the actuators, the robot enacts a behavior that interacts with the environment. The resulting new situation can be characterized by a new Markov state. If we feed this new state to the same critic, it should give us the expected total future rewards from the next point of time denoted by $V(t + 1)$. Since we obtain a real reward value at this point, we can compute a temporal difference of prediction induced by the disturbance. We denote this by $\Delta_{dist} = (r(t) + \gamma \hat{V}(t+1)) - \hat{V}(t)$. Please note that we assume that the critic is well trained so that this temporal difference of prediction would be zero if the disturbance was not given [10].

Now, let us analyze the physical meaning of Δ_{dist}. If Δ_{dist} is a positive value, that means the modified action and the new state are better than what was expected when the critic predicted $V(t)$. That is the only way for $(r(t) + \gamma V(t+1)) > \hat{V}(t)$ to be true. This implies that the controller or the policy should change its internal functionality to be one that would produce an action closer to the modified action given the state $S(t)$. If the modified action was detrimental in the eyes of the reward function, this would be reflected by a negative Δ_{dist}. In this case, the policy should try to move away from generating an action in the direction of the modified action given the state $S(t)$. This can be seen in Figure 9.8.

According to Figure 9.8, the policy-updating rule can be generalized as given by

$$\pi' = \pi + \eta \Delta_{dist}(u'(t) - u(t))$$

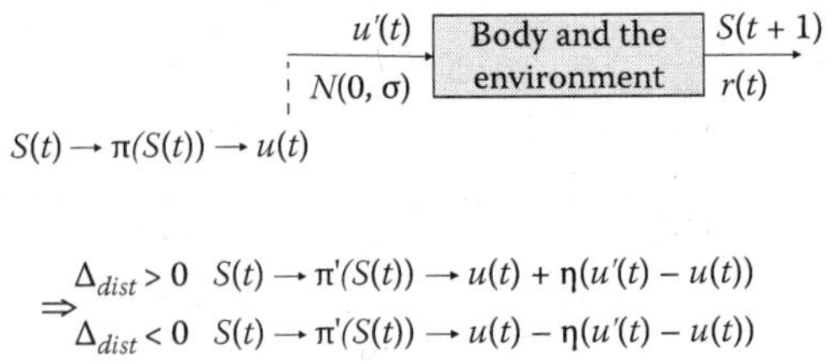

$$S(t) \to \pi(S(t)) \to u(t) \quad \begin{array}{c} u'(t) \\ N(0,\sigma) \end{array} \boxed{\begin{array}{c}\text{Body and the}\\\text{environment}\end{array}} \begin{array}{c} S(t+1) \\ r(t) \end{array}$$

$$\Rightarrow \begin{array}{l} \Delta_{dist} > 0 \quad S(t) \to \pi'(S(t)) \to u(t) + \eta(u'(t) - u(t)) \\ \Delta_{dist} < 0 \quad S(t) \to \pi'(S(t)) \to u(t) - \eta(u'(t) - u(t)) \end{array}$$

FIGURE 9.8
How the temporal difference can be used to improve the policy.

9.4 Extension to Q Learning

Temporal difference-based learning can be extended to Q learning, where the value function is an action-value function and the action space is discrete [11]. In this case, the disturbance is a quantum value, as shown in Figure 9.9, where the bound $L < N$ of the uniform random distribution $U(-L,L) = \{-L,-L+1,\ldots,-2,-1,,0,1,2,\ldots,L\}$ such that any disturbance that falls outside the action space $\{a_1, a_2, \ldots, a_N\}$ is neglected. In this case, the control policy is given by

$$\pi(s,a) = P\{a = a_k \mid s = S(t)\} = \frac{e^{p(S(t),a_k)}}{\sum_{\forall a} e^{p(S(t),a)}}$$

where $p(s(t), a_k)$ is the probability of taking action a_k at state $S(t)$.

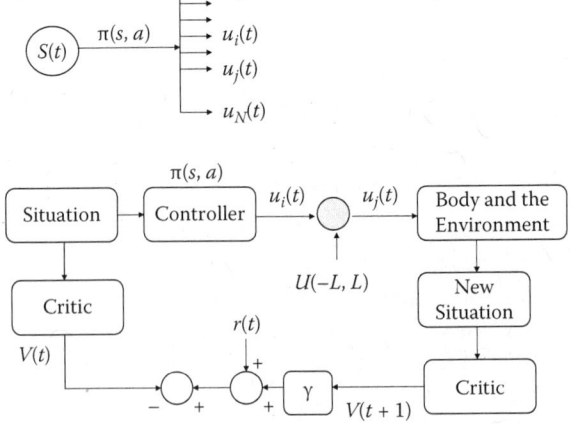

FIGURE 9.9
Temporal difference-based policy improvement in a Q learning paradigm.

The temporal difference of value prediction Δ_{dist} induced by the disturbance is used to update the probability of taking each action given by

$$p(S(t), a_i) = p(S(t), a_i) - \eta \Delta_{dist}$$

$$p(S(t), a_j) = p(S(t), a_j) + \eta \Delta_{dist}$$

The argument is the same as in the continuous-action case. If Δ_{dist} is positive, it means that modified action a_j is better than action a_i being recommended by the internal policy at state $S(t)$. Therefore, we should increase the probability of taking action a_j at state $S(t)$ and decrease that of taking a_i at state $S(t)$.

9.5 Exploration versus Exploitation

In any learning process, the initial fidelity of the internal policy is very low. This is reflected in our confidence in using the internal policy to drive us. Take, for example, a man learning to drive a car. In the beginning, he is very nervous to drive alone. He tends to follow what an instructor advises from the side. The actions are not very repeatable given the same situation. He wants to try out slightly different behaviors to see if they improve his driving performance. This is known as initial exploration. The amateur driver is a little scared to exploit the knowledge he has gathered so far until he becomes very thorough with his skills. Yet, over time, the degree of exploration dies down with the rise of the degree of exploitation of the internal skills. This natural phenomenon is also mimicked in our reward-based learning paradigm. Usually, the width of the disturbance function σ in the continuous case and L in the discrete case reduces over time, rendering maximum exploration at the beginning with a gradual transition to more exploitation of the learned policy at the end. However, if the environment is very uncertain, intermittent expansion of exploration can lead to good results. In our driver's case, it may be that suddenly he realizes that one of his tires skids. This is a sudden change in the external world with which his internal policy interacts. In such a situation, he has to readapt his skills to suit the new situation within a limited amount of time. He does this by exploring for new behaviors by disturbing what his internal models of the car recommend. This is also known as annealing of learned skills. In annealing, a metal object is reheated to allow its molecules to settle in desired configurations so that the object, such as a knife, will gain strength. Once reshaped and cooled down, the object is reheated to expand the spin of molecules and allowed to cool down again to do away with uneven stress distributions, brittle areas, etc. The expanded spin of molecules resembles expanded exploration due to larger σ or L values at the beginning. When σ or L values drop gradually, the size of exploration around actions recommended by the internal policy drops, mimicking the reducing spin of molecules when the metal object cools down.

9.6 Vector Q Learning

The idea of vector Q learning was first proposed by Kiguchi et al. [12]. The idea is to make an array of behaviors learn in parallel when one behavior is being implemented. This can happen within an individual agent or in a society of agents. Consider, for instance, that in a society, we do not have to experience everything ourselves to learn how to behave in the world. In many instances, we learn by watching how others react to various situations. This happens in sports very often. Amateurs learn by watching how experts react to various situations. Therefore, other individuals create situations for us to imagine how we will react if we are to face the same situations. In such a case, the internal reward functions or critics can evaluate our imaginary actions and compare with what we saw in the real enactment. If the person who was reacting to the real-world situation did better than what we would have done in the same situation, we tend to change our internal models so that our behavior will try to follow the good example. We learn even if the example was worse than what we would have done in the same situation. In that case, the example works as a negative teacher that reinforces our "would be" behavior as a good behavior.

This is true in the case of an array of behaviors that are implemented with different objectives by the same agent. In the study of Kiguchi et al. [12], three mutually contradicting behaviors of a mobile robot were considered.

Navigation from a start point to a goal with the objective of minimizing energy spent

Navigation from a start point to a goal with the objective of minimizing time spent

Navigation from a start point to a goal with the objective of minimizing the danger to which the robot was exposed

Although the robot has three distinct behaviors, it can implement only one at a given time. The question is whether the robot can learn to improve all three behaviors simultaneously while implementing only one of the behaviors at any given time. This was addressed by designing a vector Q network. The vector Q network maps a Markov state vector summarizing

Reward-Based Behavior Adaptation

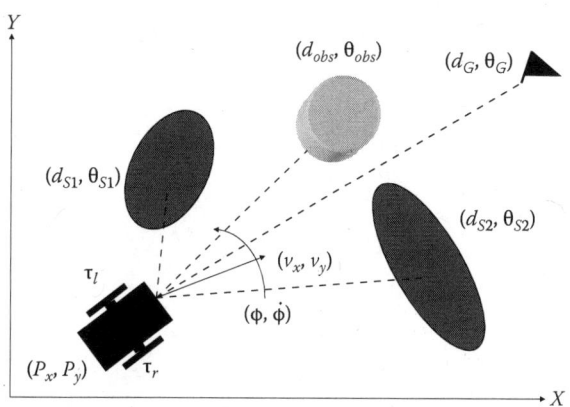

FIGURE 9.10
Environment of the robot.

the information needed to take an action at a given time to a vector of Q values that estimate the total expected future rewards for all three behaviors. Furthermore, the internal reward functions produce instantaneous rewards for all three behaviors regardless of what behavior is actually being implemented. However, only one behavior influences the transition of the state vector. The vector Q network facilitates the agent in asking the question: "What if the other behavior is implemented in this situation?"

Figure 9.10 shows the scenario being considered. There are two slippery areas d_{S1} and d_{S2} distances away from the robot so that the lines connecting the robot with the centers of these two areas make θ_{S1} and θ_{S2} angles. Similarly, there is one obstacle with (d_{obs}, θ_{obs}) and a goal with (d_G, θ_G). Moreover, the robots positions in X and Y coordinates are given by (P_x, P_y), the corresponding velocities are (v_x, v_y), the azimuth and rate of change of azimuth are $(\phi, \dot{\phi})$, and the left and right wheel torques are given by (τ_l, τ_r).

Therefore, the Q net gets these 16 valued state vectors as inputs and computes three Q values corresponding to the three behaviors. Figure 9.11 shows how the Q net is organized.

The output of the Q net is a vector of three Q values corresponding to the three behaviors given by

$$Q_v = \sum_{j=1}^{50} w_{jo} y_j$$

where $w_{j0} = \{w_{1jo}, w_{2jo}, w_{3jo}\}$ is the vector of weights connecting the hidden layer to the output layer. There are 50 hidden neurons in this Q net. Each uses a primitive function given by

$$y_j = \frac{1}{1+\exp(-s_j)}, \quad j=1,2,\ldots 50$$

$$s_j = w_{oj} + \sum_{j=1}^{16} w_{ij} x_i$$

$$w_{oj} = \{w_{1oj}, w_{2oj}, w_{3oj}\}$$

$$w_{ij} = \{w_{1ij}, w_{2ij}, w_{3ij}\}$$

where w_{oj} is a vector of weights adding a bias to the primitive functions and w_{ij} is a vector of weights scaling the 16 input variables to the networks.

Therefore, each connection between layers shown in Figure 9.11 contains a vector of three weights. This makes the Q network produce three Q values corresponding to three behaviors for a given input state simultaneously.

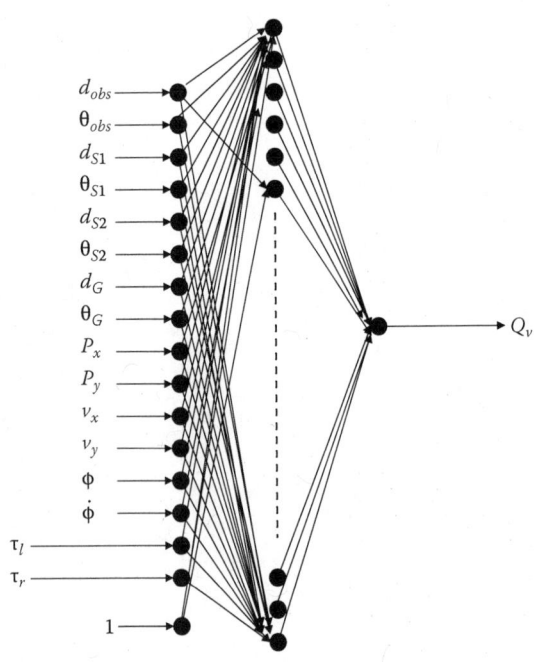

FIGURE 9.11
Vector Q net structure.

There is a vector reward function that assigns to each behavior a scalar reward value at every sampling step. The vector reward function is given by

$$R = \begin{cases} R_1 = r_\tau + r_{obs} + r_G + P \\ R_2 = r_{reach} + r_{obs} + r_G \\ R_3 = r_{safety} + r_{obs} + r_G \end{cases}$$

$$r_\tau = \frac{4}{1 + 100\exp(|\tau_l| + |\tau_r|)}$$

$$P = -10 \text{ if } |\tau_l|, |\tau_r| > 0.04$$

$$r_{obs} = -100\exp(-5(d_{obs} - 0.5))$$

$$r_G = \exp(-d_G)$$

$$r_{reach} = \sqrt{v_x^2 + v_y^2}$$

$$r_{safety} = 100 r_{obs}$$

Now, let us have a closer look at the design of this vector reward function. We can see that all three elements of the vector have two common components to reflect how the robot avoids the obstacle and how it reaches the goal given by $r_{obs} = -100\exp(-5(d_{obs} - 0.5))$, $r_G = \exp(-d_G)$. The reward component for obstacle avoidance r_{obs} punishes the agent for going closer to the obstacle. It starts to increase this punishment dramatically if the robot goes within a radius of 0.5 m. The user can define this value to make the robot maintain a safe distance from the obstacle. The goal-reaching component r_G rewards the agent for bringing the robot closer to the goal. Other than these common components, what make each element distinct are the unique components they have. The energy conscious behavior rewards the agent for keeping low torque values by the component $r_\tau = \frac{4}{1+100\exp(|\tau_l|+|\tau_r|)}$. It goes on to give a special penalty $P = -10$ if $|\tau_l|, |\tau_r| > 0.04$ to ensure that the torques remain within limits. The hasty behavior that makes the robot reach the goal as fast as possible rewards the agent for keeping high velocities given by $r_{reach} = \sqrt{v_x^2 + v_y^2}$. The safety conscious behavior discourages the agent from bringing the robot close to the obstacle by using $r_{safety} = 100 r_{obs}$.

Each of these behaviors may be implemented for a physical reason. For instance, if the internal battery level is low, it may be good idea to avoid slippery areas as much as possible, avoid colliding with obstacles, and keep the torque levels low to ensure that the battery is not drained too soon. Therefore, components in the reward function for the energy conscious behavior try to reinforce favorable behavior. In a case where the robot has run out of time to reach the goal, moving faster toward the goal should be given high priority.

Therefore, components in the reward function for the hasty behavior reinforce the agent for behaviors favorable to going fast. Similarly, if the robot has suffered some body damage, or if the sensors and actuators have some loose connections, the most important factor for survival is to avoid collision with obstacles. This behavior is reinforced by the element in the reward function for safe navigation.

The steps for Q net learning are as follows.

Step 1: Initialize the weights of the Q net and set time $t = 1$.

Step 2: Sense the environment of the robot that is represented by a 14-valued Makov state vector shown in Figure 9.11, and find the left and right wheel toques that will give the best Q value depending on the particular behavior to be implemented.

Step 3: Feed the chosen torque pair to the Q net and evaluate all three elements of the Q_v vector.

Step 4: Run the robot for one sampling step and obtain a vector of rewards $R(t + 1)$.

Step 5: At time $t + 1$, evaluate the best Q value $Q_{v\_max}(t + 1)$ for a specific pair of torques, for a given behavioral objective.

Step 6: Calculate the temporal difference vector for all three behaviors given by

$$\Delta(t+1) = \begin{cases} R_1(t+1) + \gamma Q_{v\_max}(t+1) - Q_{v1}(t) \\ R_2(t+1) + \gamma Q_{v\_max}(t+1) - Q_{v2}(t) \\ R_3(t+1) + \gamma Q_{v\_max}(t+1) - Q_{v3}(t) \end{cases}$$

$$0 < \gamma < 1$$

Step 7: Use the temporal difference vector to update the parameters of the Q net. Go to step 2.

9.6.1 Summary of Results

Energy conscious behavior. The robot moved to the goal in 10.31 s without going through slippery areas or colliding with the obstacle. The average torques remained about 0.01 N m and the torque increased to 0.04 N m when the robot approached the goal.

Hasty behavior. The robot moved to the goal in 5.99 s. It achieved this at the expense of elevated energy consumption and going very close to adverse regions. The average torque remained about 0.02 N m and reached about 0.1 N m when the robot reached the goal.

Safe behavior. The robot took 21.79 s to reach the goal. It did not collide with obstacles or go close to adverse regions. The torques remained very low at 0.001 N m, but they reached 0.04 N m when the robot reached the goal.

There are many opportunities to improve the above approach. Clearly, the size of the state vector is very large (16), and the resulting number of neurons in the hidden layer is 50. This has resulted in 150 unknown parameters for each behavior. This means there are 550 total unknown parameters to be optimized, making the learning task a very daunting prospect. Typically, this is an open problem to be solved, that is, to construct a very compact state vector so that the Q net will have a manageable number of parameters. Another disadvantage of having a large number of parameters is that the gradient descent-based learning converges to a local optimum due to the complex cost landscape made by a large number of learning parameters.

One possible approach is to split the Q net into finer modules so that one module is activated only by a smaller group of sensors organized to perceive a given environmental condition such as slippery areas, obstacles, the goal, etc. For instance, the goal-reaching behavior will be identified as one primitive behavior connected to the goal-seeking sensor only. Learning the parameters of the module is active only when the goal-seeking sensor starts to give some meaningful information. The output of such a module may be shared by other modules to compute the final Q value for a given behavior defined with a larger meaning, such as hasty behavior. This is the idea behind finite state acceptors (FSA) shown in Figure 9.12.

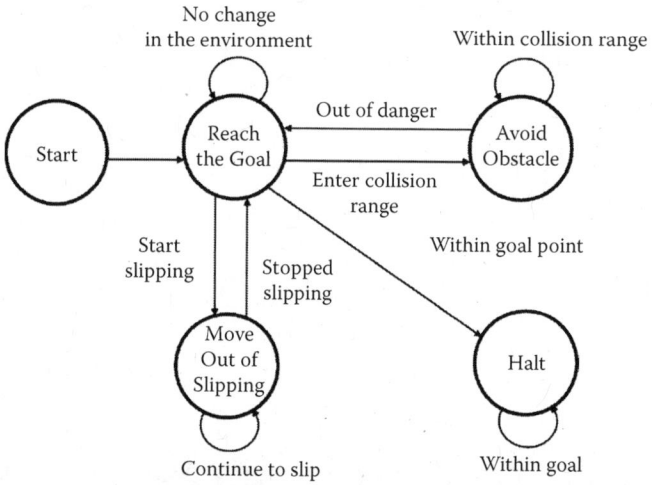

FIGURE 9.12
A finite-state acceptor diagram for robot navigation through adverse regions.

Each subbehavior in this FSA, such as "reach the goal," "avoid obstacle," and "move out of slipping," will have vector parameters to give different Q values for different overall behaviors, such as hasty, safety conscious, and energy conscious movements.

Project. A kitchen guard complains that he has been seeing rats in the night during the last f weeks. He planned various methods to keep them away. However, he cannot use any toxic or poisonous materials to kill the rats because they are in a kitchen. Because this is a health hazard, the best approach would be a noninvasive one that scares the rats away. Therefore, you are required to design a robot that can learn to move around in the kitchen in the darkness by avoiding tables and chairs, move through 1-m gaps, reach a maximum speed of 0.5 m/s, and generate the sound of a cat if it detects a rat.

> Draw an FSA diagram showing the essential behaviors it should have to accomplish the goals.
>
> Decide on the type of sensors it should have and how they should be physically organized on the body of the robot to give it a meaningful perception.
>
> Design a reward function that takes into account all the requirements, such as maintaining a speed, preserving onboard energy, avoiding collisions, detecting rats, scaring rats, etc., and show how a Q network can be used to enable the robot to learn over time.

Use the dynamics given below to simulate your design.

The state-space equation of the robot is given by

$$x = [v \quad \phi \quad \dot{\phi}]^T$$

$$\dot{x} = \begin{bmatrix} \dfrac{-2c}{(Mr^2 + 2I_w)} & 0 & 0 \\ 0 & 0 & 1 \\ 0 & 0 & \dfrac{-2cl^2}{(I_v r^2 + 2I_w l^2)} \end{bmatrix} x + \begin{bmatrix} \dfrac{kr}{(Mr^2 + 2I_w)} & \dfrac{kr}{(Mr^2 + 2I_w)} \\ 0 & 0 \\ \dfrac{-krl}{(I_v r^2 + 2I_w l^2)} & \dfrac{krl}{(I_v r^2 + 2I_w l^2)} \end{bmatrix} u$$

$$u = [\tau_l \quad \tau_r]^T$$

where v is the velocity of the robot, ϕ is the azimuth, M is the mass, l is the distance between the two wheels, r is the radius of a wheel, I_v moment of inertia of the robot around the center of gravity, I_w is the moment of inertia of a wheel, k is a gain constant, c is the friction coefficient, and τ_r and τ_l are the torques of the right- and left-hand side wheels, respectively.

References

[1] Dayan, P., and Balleine, B., "Reward, motivation, and reinforcement learning," *Neuron* 36(2), 285–298, 2002.

[2] German, P. W., and Fields, H. L., "Rat nucleus accumbens neurons persistently encode locations associated with morphine reward," *Journal of Neurophysiology* 97(3), 2094–2106, 2007.

[3] Hollerman, J. R., Tremblay, L., and Schultz, W., "Influence of reward expectation on behavior-related neuronal activity in primate striatum," *Journal of Neurophysiololgy* 80(2), 947–963, 1998.

[4] Montague, P., and Berns, G., "Neural economics and the biological substrates of valuation," *Neuron* 36(2), 265–284, 2002.

[5] Arkin, R. C., *Behavior Based Robotics*, MIT Press, Cambridge, MA, 1998.

[6] Mataric, M. J., "Reward functions for accelerated learning," in *Proceedings of the Eleventh International Conference on Machine Learning*, pp. 181–189, Morgan Kaufmann, San Francisco, CA, 1994.

[7] Mataric, M. J., "Sensorymotor primitives as a basis for imitation: linking perception to action and biology to robotics," in *Imitation in Animals and Artifacts* (C. Nehaniv and K. Dautenhahn, Eds.). MIT Press, Cambridge, MA, 2000.

[8] Ng, A. Y., Harada, D., and Russell, S., "Policy invariance under reward transformations: Theory and applications to reward shaping," in *Proceedings of the Sixteenth International Conference on Machine Learning*, pp. 278–287, Morgan Kaufmann, San Francisco, CA, 1999.

[9] Schultz, W., Dayan, P., and Montague, P. R., "A neural substrate of prediction and reward," *Science* 275(5306), 1593–1598, 1997.

[10] Sutton, R. S., and Barto, A. G., *Reinforcement Learning*, MIT Press, Cambridge, MA, 1998.

[11] Watkins, C. J. C. H., "Learning from delayed rewards," PhD dissertation, Cambridge University, 1989.

[12] Kiguchi, K., Nanayakkara, T., Watanabe, K., and Fukuda, T., "Multi-dimensional reinforcement learning using a vector Q-net—Application to mobile robots," in *Proceedings of 2002 FIRA Robot World Congress*, pp. 405–410, 2002.

10

An Automated System to Induce and Innovate Advanced Skills in a Group of Networked Machine Operators

10.1 Introduction

This chapter expands on the work presented by Nanayakkara et al. (2007). The objective of this project was to evolve an expert skill among a group of machine operators in a garment factory by inducing one individual's skill on others. We take advantage of the fact that individuals innovate better skills while trying to mix their own skills with those of an elite to emerge as a new elite. Here, we discuss how such a group innovation process can be automated.

Let us observe some of the salient features characterizing the skill innovation context:

1. Every individual in the group of machine operators is tasked with making a given product according to a given set of specifications.
2. Each machine operator works on the same type of machine. Therefore, each operator is independent from others. However, the group as a whole works to achieve a common production target.
3. The process continues even if one or several machines break down. It will only increase the workload of the remaining workers.
4. The skill of a machine operator is evaluated by how well he/she meets a predefined set of performance standards. We define the machine operating skill using three criteria: maximization of smoothness of the machine speed profile, minimization of time taken to finish the job, and minimization of non-value-adding time during the job. Data needed to evaluate these criteria were obtained by attaching a data acquisition module to each machine. All data were transmitted to a remote console where experts could evaluate each machine operator and compare one against another.

5. Each machine operator has his/her own body and mental dynamics, emphasis on different details of the performance standards, and interpretation of the feedback given by process supervisors.
6. Each machine may have its own slight variation of dynamics due to uncertainty of friction, electrical resistance, etc.

Although this context looks very simple, complexity emerges when the number of workers is large. In the factory where we conducted this study, certain modules were run by as many as 400 machine operators. In such cases, there was more than one process supervisor. Therefore, different supervisors may tend to evaluate the same performance in slightly different ways. Furthermore, different machine operators may give emphasis on different details of the performance evaluation criteria. Therefore, what is perceived as best by each machine operator may not be the best in the eyes of a supervisor.

This chapter presents a method that can be adopted to automate the evolution of an elite skill in such a context. We will also discuss a simple model that can be used to explain complex phenomena that cannot be explained by the conventional human learning models.

10.2 Visual Inspection and Acquisition of Novel Motor Skills

Operating a sewing machine to manufacture a piece of a garment involves continuous generation of corrective motor commands that interact with the dynamics of the muscles and the machine to make a product according to certain specifications. Since the motor system in the brain is predominantly responsible for learning such skills, we call it a motor skill. A motor skill is synthesized by a collection of motor primitives that map perception to suitable motor commands (Bizzi et al., 1991; Arbib et al., 1981; Thoroughman and Shadmehr, 2001; Shadmehr and Mussa-Ivaldi, 1994). Perception is an interpretation of the situation based on various sensory inputs such as vision, touch, audition, etc. A motor primitive is a neural organization in the brain that maps a given locality in the perception space to a simple motor command. A complex motor skill seems to be constructed by the adaptive combination of these motor primitives (Thoroughman and Shadmehr, 2001). Therefore, the skill to operate a machine to produce the desired results largely depends on the basis set of motor primitives, their shapes both in the perception and action spaces, and the manner in which they are combined given the need to take a corrective action. Humans acquire a fair portion of motor skills through inspection. A good elaboration on how players may be acquiring the skill to play squash through inspection of expert players is given by Wolpert et al. (1995). It seems that a significant part of a motor skill can be constructed by combining one's own first-hand experience with visual inspection of a demonstration of a coach. Therefore, visual inspection must be a major contributing factor to the brain to construct new motor primitives. A model that translates visual extraction of movement features to internal motor primitives is suggested by Abernethy (1990). Work performed on the relationship between saccadic eye movements to track a hand movement and the adaptation of the internal model that maps a desired trajectory to motor commands to control the muscles shows that eye movements change with the adaptation of the internal motor primitives, suggesting that visual inspection is somehow yoked to the internal motor models (Mataric et al., 1998; Arrif et al., 2001; Nanayakkara and Shadmehr, 2003).

Visual inspection of a motor skill can extract two types of information: one is the exact position and speed profiles of the limb movement, and the other is hidden information about the optimality criteria. For instance, when one watches how a football player kicks a ball, one could try to remember the exact posture and the way the leg moved, as well as extracting optimality criteria such as smoothness of the movement at different critical special segments, the trajectory of the ball after kicking, stability of the body, etc. Therefore, a second person could try to focus on one or many such extracted information to develop the skill of kicking a ball. In this chapter, we look at several experimental results of how people set about developing or improving a skill by watching a demonstration.

10.3 Experimental Setup

A data terminal was attached to each machine that sensed the speed profile and plotted it on a screen as shown in Figure 10.1. The data terminal transmitted these data to a remote data server through a wireless data link. The speed of the machine was sensed using an optical encoder and then sent to the graph module as shown in Figure 10.2. This pulse train is counted using the internal hardware counter of the microcontroller. The sampling time is set proportional to the standard minute value (SMV), which is the standard time taken by an average operator to complete the operation.

Before starting the operation, the operator has to log in to the database server. The operator can use either the keyboard or the I-button input to enter job-related information such as style number, operation number, and the employee provident fund number. This information is then sent to the main database server via a wireless link, which then resends the SMV corresponding to the job concerned back to the data terminal. This SMV is then used to scale the X axis, and the Y axis of the display.

As the operator logs in to the system, data points of the standard speed profile chosen beforehand by the manager as the best available graph for the current operation were sent to the graph module via the wireless link. Once the machine operator has learned to match this standard speed profile

284 *Intelligent Control Systems*

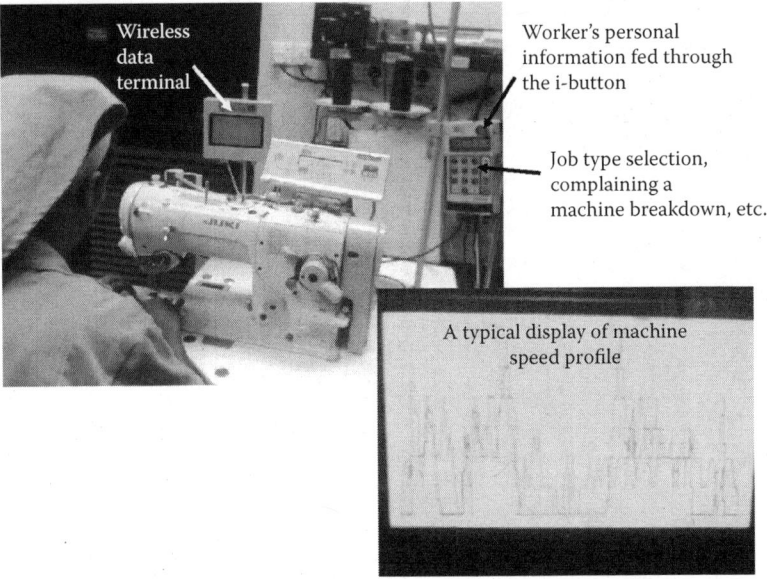

FIGURE 10.1
Wireless data terminal attached to the machine.

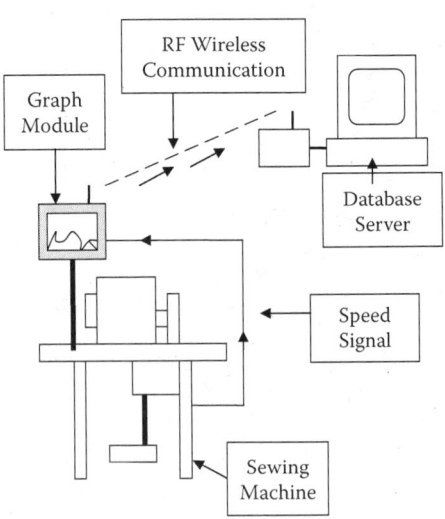

FIGURE 10.2
Data communication network.

continuously ten times, the learned speed profile of the worker was sent to the main database server and saved until it was studied by a technical expert. The technical expert then picked what he/she perceived to be the best profile out of many new speed profiles sent to the server by individual data terminals. Once an elite profile was chosen, it was transmitted back to all the data terminals of the workers to be plotted on top of each of the corresponding machine speed profiles.

10.4 Dynamics of Successive Improvement of Individual Skills

In the garment factory considered in this study, supervisors used three evaluation criteria to assess workers. Although the supervisors express these evaluation criteria in linguistic terms, we estimate a quantitative model given by:

$$J_1 = \max(t_i), \nabla \omega(t_i) > 0$$

$$J_2 = \ddot{\omega}(t_i)$$

$$J_3 = \sum t_i, \nabla \omega(t_i) < \varepsilon$$

where $\omega(t_i)$ is the speed of the machine at time t_i and ε is a speed threshold for the machine speed. Therefore, J_1, J_2, and J_3 estimate the time taken to finish the task, the jerk of the machine movements given by the double derivative of the speed, and the total idle time, respectively. Since the workers had been verbally briefed about the evaluation criteria, visual feedback of a template machine speed profile helped them to identify the details of mismatches between their individual machine operating behaviors and that of the given template. Apparently, the rich feedback information helps the workers to plan more comprehensive learning strategies to improve their performance. It was observed that, while trying to match the elite profile, some workers emerged as new elite workers due to the emergence of new behaviors. This process of skill innovation through an interaction between elite demonstrations and mechanisms of human motor skill acquisition continued till the live system of systems converged to a high standard of performance.

Two stitching tasks, T_A (speed profile shown in Figure 10.3) and T_B (speed profile shown in Figure 10.4), were selected from among a set of actual operations handled by a particular group.

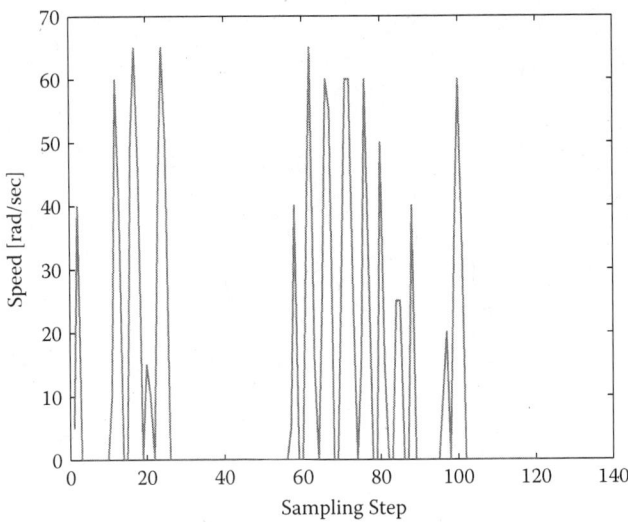

FIGURE 10.3
Elite speed profile for task T_A.

Figure 10.5 shows how two workers responded to such an elite profile across two trials. It is evident in Figure 10.5 that worker 1 (subject 1) has taken an innovative approach to optimize the evaluation criteria, whereas worker 2 has made an attempt to match the pattern given by the elite. Therefore,

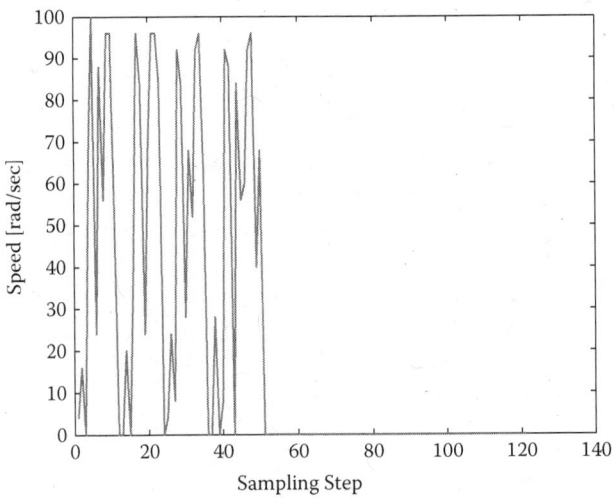

FIGURE 10.4
Elite speed profile for task T_B.

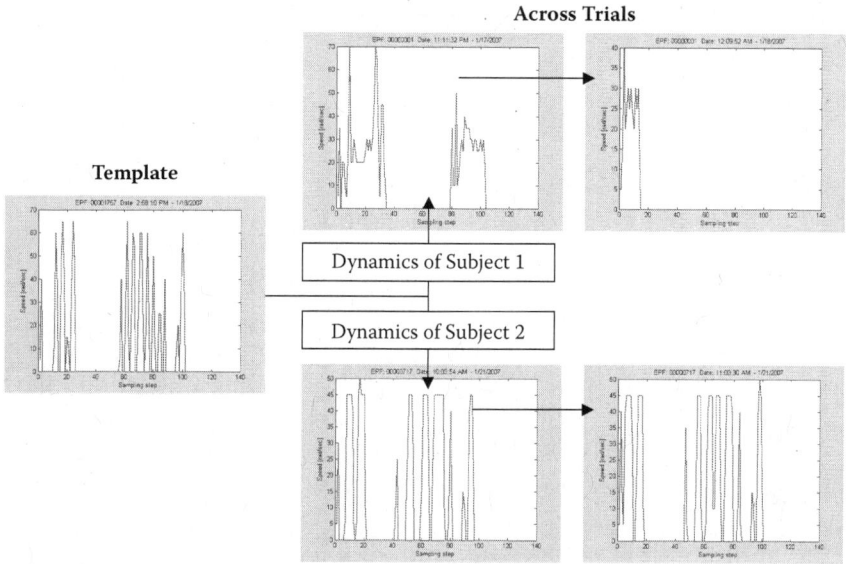

FIGURE 10.5
Example of emergence of new machine operating skills.

both approaches may achieve the same evaluation under the above cost functions.

Therefore, this way of evolution of elite skills in a group of diverse humans is quite different from that of a colony of robots or machines that tries to mimic the salient features of an elite behavior selected or demonstrated by a human. In contrast, the dynamics in a human group can be very explorative, although the trigger to change is given by a particular machine speed profile chosen by another human technical expert. For instance, worker 1 had been operating the machine in a different manner until the particular elite profile was displayed on his/her screen. Therefore, the crossover of new information with the existing background training, attitudes, creativity, and, of course, body dynamics and that of the machine itself could produce a new elite that is much better than what is available at a given time.

Figure 10.6 shows the strategy of worker W_2 to improve his/her performance once an elite speed profile for task T_A was displayed on his/her screen. The increasing degree of match suggests that the worker tried to follow the elite profile. This involves adaptation of the internal motor primitives to match those of the worker who produced the elite speed profile. This is also confirmed by the evolution of cost components shown in Figure 10.7. Obviously, they converge toward the elite cost components. Yet, it is unlikely that this type of a strategy will lead to a new elite.

In contrast to worker W_2, W_1 has adopted a strategy of deviating from the elite profile for task T_A. This is elicited by the fact that the degree of match

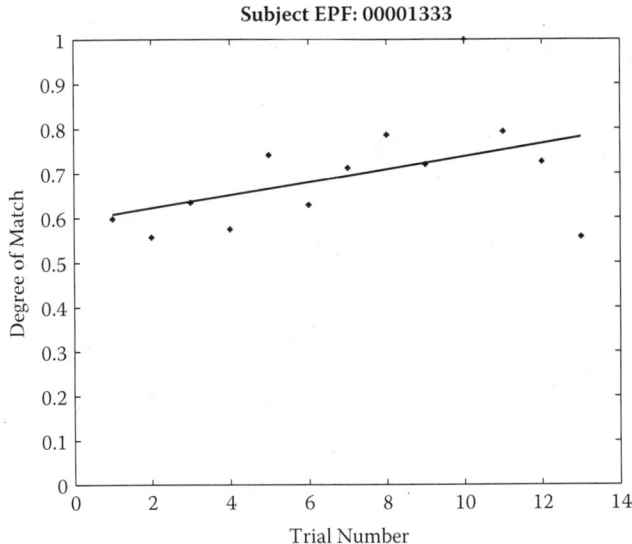

FIGURE 10.6
Evolution of the degree of match between the successive attempts of W_2 and the elite machine speed profile for task T_A shown in Figure 10.7.

shown in Figure 10.8 reduces across trials. Figure 10.9 suggests that worker W_1 seems to have given top priority to the reduction of idle time during the task, followed by minimization of total time spent to complete the task and reduction of the jerk of the machine speed profile. This is a creative move in response to the presence of the elite speed profile on the screen of worker W_1.

An interesting phenomenon can be seen in Figures 10.10 and 10.11, where subject W_3 tried to innovate a new machine speed profile to reach the cost of an inferior reference for task T_B shown in Figure 10.4. The fact that the degree of match between the speed profile of the reference and that of worker W_3 reduces across trials depicts that worker W_3 was deviating from the reference pattern on the screen.

Yet, the evolution of the cost components shown in Figure 10.11 clearly shows that the worker was reaching the levels of the cost components attributed to the given reference. This phenomenon cannot be explained by traditional learning theories because the fact that the cost of the speed profile of the worker concerned converged toward that of the reference should traditionally imply that the speed pattern also converged toward that of the reference. Therefore, this scenario suggests that there is a learning mechanism where the subject gives emphasis to the minimization of error between internal evaluations between two speed profiles without having to worry about the pattern mismatch seen on the screen.

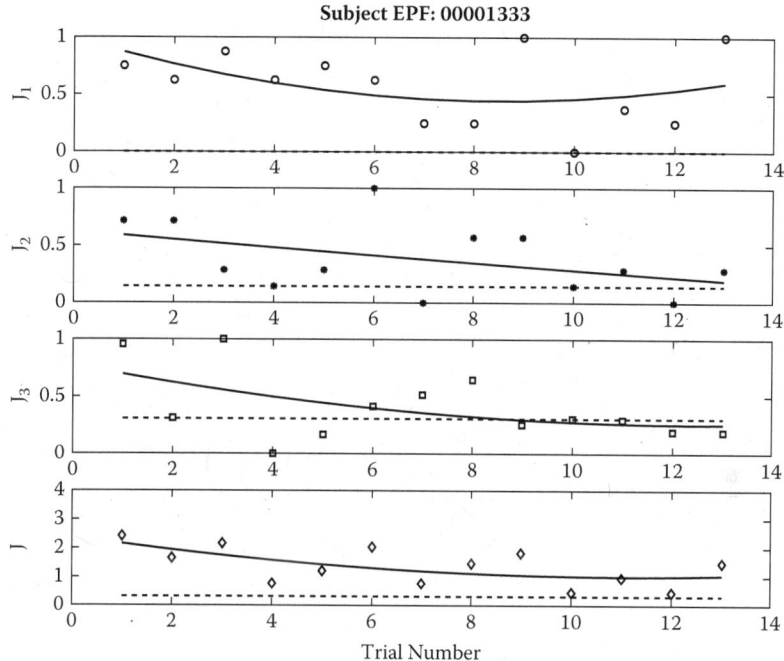

FIGURE 10.7
Variation of normalized individual cost components (top three plots) and the total cost for W_2 (bottom plot). Dotted line shows the respective cost of the elite speed profile for task T_A normalized for the data of worker W_2.

10.5 Proposed Model of Internal Model Construction and Learning

Although the factory supervisors do assess the skill of the workers using the criteria described in Section 10.4, they find it very difficult to give specific instructions to the workers as to where they should improve and exactly what they should do to achieve that improvement because there is no visual comparison among alternative solutions. However, when a proper visual feedback was given, the workers could

1. Better interpret the physical meaning of the above evaluation criteria
2. Better combine the individual criteria into one mental goal
3. Identify detailed mismatches between the expected skill and the current level of skill and come up with self-motivated strategies to improve performance

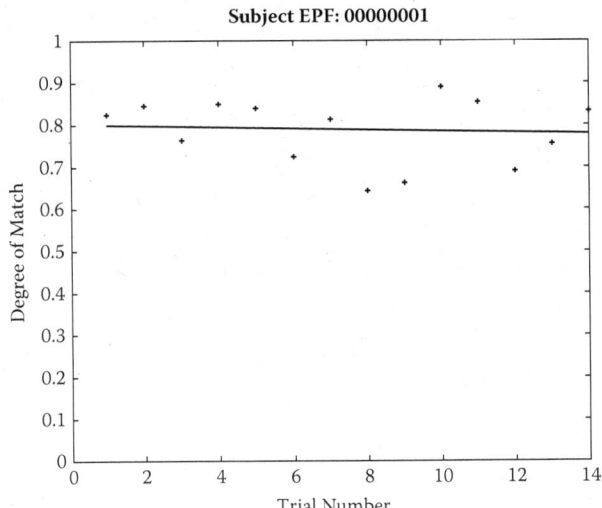

FIGURE 10.8
Evolution of the degree of match between the successive attempts of W_1 and the elite machine speed profile for task T_A.

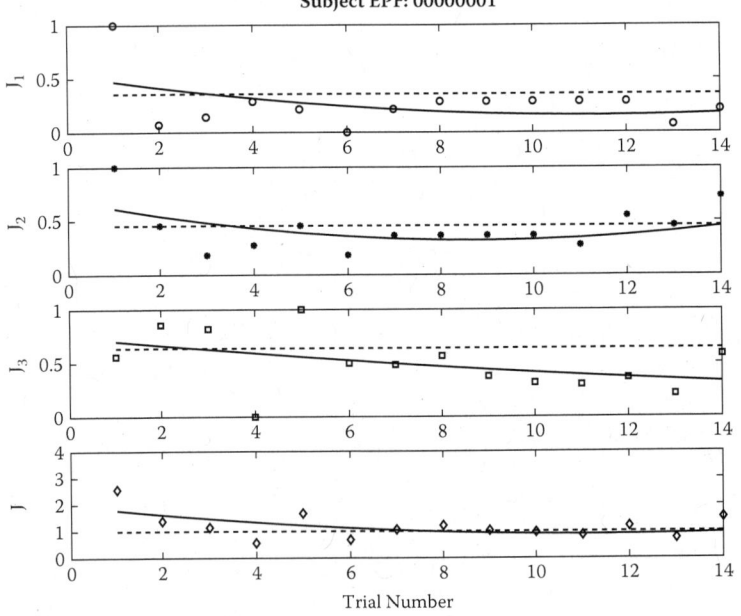

FIGURE 10.9
Variation of normalized individual cost components (top three plots) and the total cost (bottom plot) for W_1. Dotted line shows the respective cost of the elite speed profile for task T_A normalized for the data of worker W_1.

An Automated System to Induce and Innovate Advanced Skills in a Group

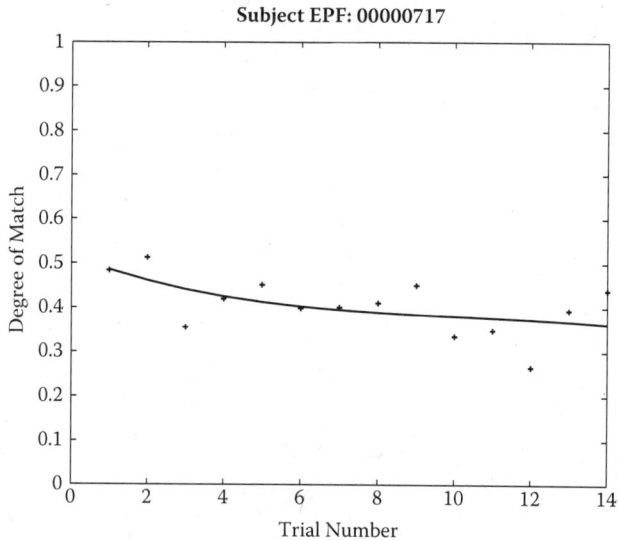

FIGURE 10.10
Degree of match between the successive attempts of W_3 and the elite machine speed profile for task T_B.

We identify this process of combining visual feedback with the supervisors' qualitative feedback as a construction of an internal model of the evaluation criteria. This internal model is known as an internal critic in a reinforcement-based learning mechanism (Sutton and Barto, 1998). As described in Chapter 9, a critic helps an individual evaluate his/her own performance without outside help. However, outside help is needed to construct a critic of one's self. It is recognized throughout the literature of reinforcement learning that the accuracy, robustness, and comprehensiveness of the internal critic decide the ultimate destination of a reward-based learning system in terms of developing a desired skill.

The internal critic helps the workers predict the evaluation he/she will earn given a machine speed profile. Here, we propose the skill innovation model shown in Figure 10.12, where ψ^* is the elite machine speed profile at a given time, ψ is the actual machine speed profile of a given worker, Δ_v is the visual error between the elite machine speed profile and the actual machine speed profile of a given worker, Δ_V is the corresponding weight given by the worker concerned, α_j is the difference between the internal evaluation of the actual machine speed profile and that of the elite speed profile, Δ_J is the corresponding weight assigned by the subject, ψ_m is the modified machine speed profile innovated by an internal model responsible for profile planning, and u is the motor command computed by the motor primitives. Since visual feedback of the elite speed profile and that of the worker concerned is given

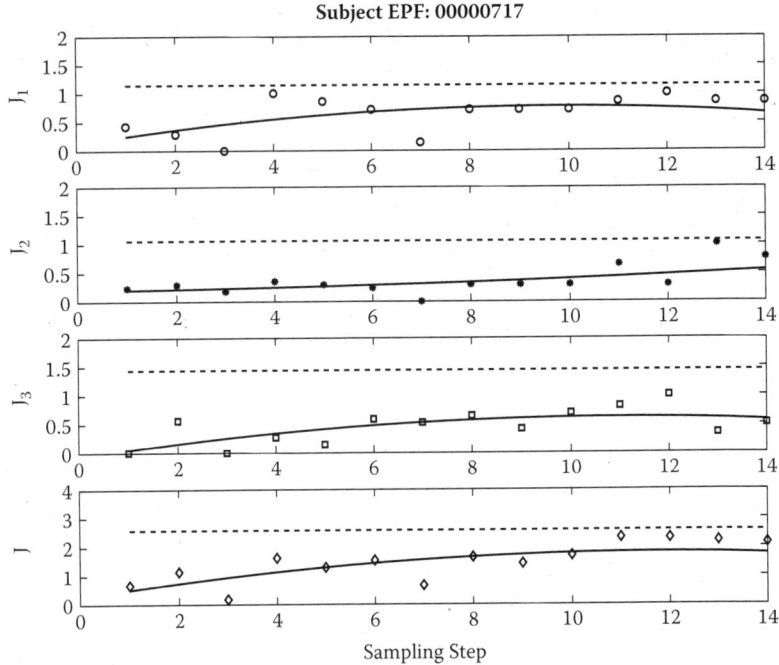

FIGURE 10.11
Variation of normalized individual cost components (top three plots) and the total cost (bottom plot) for W_3. Dotted line shows the respective cost of the elite speed profile for task T_B normalized for the data of worker W_3.

on the screen, the internal critic can calculate the difference between the evaluations of the two profiles given by Δ_J. The subject also notices the visual disparity Δ_V between the elite reference speed profile and the speed profile of the worker. Both of these pieces of information can be used to update an internal model that explores for novel machine speed profiles.

The memory of the modified or innovated speed profile and the visual feedback of the resulting actual speed profile of the worker concerned can be used to update the parameters of the internal motor model controlling the muscles in a supervised learning sense.

However, one should note that there are many possibilities for the learning outcome in the proposed learning mechanism. The worker concerned can decide to follow a conservative approach by trying to use the visual disparity between ψ^* and ψ to maximize the evaluation he/she receives. The worker could also use the internal evaluations of ψ^* and ψ and use the difference between the evaluations of the two profiles given by Δ_J to update the internal model that plans innovative speed profiles. The worker may also decide to use a combination of the two given by $\alpha_v \Delta_V + \alpha_j \Delta_J$. Although the choice to go for an innovation is risky for the worker concerned, it can lead

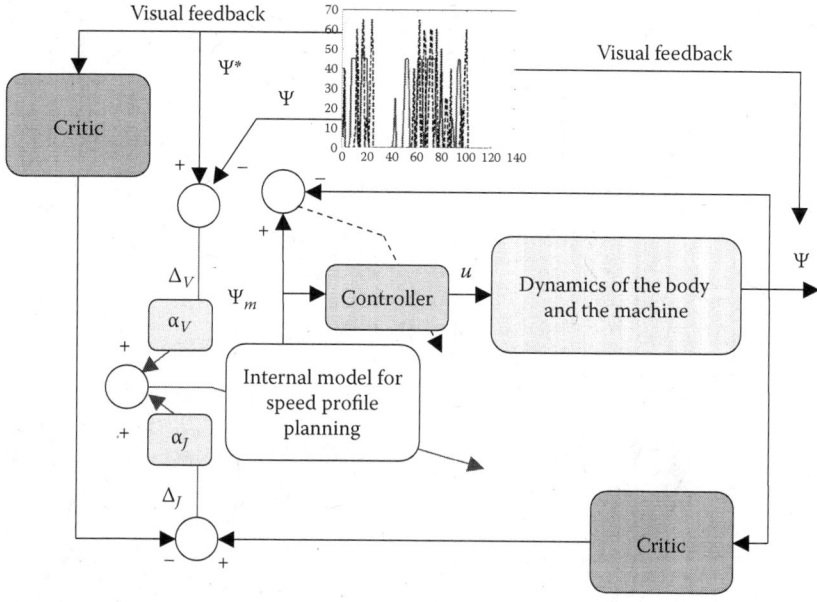

FIGURE 10.12
Proposed reinforcement-based innovation mechanism.

to revolutionary improvements in the whole factory. The innovation process can continue till the worker is satisfied with the internal evaluation he/she receives.

Let us refer to the model shown in Figure 10.12 to explain the experimental outcomes discussed in Section 10.4. The proposed model suggests that a subject has the choice to compare two speed profiles displayed on the screen by the pattern itself or by the evaluation assigned to such patterns by the internal critic. The error between the cost estimated for the reference and that for the actual speed profile of the worker is given by Δ_J, and the visual disparity between the two speed profiles is given by Δ_V. The worker has few options for the strategies he/she can take based on this internal signal. A normal subject would decide to modify the speed profile so that $\|\Delta_V\|$ is minimized. Yet, a subject could also take the decision to give more weight to minimize the norm of the error of the cost given by $|\Delta_J\|$. Another subject might even take the decision to maximize $\|\Delta_J\|$, so that a completely innovative elite solution emerges. The choice depends on the personality and attitudes of the subjects. The results shown in Figures 10.10 and 10.11 can be explained only by assuming that the subject W_3 took a strategy where $\alpha_V \ll \alpha_J$ and wanted to minimize $\|\Delta_J\|$. This led to the emergence of a machine speed profile that was different from the reference but, as far as the cost is concerned, was similar to the reference speed profile.

10.6 Discussion and Conclusion

The orchestration of skills in a system of live systems differs from that of a system of mechanized systems in that the results cannot be predicted due to many possibilities of the strategies the individual systems could take to interact with other systems. This chapter proposes a simple model to explain this complex phenomenon. The experiments were conducted for human subjects in a garment factory for two tasks. Completely noninvasive techniques were adopted to collect data, and all data communications were done using a wireless network without adding constraints to the positioning of the machines.

A simple model like the one proposed in this paper can be very useful in guiding a team of workers to evolve an advanced skill through the exchange of visualized information. This can also be useful in training players in any sport where intricate details of a speed profile of an arm or leg movement could carry the biggest secrets of the talents.

Our results showed that the plasticity of the brain could give room for a wrong elite speed profile to wash away some of the skills learned so far. Therefore, when applying this technique in a factory environment, care should be taken to plan the training sessions well and give clear instructions to the workers and supervisors.

References

Abernethy, B. "Expertise, visual search, and information pick-up in squash," *Perception* 19, 63–78, 1990.

Arbib, M. "Perceptual structures and distributed motor control" in *Handbook of Physiology: Motor Control*, V. B. Brooks, Ed., MIT press, Cambridge, MA, pp. 809–813, 1981.

Arrif, G., Donchin, O., and Shadmehr, R. "A real time state predictor in motor control: study of saccadic eye movements during unseen reaching movements," *Journal of Neuroscience* 22, 7721–7729, 2001.

Bizzi, E., Mussa-Ivaldi, F. A., and Giszter, S. "Computations underlying the execution of movement: A biological perspective," *Science* 253, 287–291, 1991.

Mataric, M. J., Zordan, V. B., and Mason, Z. "Fixation behavior in observation and imitation of human movement," *Cognitive Brain Research* 7(2), 191–202, 1998.

Nanayakkara, T., Piyathilaka, L., Subasingha, A., and Jamshidi, M. "Development of advanced motor skills in a group of humans through an elitist visual feedback mechanism," in *Proceedings of IEEE International Conference on Systems of Systems Engineering*, San Antonio, 2007.

Nanayakkara, T., and Shadmehr, R. "Saccade adaptation in response to altered arm dynamics," *Journal of Neurophysiology* 90, 4016–4021, 2003.

Shadmehr, R., and Mussa-Ivaldi, F. A., "Adaptive representation of dynamics during learning of a motor task," *Journal of Neuroscience* 14, 3208–3224, 1994.

Sutton, R. S., and Barto, A. G. *Reinforcement Learning*, MIT Press, Cambridge, MA, 1998.

Thoroughman, K., and Shadmehr, R. "Learning of action through adaptive combination of motor primitives," *Nature* 407, 742–747, 2001.

Wolpert, D. M., Ghahramani, Z., and Jordan, M. I. "An internal model for sensory motor integration," *Science* 269, 1880–1882, 1995.

11

A System of Intelligent Robots–Trained Animals–Humans in a Humanitarian Demining Application

11.1 Introduction

In this chapter, we discuss an animal-robot-human integrated system. Parts of the work were covered in two undergraduate group projects. The legged field robot discussed here was developed by the team of Lloyd et al. (2002), and the mongoose training and animal-robot integrated system was developed by the team of Dissanayake et al. (2007) at the University of Moratuwa, Sri Lanka.

Antipersonnel landmines buried during armed conflicts to maim or kill opponents are an indiscriminate weapon because they have no particular target. Today, about 60 countries have been affected by landmines. Once a conflict comes to an end, the affected areas have to be cleared to expedite human resettlement. Usually, the first resettlement programs are launched in the lands best for farming. Therefore, it is preferred to have techniques that do not cause damage to the top soil layer or the vegetation. However, current demining approaches require a lot of vegetation cutting and top soil removal in the area-reduction phase. Area reduction is the first phase of any demining project, where a large suspected area is segmented into several blocks of varying priority levels. High-priority areas are those with the highest probability of finding mines. Various techniques, which range from identifying former landmarks of a battlefield, such as trenches, forward defense lines (FDL), etc., to scraping the top soil layer in random blocks, are used today. The most frequently used techniques, such as burning the vegetation and raking or scraping the top soil layer using heavy earth-moving vehicles, cause long-lasting damage to the environment that render the area unsuitable for the resettling of farmers. It is a severe problem in areas where the fertile soil layer is very thin and vulnerable to erosion. Therefore, the availability of a lightweight field robot that burrows through an unstructured vegetated environment to carry a sensor or to guide a trained animal without setting off mines can improve the efficiency and effectiveness of an area reduction phase. Moreover, it removes the requirements of sending humans into a minefield. Naturally, manual deminers tend to be extremely slow due to the danger associated with demining. In addition, manual demining is boring and repetitive.

The purpose of building an animal-robot integrated system is to derive a synergy in a combination of individual strengths of an animal and a robot to navigate in a forest environment looking for landmines. In this case, we use a rodent called a mongoose to detect landmines.

The strengths of the rodent are as follows:

1. Powerful olfactory capability enabling the animal to walk along smell gradients.
2. Dexterous navigation skills in forest environments.
3. Weight below the threshold to set off a mine. Furthermore, this allows a lightweight robot to restrict its movements.

The strengths of a walking robot are as follows:

1. Efficient locomotion in soft soil conditions (muddy, sandy, grassy, etc.)
2. Can communicate data and images to a remote location
3. Can restrict the animal's movements to desired paths
4. Can learn from the animal to navigate in cluttered environments

11.1.1 Mine Detection Technology

Modern antipersonnel landmines, as shown in Figure 11.1, are made of plastic. There is only one metal pin attached to the knob at the top center that is pressed against a capsule of explosives when somebody steps on it. Therefore, this makes it very difficult to detect a mine using conventional metal detectors. Various other detection technologies, such as ground-penetrating radar (GPR), seismic sensors, biosensors, trace explosive detection systems, nuclear quadruple resonance sensing, etc., have been tested. However, they have limited or no application due to the extremely high cost involved. We will discuss these techniques later in this chapter.

According to a survey performed by Bunyan and Barratt (2002), different technologies available for landmine detection can be classified into nine classes. Those technologies with only basic principles observed and reported go into class 1. Class 2 is for those with a technology concept and/or application formulated. Class 3 covers those with analytical and experimental critical functions and/or given characteristic proof of concept. Class 4 is composed of those with a technology component and/or basic technology subsystem validation in a laboratory environment. Class 5 includes those with a technology component and/or basic subsystem validation in the relevant environment. Class 6 includes those with a technology system/subsystem model or prototype demonstration in a relevant environment. Class 7 covers those with a

FIGURE 11.1
An antipersonnel landmine (courtesy of the United Nations Development Program).

technology system prototype demonstration in an operational environment. Class 8 refers to those with an actual technology system completed and qualified through test and demonstration, whereas class 9 applies to those with a technology system "accredited" through successful mission operations.

11.1.2 Metal Detectors (Electromagnetic Induction Devices)

A time-varying current in a transmitting coil induces an "eddy" current in nearby metallic objects. The magnetic field created by this induced current develops a voltage in a receiving coil of the metal detector. This voltage signal is filtered and amplified to generate an acoustic signal that works as the alarm (Collins et al., 2002). However, the false alarm rate can be as high as 98% because it gives alarms for shell fragments, bullet cases, and other metal debris usually found in minefields. Moreover, it is difficult to tune in Laterite-rich soil or conductive soil (red soil, sea beaches, etc.), and it is difficult to detect mines if soil deposits over it exceed 1 ft. There can also be false alarms due to electromagnetic interference.

11.1.3 Ground-Penetrating Radar

GPR works by detecting the dielectric properties of the medium by emitting a microwave signal into the soil. If an object, such as a landmine, that has a different dielectric property from the surrounding soil is present, the sensor can use the reflected signal to differentiate the object from the soil

(Sun and Li, 2003; Savelyev et al., 2007). Therefore, GPR technology is much more advanced than metal detection. However, it is an expensive technology. Therefore, it is very rarely used in the affected regions of the world. Microwaves attenuate in conductive and wet soils. Because they give false alarms for roots, rocks, water pockets, etc., there is a need to compromise between the resolution and penetration. Resolution is high at elevated frequencies, and penetration is high at low frequencies. Sometimes, it is difficult to train even a human to look at an image given by the GPR to distinguish between a mine and the surrounding clutter.

A typical application is described by Sun and Li (2003), where a GPR sensor is mounted in front of a field vehicle so that the GPR detector is directed forward when the vehicle moves. When the vehicle moves forward, the data are collected in time domain. The frequency analysis of these data has been done. Two-dimensional (2-D) images have been obtained by plotting the real part of the Fourier transformed data on the X axis and the frequency spectrum on the Y axis. The corresponding image gives different patterns for different mines buried at different depths. This could be considered an effective method to detect suspicious items by visual inspection. However, vehicle-mounted sensors can misguide the human interpreter due to false readings given by sensor vibrations and the jolt of the vehicle in rough terrain.

11.1.4 Multisensor Systems Using GPR and Metal Detectors

Several multisensor systems have been field-tested so far.

1. Handheld Standoff Mine Detection System (HSTAMIDS or acronym AN/PSS-14) (Xu et al., 2002). This U.S. army project has developed a multisensor system containing a GPR and a metal detector that are used by a sensor fusion algorithm for feature extraction.
2. MINEHOUND (Doheny et al., 2005) is claimed to be simple and effective compared to other GPR-based technologies. The project is sponsored by the UK Department for International Development and developed by ERA Technology.
3. Advanced Landmine Detection System (Daniels et al., 2005). Similar to HSTAMIDS, this system uses GPR and pulse-induced metal detectors. The limitations are that it is very expensive for normal use and needs a large platform to carry it, which may cause problems in a tropical minefield.

There has been a considerable amount of effort given to improve the sensor fusion algorithms (Sato et al., 2005; Ferrari and Vaghi, 2006). In their study, Sato et al. (2005) proposed a sensor fusion technique based on Bayesian networks. The method is tested on a system of GPR, electromagnetic induction, and infrared sensors. The data have been processed as batches. Therefore, there is no guarantee that this method can be deployed in an unstructured environment with no prior information to detect landmines online.

11.1.5 Trace Explosive Detection Systems

Samples of the environment have to be obtained and the detection of explosives is done based on a chemical reaction or a mass-spectroscopy measurement (Perrin et al., 2004; George et al., 1999; Cumming et al., 2001). For instance, one technology is to use amplifying fluorescent polymers (George et al., 1999). In the absence of explosive agents, the polymer fluoresces when exposed to light. With the presence of nitro-aromatic compounds, the fluorescence decreases. The limitations are that it is slow to respond and expensive, and requires samples of the environment and, therefore, cannot detect efficiently while on the move.

11.1.6 Biosensors

Biosensors use a quartz crystal microbalance (QCM) together with antibody and antigens. A sample collector sucks in the vapor. If there are substrates of TNT/DNT, RDX, PETN, and tetryl, the oscillating frequency of the QCM changes due to the migration of the microorganisms (Fisher and Sikes, 2003; Crabbe et al., 2005; BIOSENS Consortium, 2004). This change in the frequency is detected to give an alarm. Limitations are that it has a slow response, can have a drift due to population changes, and is very expensive.

11.1.7 Magnetic Quadrupole Resonance

A method is proposed to observe the quadrupole resonance signals of explosives that are often interfered by background radio signals (Crabbe et al., 2004). For instance, in TNT, there are 18 resonant frequencies, 12 of which are between 700 and 900 kHz, which is susceptible to interference from AM radio. Since the resonant signal is weak, filtering and observation is necessary. This method proposes to use the noise as the state and the signal as the measurement noise of the Kalman filter. If the interference from AM radio is handled with improved signal processing methods, this method can be useful as a remote sensing technique. However, the technology is not yet mature.

11.1.8 Seismoacoustic Methods

Seismic waves are generated from one end, and from the other end, a non-contact (acoustic) transducer and a contact (seismic) transducer pick up the signal (Tan et al., 2005; Xiang and Sabatier, 2004; Scott et al., 2001). If there are hollow objects such as landmines, they will be reflected in the received signal because they are different from the mechanical properties of the surrounding soil. Attempts have been taken to identify the vibration signature of a mine. However, aging and changes in the mechanical properties of the mine may lead to false alarms.

In essence, given the scale of the need to have technology to detect landmines and explosives, the current state-of-the-art is very expensive, computationally cumbersome, and gives high false alarm rates. Therefore, there is an open need to have a simple, cost-effective, and powerful technique that can be applied in most practical cases. We believe the answer lies in a system that combines the strengths of animals to detect chemical agents, semiautonomous robots that can navigate in a cluttered environment to guide the animal, and a human's ability to control a robot and to analyze the visual feedback of an animal's behavior.

11.2 A Novel Legged Field Robot for Landmine Detection

Almost all robots available today are designed to suit desert environments or the insides of buildings. Many of them will find it difficult to move in a forest environment. To suit a forest environment, the robot should be able to adapt to the uneven ground profiles, be stable in an unstructured environment, and be able to deal with typical obstacles such as plants, tall grasses, and rubble in an efficient manner. In addition, to ensure safety in a minefield, the contact forces should fall within certain limits. Many robots found today are too heavy to ensure safety if the detectors miss a mine. Moreover, the robot platforms are too expensive to justify deployment in many of the developing countries affected by landmines.

11.2.1 Key Concerns Addressed by the Moratuwa University Robot for Anti-Landmine Intelligence Design

Cost. Most landmines are found in developing countries such as Cambodia, Mozambique, Afghanistan, Kosovo, Egypt, Somalia, Sri Lanka, etc. One of the reasons as to why robotic solutions have not yet been successful in these countries is the cost involved. Therefore, one of the major design goals is to reduce the cost of the robot. This can be done by reducing the need for expensive sensors, body parts, and actuators. The cost of repair should also be kept low.

Adaptation to the working environment. Moratuwa University Robot for Anti-Landmine Intelligence (MURALI) has been designed to work in vegetated environments found in tropical countries such as Sri Lanka. In the affected regions of Sri Lanka, the abandoned minefields are often covered with bushes and tall grass. Moreover, the fertile soil layer is very thin (sometimes 6 in.).

Therefore, vegetation cutting should be avoided as much as possible to avoid erosion. This is a very important concern because the final goal is to clear the land to resettle farming communities. This implies that the robot has to move through vegetation. Therefore, conventional obstacle avoidance algorithms do not suit this application because the robot is always surrounded by obstacles. Therefore, in this case, the meaning of an obstacle is redefined as an object that cannot be bent or pushed by the robot to move forward. In other words, the mechanical impedance of the obstacle is the only important information. There are objects such as grass with low mechanical impedance, and there are other objects such as large trees and vines with higher mechanical impedance. The robot has to use tactile sensors to estimate the impedance in the surrounding objects and classify them into those objects that can be bent or pushed easily, and those that should be avoided.

Critical safety constraints. Since the robot has to move on a minefield, the weight on a given ground contact point should not exceed the critical pressure needed to trigger a mine. For an antipersonnel mine, this limit is generally considered to be 7 kg. Therefore, the body weight has an upper constraint. Moreover, the sensor should scan the environment at an acceptable speed to make sure no mine is left undetected. Therefore, the speed of the robot should be within an acceptable range. In this case, we considered a speed range of 0.1 to 0.2 m/s.

Exploitation of passive dynamics. Most tropical minefields have been abandoned lands for decades. Due to a long period of soft deposits and growth of weeds and grasses, the robot has to walk on a soft terrain. Usually, soft terrains are springy and readily deformable. Therefore, the robot should be designed to make the best use of the passive dynamics of locomotion on soft terrain conditions. Out of many possible locomotion technologies, such as traction, wheels, legs, gliding, slithering, etc., one has to decide the best method to suit the typical environments.

Stability. The robot has to move in an unstructured environment. Sometimes, the exact terrain conditions of the minefield are not known due to limited accessibility. Therefore, the static and dynamic stability of the robot is of paramount importance. Since we have limitations on the weight, complexity, and cost of the robot, the best solution would be to increase the passive compliance of the robot to adapt to various terrain profiles.

11.2.2 Key Design Features of MURALI Robot

The physical features of a robot, such as the shape, degrees of freedom, forces it can exert, locomotion strategy, the number of ground contact points, the type of sensors and actuators, mechanical impedances of body parts, etc., and algorithms that map sensory information to robotic behaviors should suit the environment with which the robot is intended to interact (Arkin, 1998; Brooks, 1986). In an ideal case, all these features should evolve to be in perfect harmony with the environmental dynamics over time. Therefore, we paid considerable

FIGURE 11.2
Front view of the robot.

attention to creatures that are normally found in the affected regions of Sri Lanka. Among them, the iguana caught our eye because its behaviors suit the bushy environment found in the northern part of Sri Lanka. Because the ground contact points are located well outside the locus of the center of gravity, static stability is guaranteed. Furthermore, due to its low height, the iguana can creep through common obstacles such as barbed wire fences.

Figures 11.2 and 11.3 show the laboratory-made robot. The legged robot consists of two independent units, each resembling the shape of an iguana. Figure 11.4 shows the basic kinematic features of one such module. Figure 11.5 shows some photos of the real implementation of the mechanism.

The key strength of this design is that four legs are driven by just one motor. By having two such independently controlled modules, we can achieve any turning behavior. Differential velocity control is done using onboard embedded processors and drivers. The robot is even capable of turning in the same position by making one module reverse while the other moves forward. Since the two modules are connected using two rods hinged to each module at the front and rear ends, a parallelogram is formed with four passive joints. This enables the two modules to passively adapt to a wide range of complex ground profiles, as shown in Figure 11.6. This ensures robustness and stability of navigation algorithms.

Exercise. Based on the simplified diagram shown in Figure 11.4, derive the relationship between the position of the tip of one leg to the angular rotation of the motor shaft for a given set of dimensions and values of the motor shaft, worm gear ratio, lengths of links in the leg, stoke length of the crank, etc.

A System of Intelligent Robots-Trained Animals-Humans

FIGURE 11.3
Rear view of the robot in a typical test environment.

Use this relationship to derive the Jacobian matrix relating the linear velocity of the robot to the angular speed of the motor shaft.

11.2.3 Designing the Robot with Solidworks

In this section, we will go through the steps of designing the robot using Solidworks. It is very important for a modern-day engineer to be able to use computer-aided design software to sketch, simulate, and visualize designs before fabricating real hardware. A good feature of Solidworks is that we get

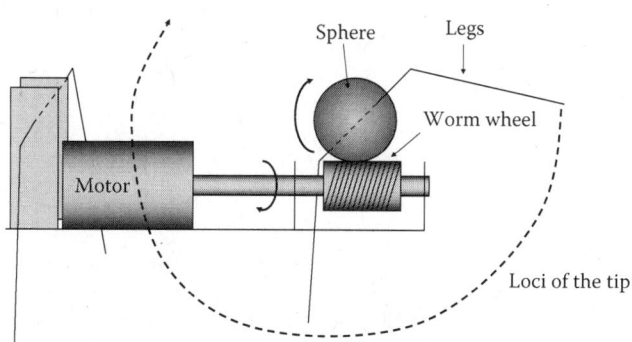

FIGURE 11.4
Basic kinematic structure of one module of the robot.

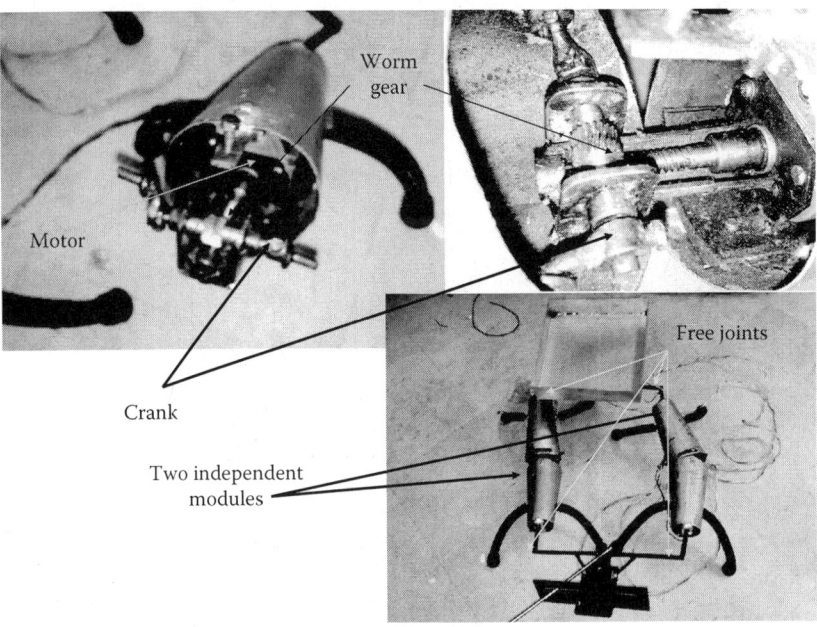

FIGURE 11.5
Internal mechanism of the robot.

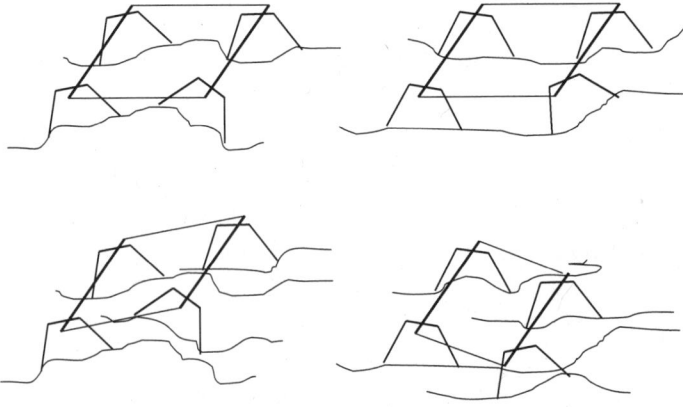

FIGURE 11.6
Passive adaptation of the robot to different ground profiles due to the parallelogram with passive joints.

the opportunity to visualize the drawings in a 3-D world, add mathematical relationships among various elements of the design, edit dimensions at any stage of the development, and, finally, run simulations to see how different components interact among each other. In the beginning, we will try to go through detailed steps. However, later on, we will skip those principles learned earlier. We will not discuss the drawing of the complete robot since the purpose is to give training on basic design using Solidworks.

First, we try to design one of the cones shown in the robot bodies (Figures 11.5 and 11.6). Figure 11.7 shows the interface obtained when we click "new" in the menu bar. Out of the three options, we select "part." Next, we are shown the three planes as in Figure 11.8. We can literally select any plane because we can rotate objects at any time in an assembly operation.

Since our objective is to make a hollow cone, we take advantage of Solidworks' facility to swipe a 2-D shape by 360° to make a 3-D solid object. In Solidworks, there are two types of menu bars we can select. One is the "sketch" menu bar that gives us normal drawing features such as circles, squares, lines, etc. The other is the "features" menu bar that gives us various facilities to add features to our sketches such as extruding a 2-D shape, drilling, adding fillets at the edges, etc. First, we go to the sketch menu bar and choose a line tool to draw the trapezoidal shape as shown in Figure 11.9.

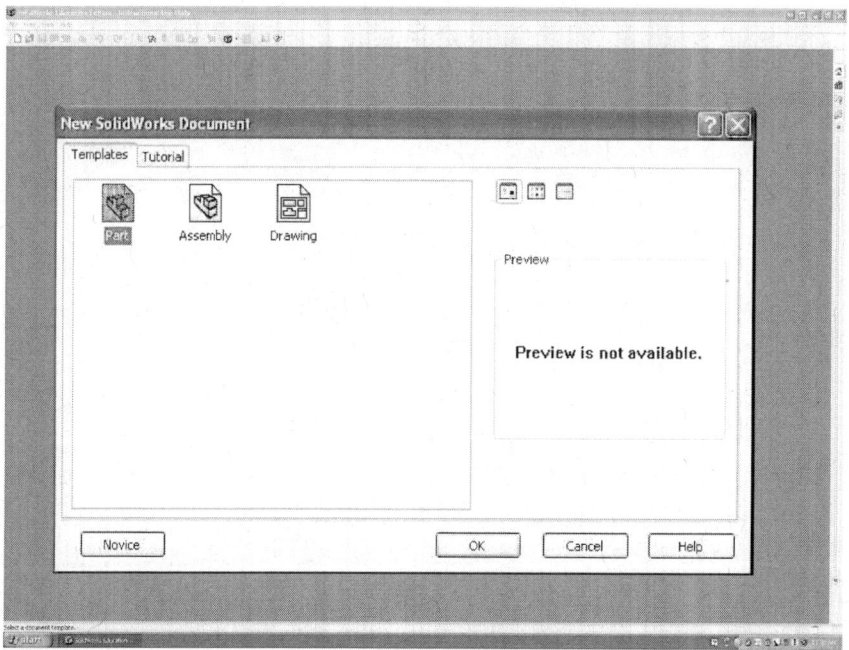

FIGURE 11.7
Drawing options—select "Part."

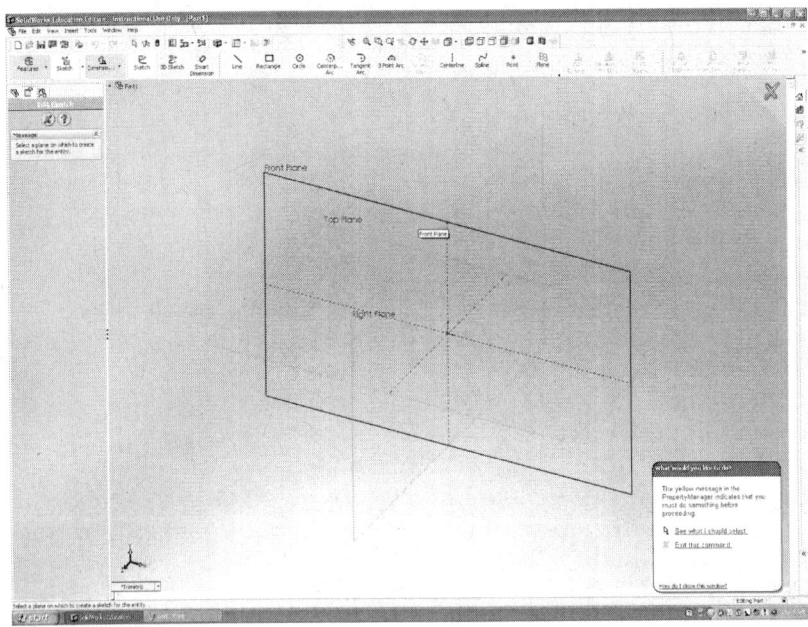

FIGURE 11.8
3-D planes.

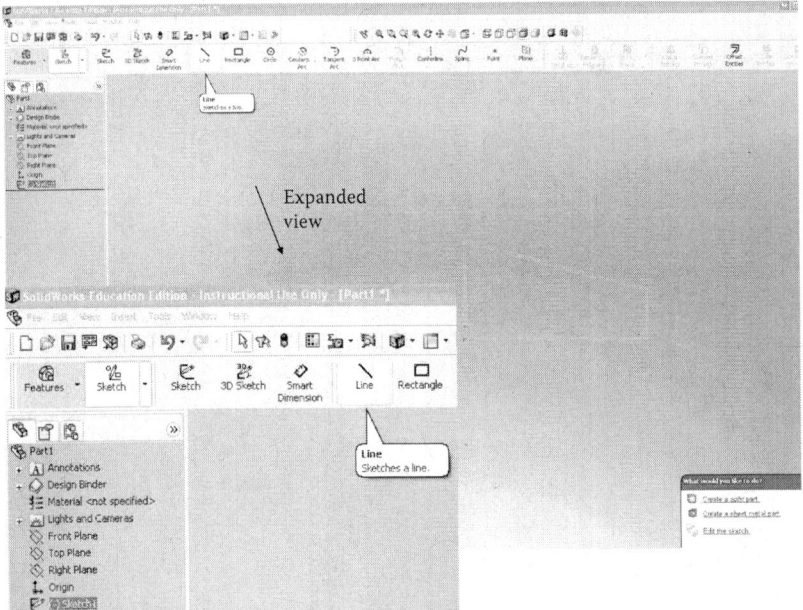

FIGURE 11.9
Trapezoidal shape.

A System of Intelligent Robots-Trained Animals-Humans 309

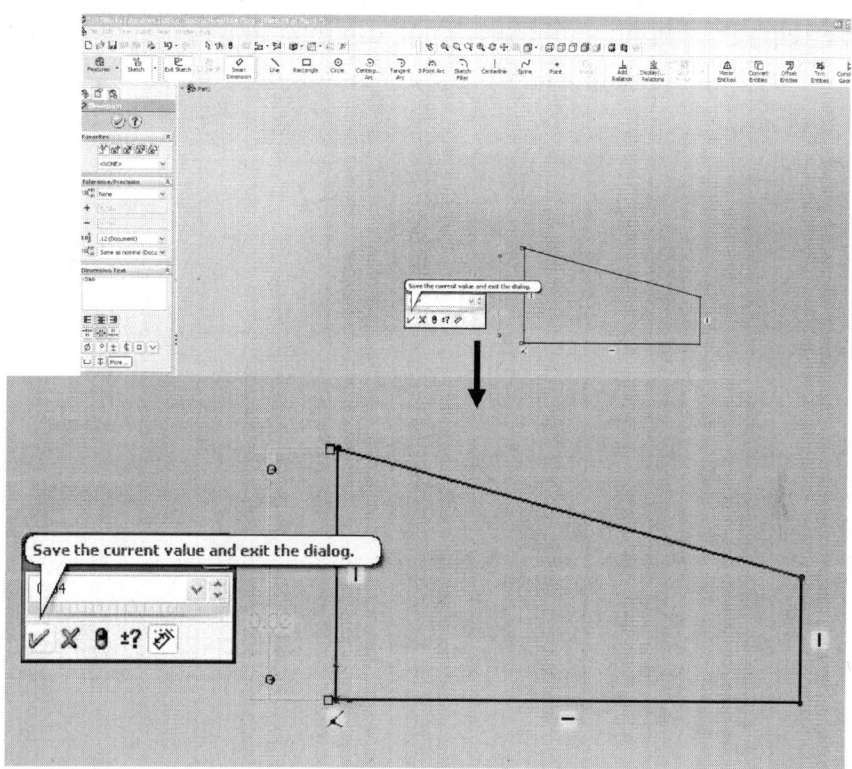

FIGURE 11.10
Adding dimensions.

Next, we go on to add dimensions. We can do this by using the "smart dimensions" tool in sketch menu bar. First, we click on the line we want to add a dimension to and then click on any point outside the line as shown in Figure 11.10. Then, in the text box, we enter the desired length of the line.

Rather than adding dimensions to each line, we can add relationships to other lines as shown in Figures 11.11 and 11.12. The result of this equation is shown in Figure 11.13. In this case, we want the small end of the cone to be half the size of the large side. Similarly, we can add the dimension for the bottom line. The complete trapezoid is shown in Figure 11.14.

Figure 11.14 shows an important property of the "features" menu bar. We choose "revolved Boss/base" and choose the bottom line as the axis around which this 2-D shape should be revolved to form a cone. The result is shown in Figure 11.15. We can accept this shape by clicking the green check mark on the left-side box, or we can reset it by clicking the red cross mark.

In the "features" menu, we select the "shell" feature and then select a face of the cone to create a shell as shown in Figure 11.16. We can rotate the 3-D

310   Intelligent Control Systems

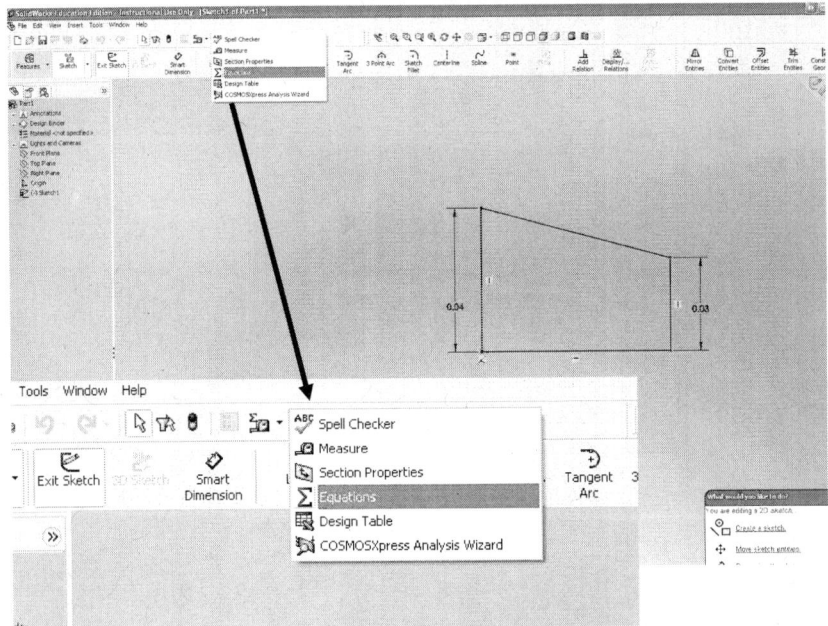

FIGURE 11.11
Adding an equation.

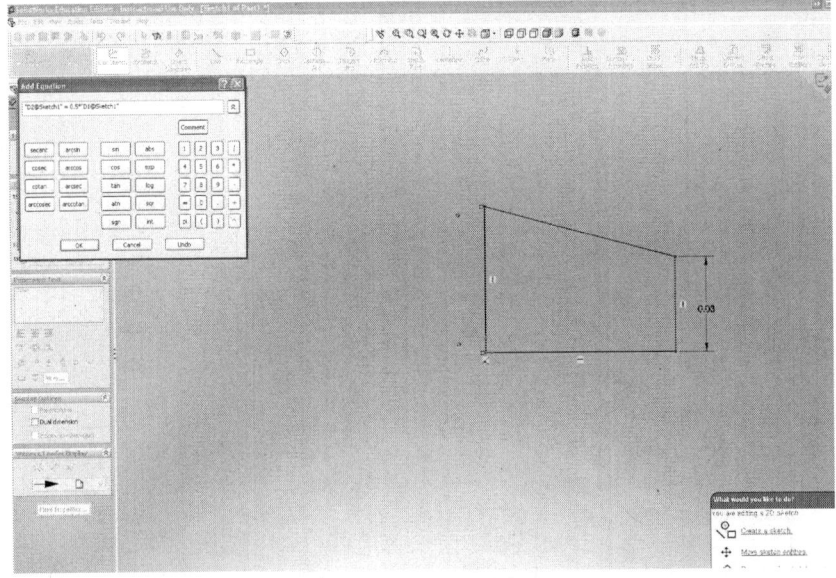

FIGURE 11.12
Adding an equation.

A System of Intelligent Robots-Trained Animals-Humans 311

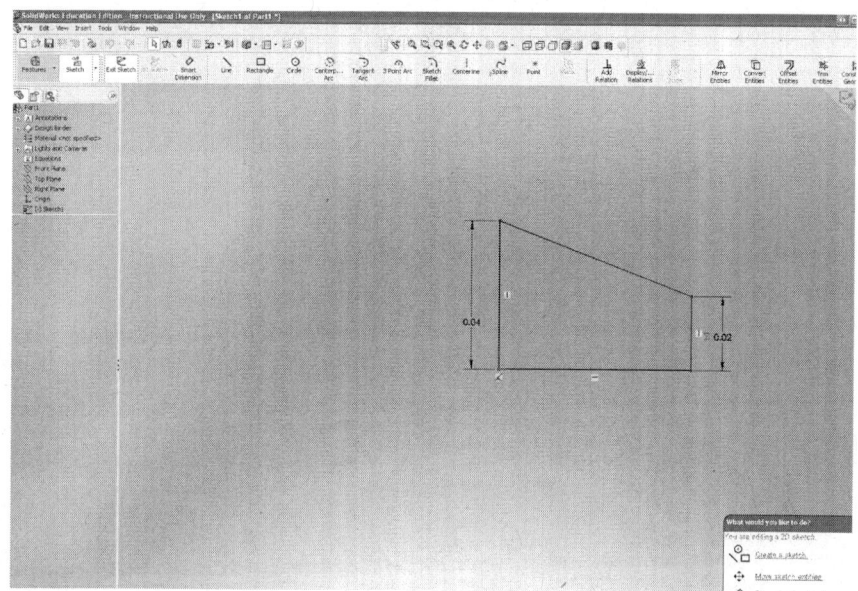

FIGURE 11.13
Effect of the equation.

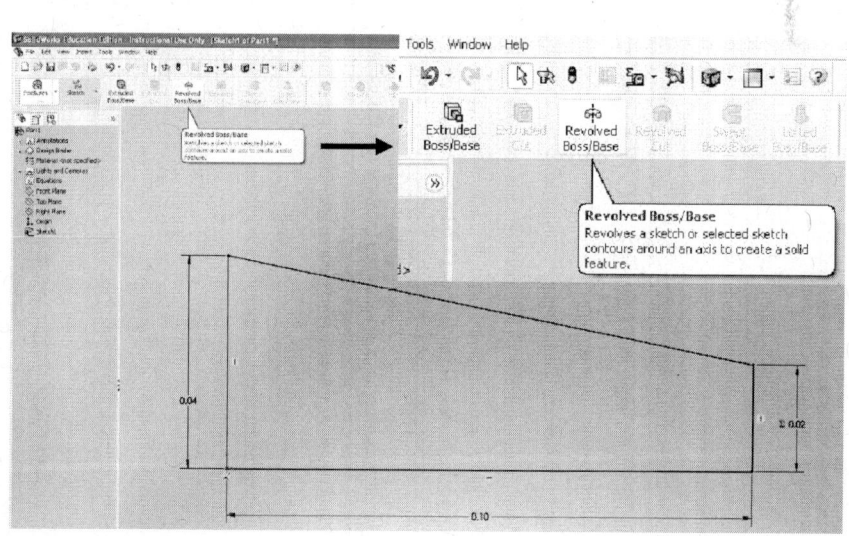

FIGURE 11.14
Revolving a 2-D shape to form a 3-D volume.

312   *Intelligent Control Systems*

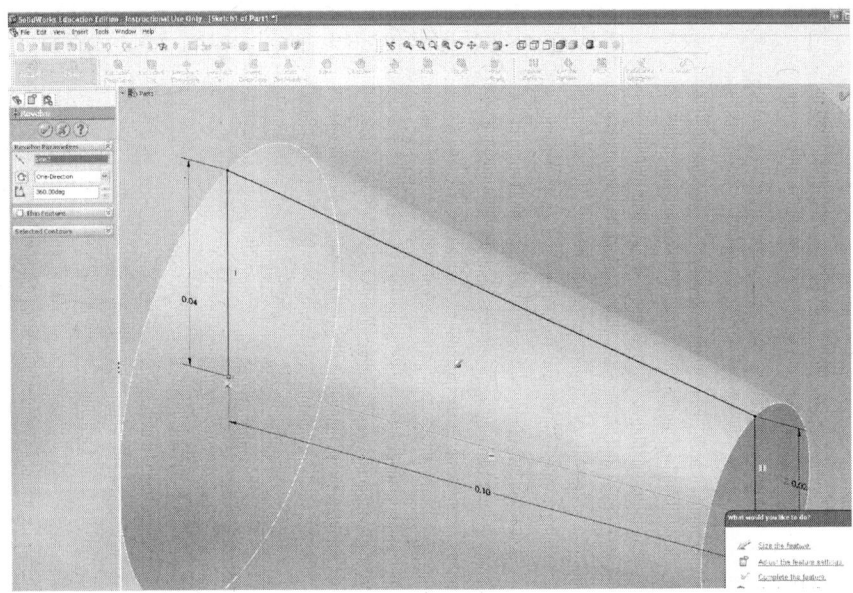

FIGURE 11.15
Revolved volume.

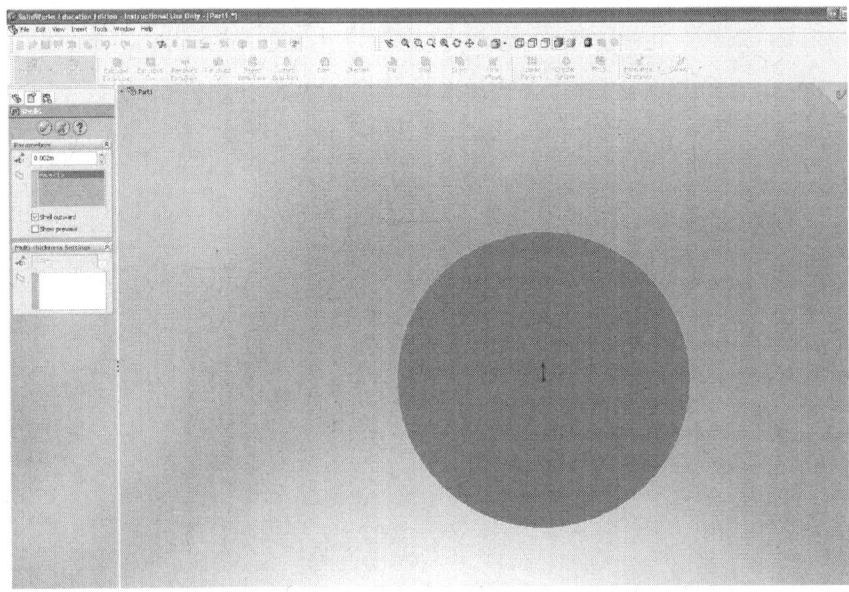

FIGURE 11.16
Select a face to create a shell.

shape to get any face using the orientation selection interface in the bottom left corner, or we can use the following key combinations:

View orientation menu: spacebar
Front view: Ctrl + 1
Back view: Ctrl + 2
Left view: Ctrl + 3
Right view: Ctrl + 4
Top view: Ctrl + 5
Bottom view: Ctrl + 6
Isometric view: Ctrl + 7

Once the face is selected and the shell feature is clicked, we get to choose the offset from the outer edge, as shown in the left-hand-side box of Figure 11.16. We have selected 2 mm. Once the green check mark is pressed, we get a shelled cone as shown in Figure 11.17.

Now, we need to drill two symmetrically located holes to fix two legs to the robot. In Solidworks, drilling a hole is done by sketching a circle on a plane perpendicular to the axis of the hole and using the extruded cut feature. Figure 11.18 shows how we can draw a plane parallel to some other

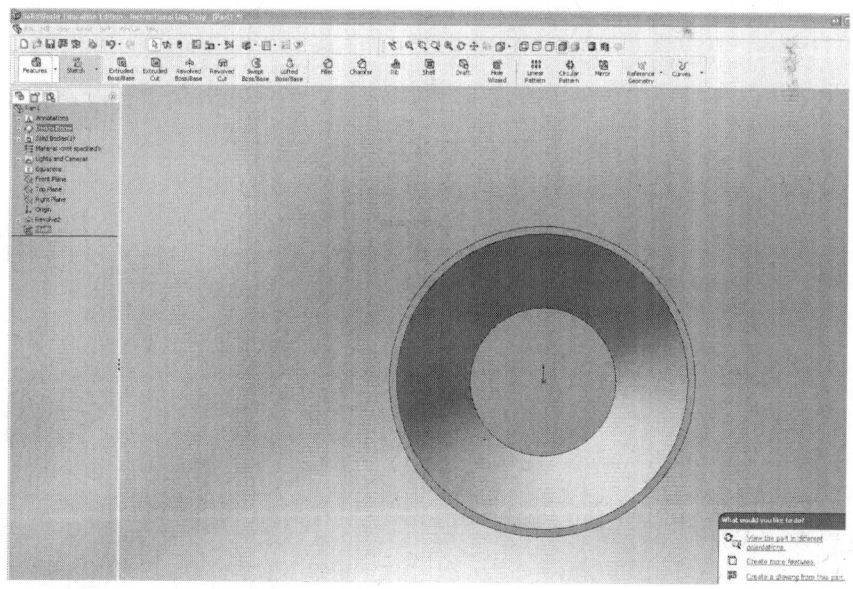

FIGURE 11.17
Shelled cone.

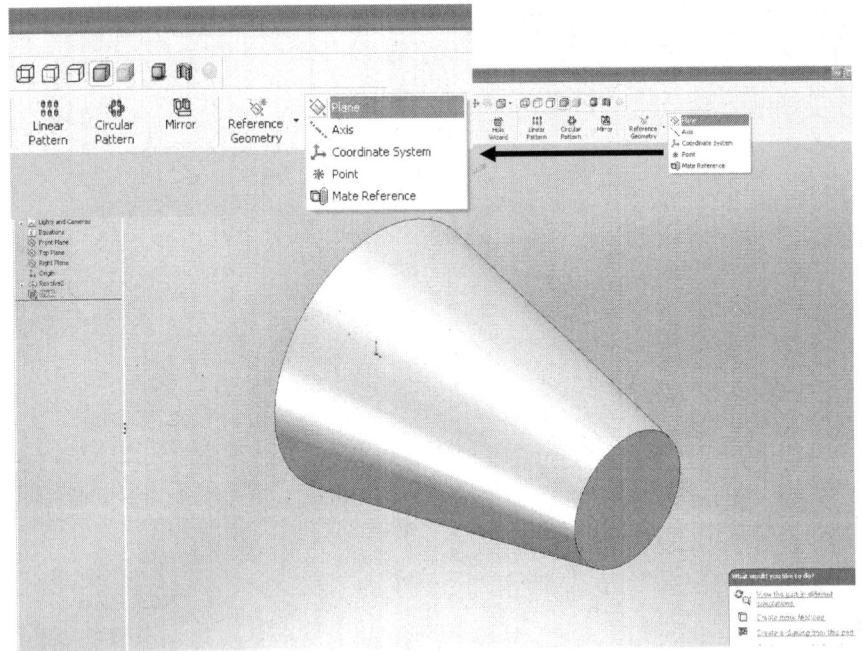

FIGURE 11.18
Adding a reference plane.

selected plane. We first go to "features" and select "reference geometry." In that pull-down menu, we select "plane." Next, we select some base plane from the design tree. You will see the reference frame in yellow color as shown in Figure 11.19. Once the distance between the two planes is given, we can accept the settings by pressing the green check mark on the left-hand-side box.

Using the sketch menu bar, we can draw a circle on the reference plane as shown in Figure 11.20. Then, we select the extruded cut feature and select the "through all" property, as shown in Figure 11.21, to make the hole go through both sides. The result is shown in Figure 11.22.

If we want to verify whether the dimensions are correct, we can choose the measurements feature and check all the added features. If we want to modify any feature at this stage, such as the extruded cuts, we can simply go to the project tree on the left-hand side and right-click on the target feature and select "edit sketch." This will put that feature into sketch mode. Then, we can change dimensions.

Next, we need to cut a smooth-edged rectangular shape on only one side parallel to the axis going through the two holes we cut earlier, to constrain the movement of the other cone of the robot module. We follow the same procedure as the one we took to drill two holes, except that we only need one rectangular slot cut here and the edges should be smooth.

A System of Intelligent Robots-Trained Animals-Humans 315

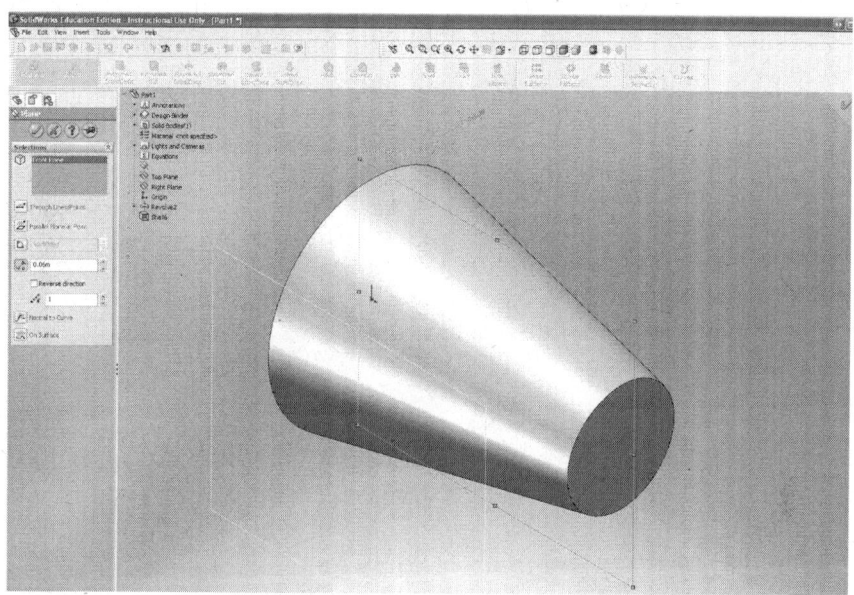

FIGURE 11.19
Adding a reference frame.

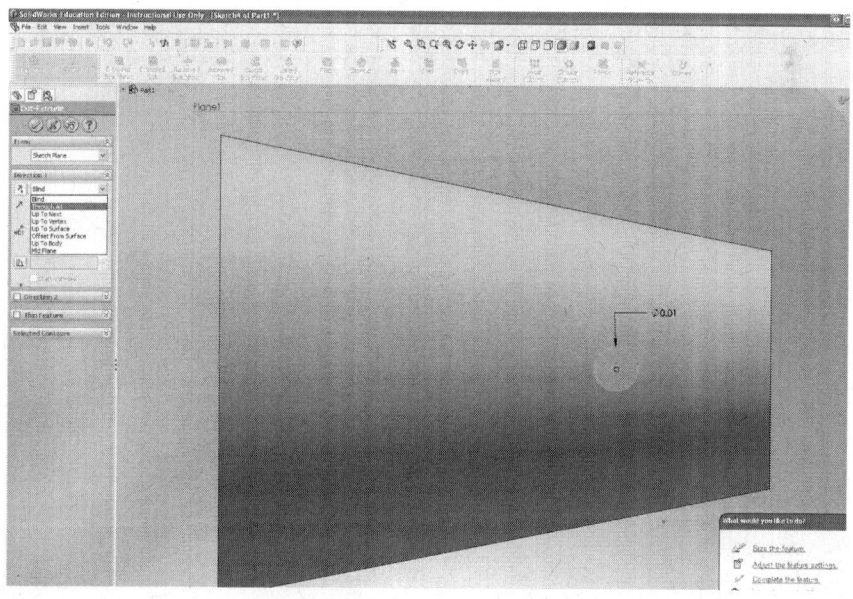

FIGURE 11.20
Preparing an extruded cut to set two holes on either side of the cone.

316   Intelligent Control Systems

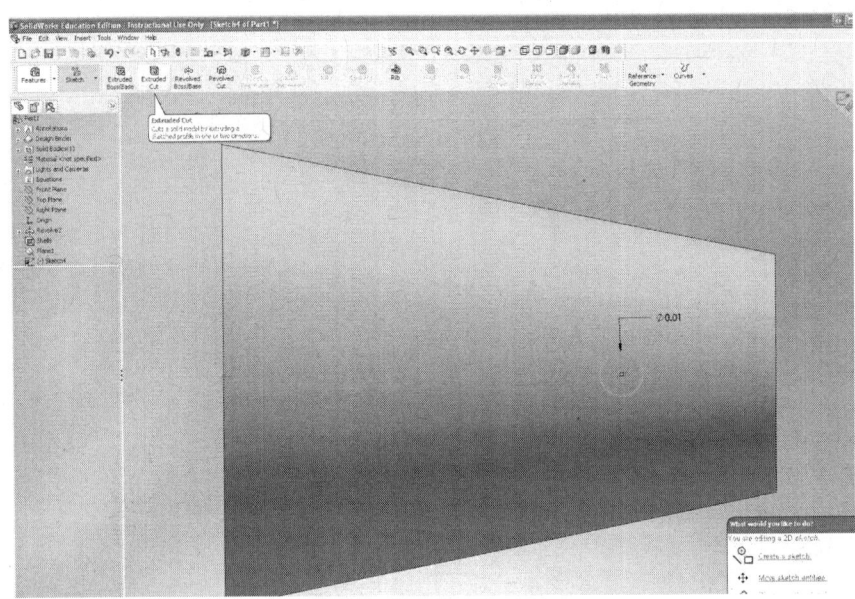

FIGURE 11.21
Extruded cutting.

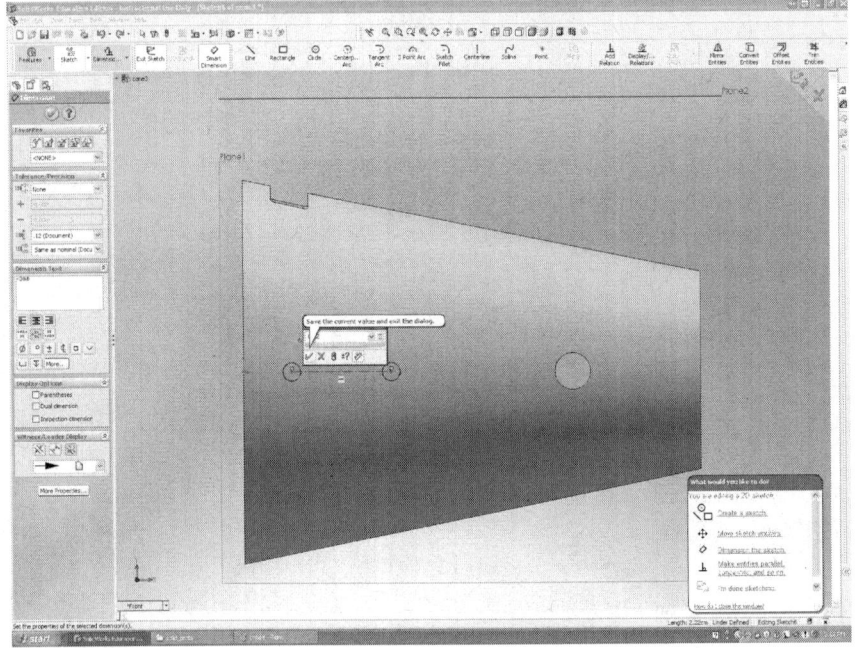

FIGURE 11.22
Result of an extruded cut.

A System of Intelligent Robots-Trained Animals-Humans

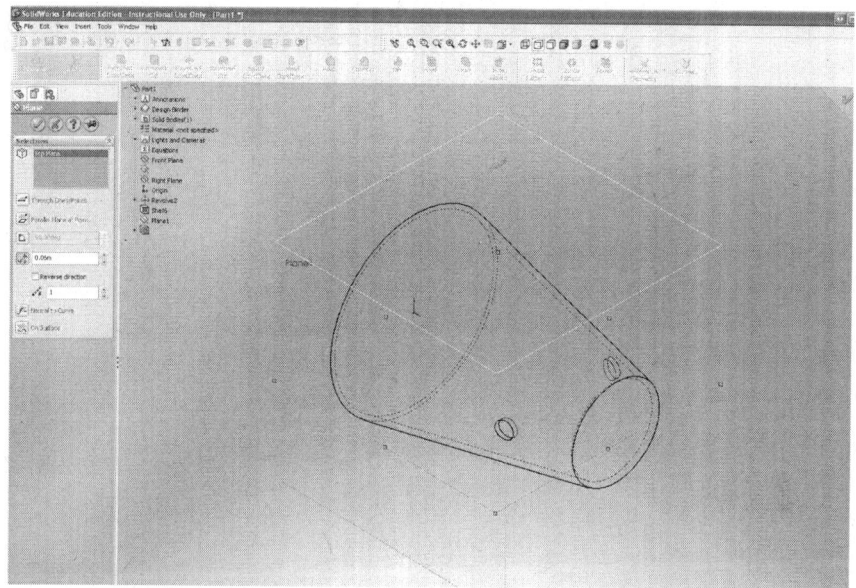

FIGURE 11.22
(*Continued*)

As we know, if we draw a rectangle, the edges are sharp. To smoothen the edges, we use a feature called "sketch fillet." Once this feature is selected, click on the corners of the rectangle and enter the required radius in the feature property dialog box on the left as shown in Figure 11.23. Then, we select the "upto next" property instead of the "through all" property we selected in the previous case. This will ensure that the cut is made only on one side (Figure 11.24).

Next, we make the legs for one module. First, we draw a line using the sketch menu as shown in Figure 11.25. Then, we add dimensions using the "smart dimensions" tool in the "sketch" menu bar as shown in Figure 11.26.

Then, we want to bring the center of the line to coincide with the center of the coordinate frame (Figure 11.27). To do this, we use the "add relations" tool in the sketch menu bar. The add relations tool wants us to select the objects that have to be related. Therefore, we select the line and then the center of the frame, and then we select the "mid point" property on the left-hand-side property selection box. When we click the green check mark, the line's center will move as shown in Figure 11.28.

We add legs that look like an arc tangential to the axle we drew. We use the "tangential curve" tool in the sketch menu bar to do that. Since we need two such arcs, we can use the "mirror entities" tool on the sketch tool bar. Then, we mirror the leg around the center of the axle as shown in Figures 11.29 and 11.30.

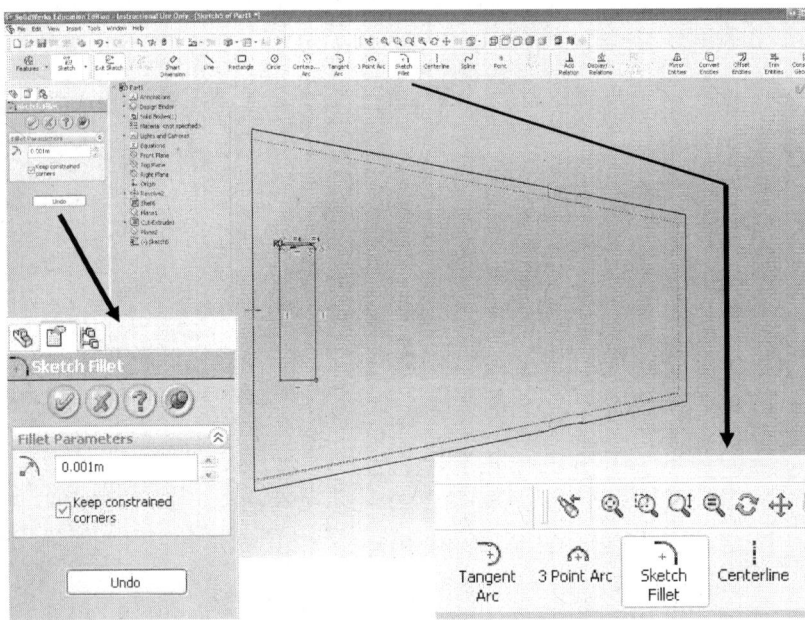

FIGURE 11.23
Cut a smooth edged rectangular shape.

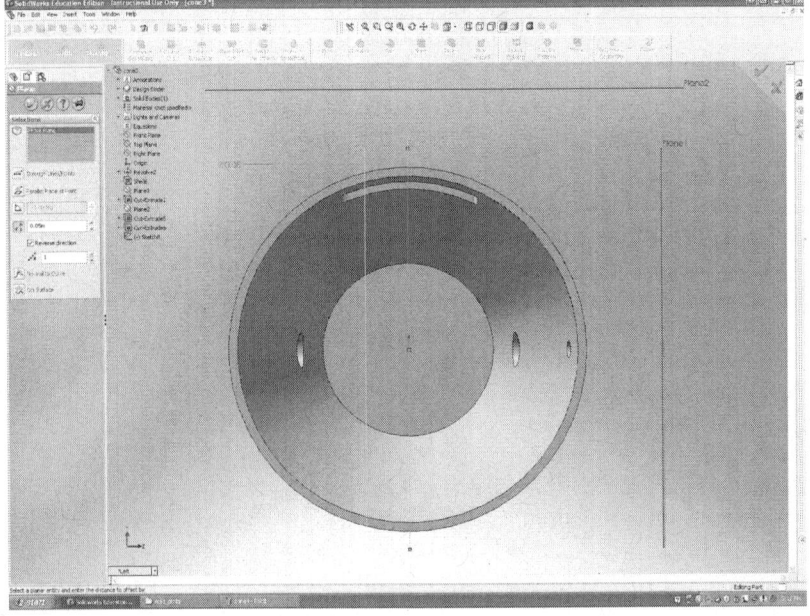

FIGURE 11.24
Drilling holes for the crank shaft.

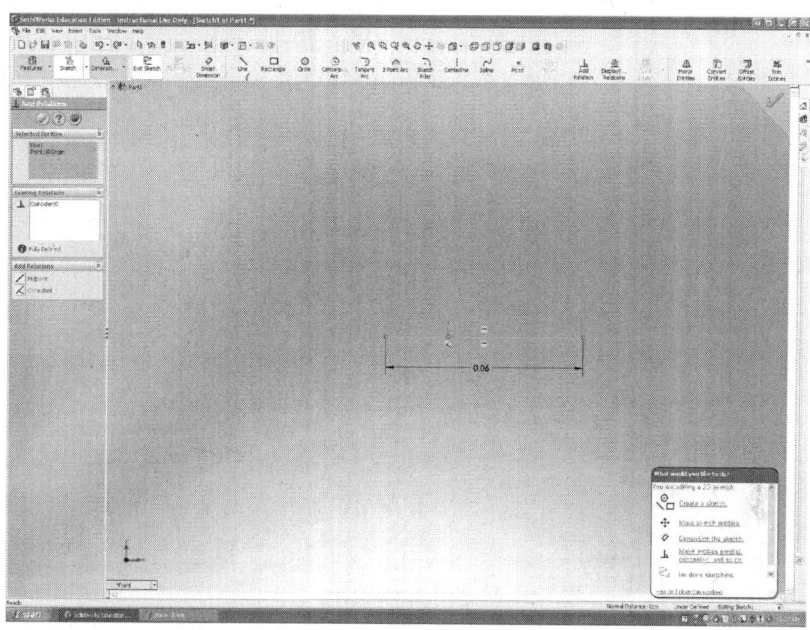

FIGURE 11.25
Sketch a line.

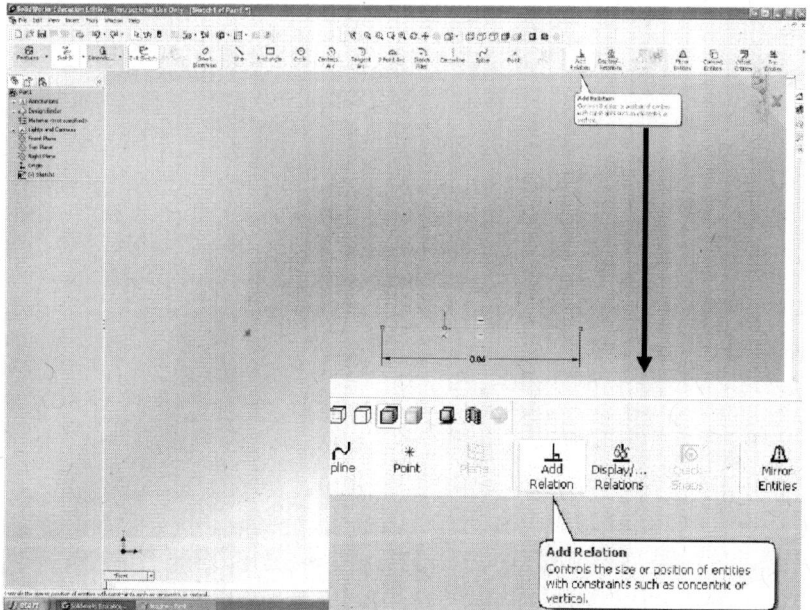

FIGURE 11.26
Dimensioning the line.

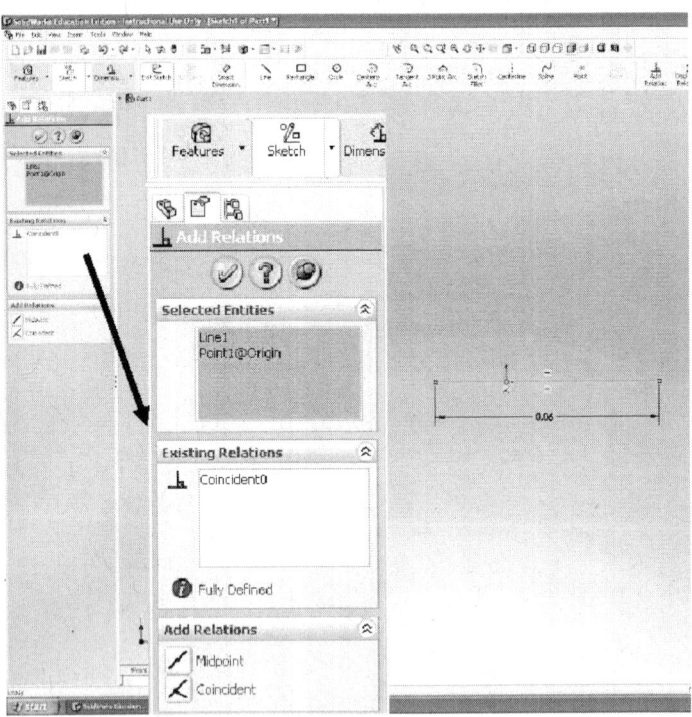

FIGURE 11.27
Adding relationships.

Next, we add flesh to the legs. We make a rod out of a line by using the "swept bose/base" tool in the features tool bar. Before sweeping the shape of a circle along the wire sketch we have in Figure 11.30, we have to draw a circle perpendicular to the tip of the leg. Out of several approaches, we first draw a centerline tangential to the arc at the tip of the leg first and draw a reference plane perpendicular to this centerline at the tip of the leg. Then, we draw a circle on this reference plane as shown in Figures 11.31 and 11.32.

Next, we use the "sweep bose/base" tool from features and select the lines along which the circular shape should be swept to get the result shown in Figure 11.33. Now, we are ready to assemble the cone and the leg we created earlier. To do this, we should start a new assembly document as shown in Figure 11.34 and import the components to be assembled.

In the assembly interface, we can add components as shown in Figure 11.35. When we add components, they come in the original orientation as shown in Figure 11.36. Then, we "mate" the components in the right orientation. Mating also gives us several options, such as "coaxial," "coincident," "parallel," "perpendicular," "at a given distance," and "at a given angle."

A System of Intelligent Robots-Trained Animals-Humans   321

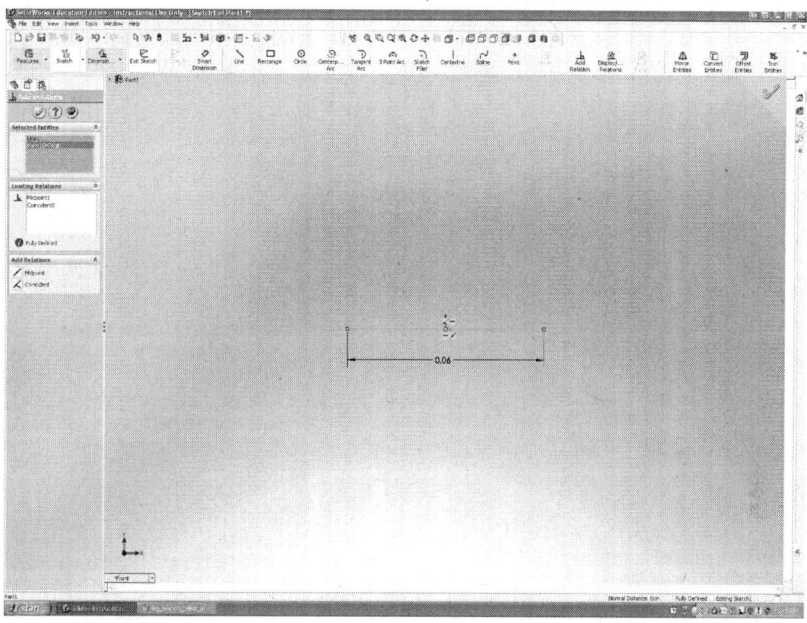

FIGURE 11.28
Centered line.

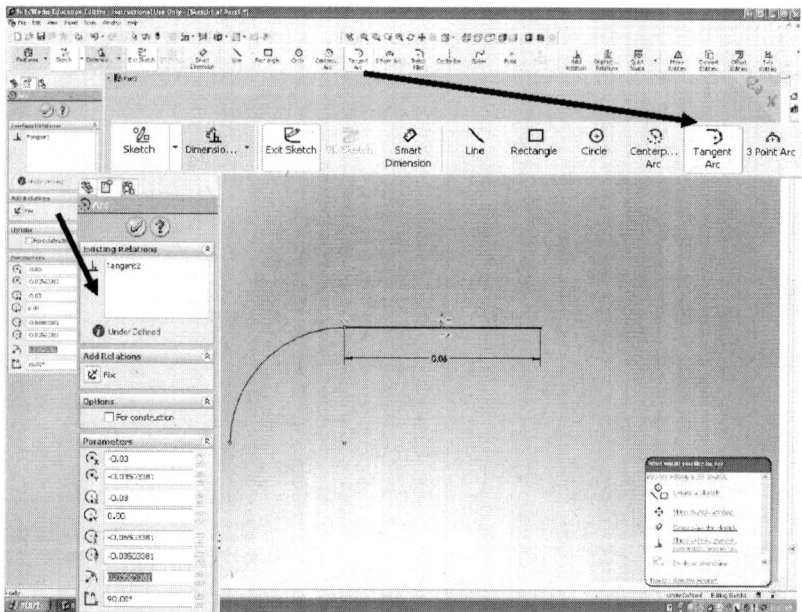

FIGURE 11.29
A tangent curve.

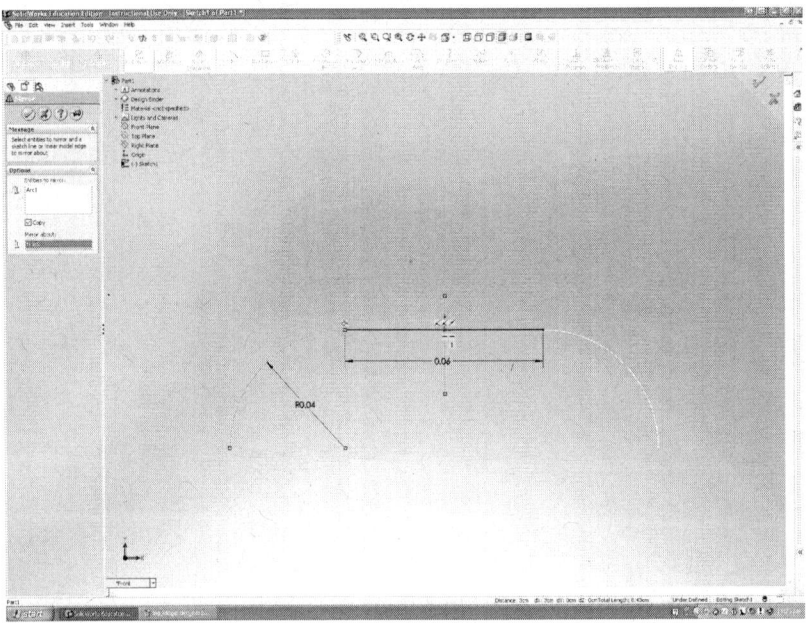

FIGURE 11.30
Mirror image of an object.

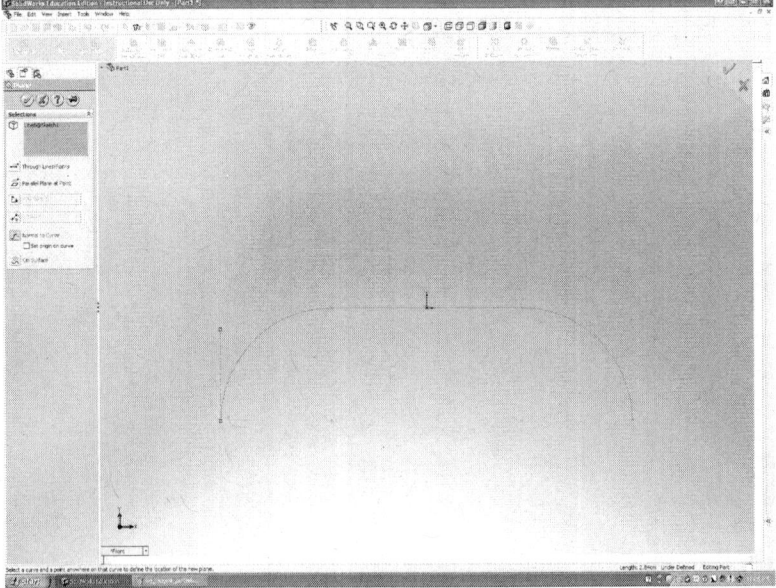

FIGURE 11.31
Adding a reference plane at the tip of a line.

A System of Intelligent Robots-Trained Animals-Humans   323

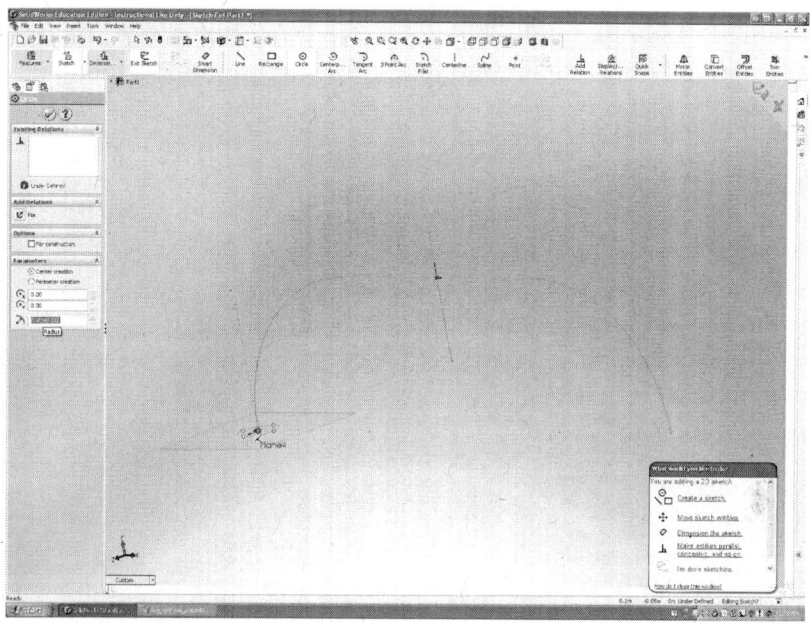

FIGURE 11.32
Circle of a reference plane.

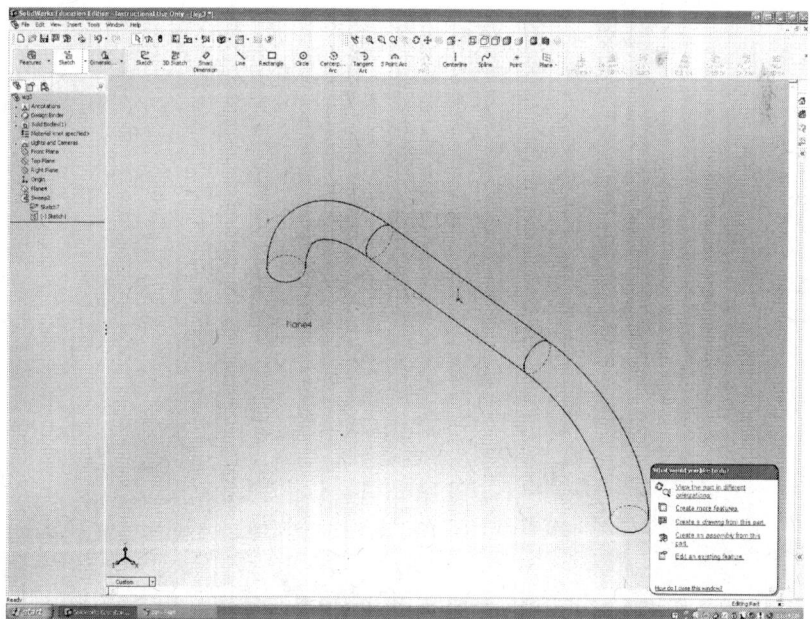

FIGURE 11.33
Finished leg.

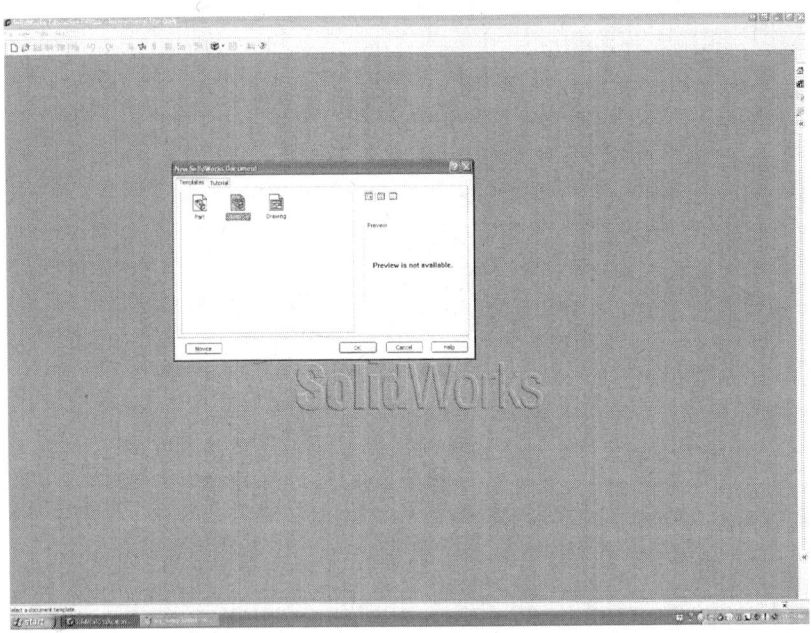

FIGURE 11.34
Components assembly interface.

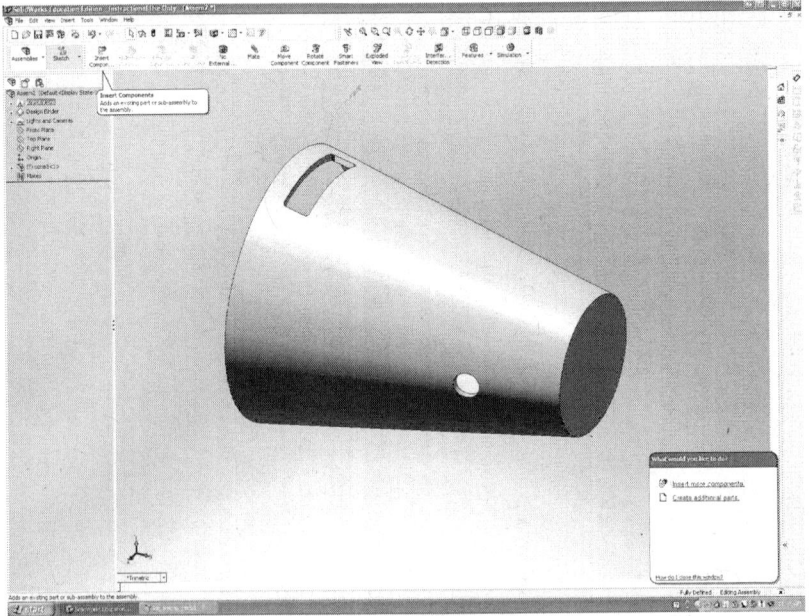

FIGURE 11.35
Adding a new component.

A System of Intelligent Robots-Trained Animals-Humans 325

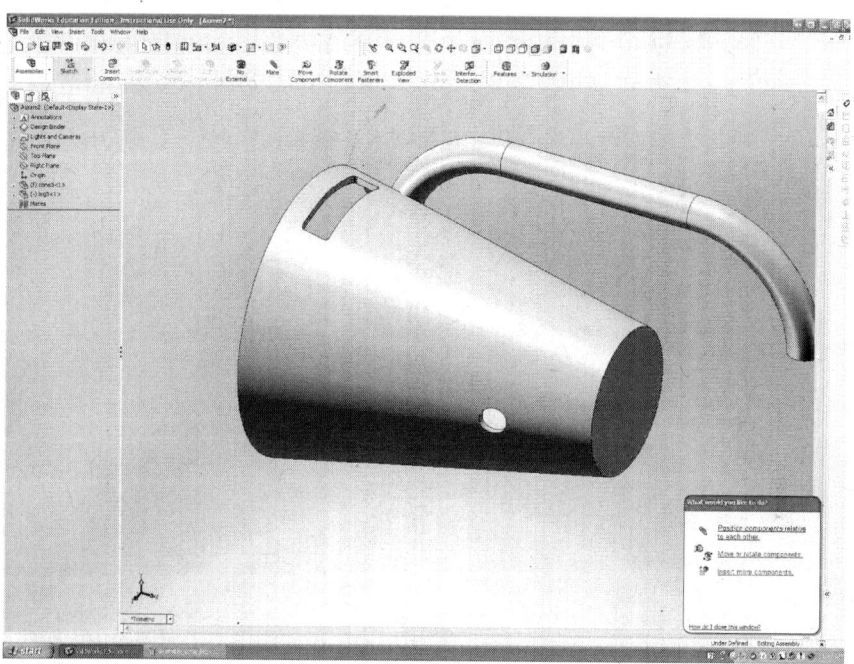

FIGURE 11.36
Leg and the cone in the assembly pool.

When we select geometrical objects to be mated, we should select matching types, for example, cylindrical surfaces, circles, planes, lines, etc. Therefore, we cannot mate a cylindrical surface with a straight line unless we have a tangent line to the surface that can be mated with the straight line (Figure 11.37). You can see how the inner cylindrical surface of the hole on the cone is mated to the outer surface of the center bar of the leg to be "coaxial" as shown in Figure 11.38.

Figure 11.39 shows the result of the mate. However, mating two surfaces does not mean that the two objects are physically constrained as shown. The leg can slide and rotate along the axis that goes through the center of the hole in the cone. To physically constrain the leg in 3-D space, we have to mate a few other geometrical objects in the two components.

To stop the leg from rotating around the axis going through the center of the hole in the cone, we mate the bottom surface of the tip of the leg with the horizontal plane of the cone to be "parallel," as shown in Figure 11.40. To fix the lateral movements of the leg, we mate the vertical plane of the leg with that of the cone to be "coincident."

Next, let discuss how we perform a motion simulation (Figure 11.45). Figure 11.42 shows a cylindrical object that we will use later to make the outer case of a motor. To insert a rotary shaft into it, we have to cut a cylindrical

326   *Intelligent Control Systems*

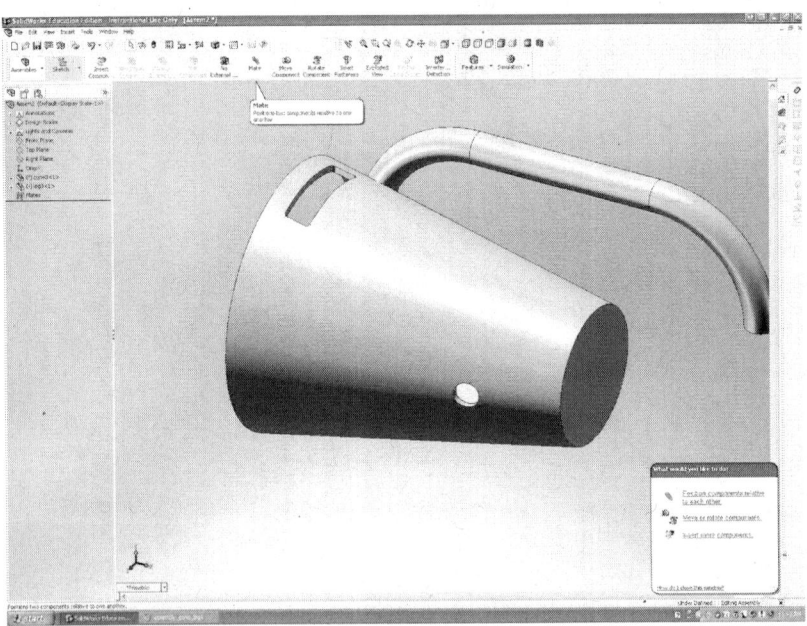

FIGURE 11.37
Mating two components.

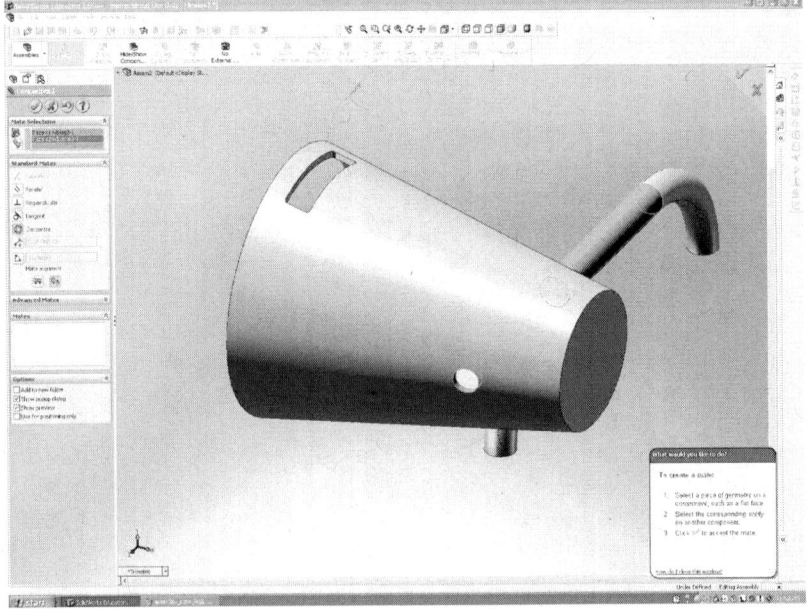

FIGURE 11.38
Selecting two surfaces to be mated.

A System of Intelligent Robots-Trained Animals-Humans 327

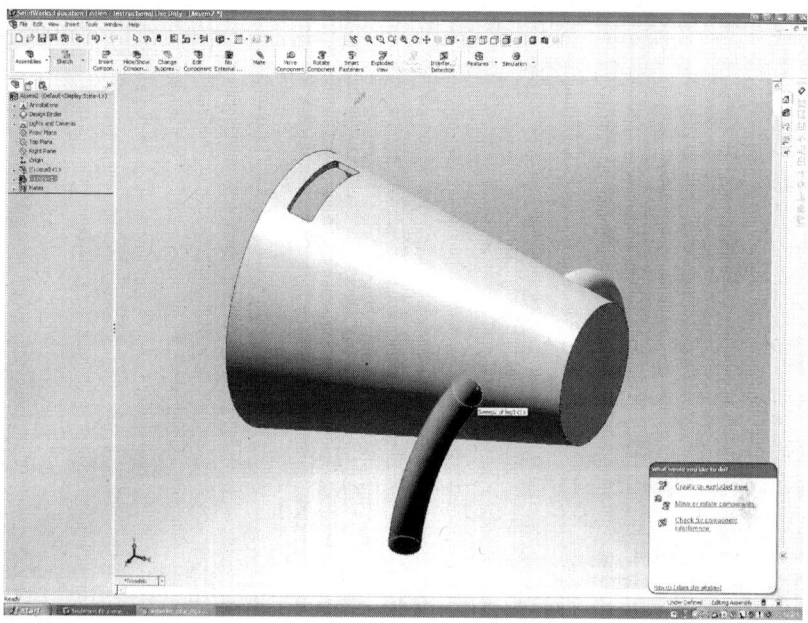

FIGURE 11.39
Result of the mate.

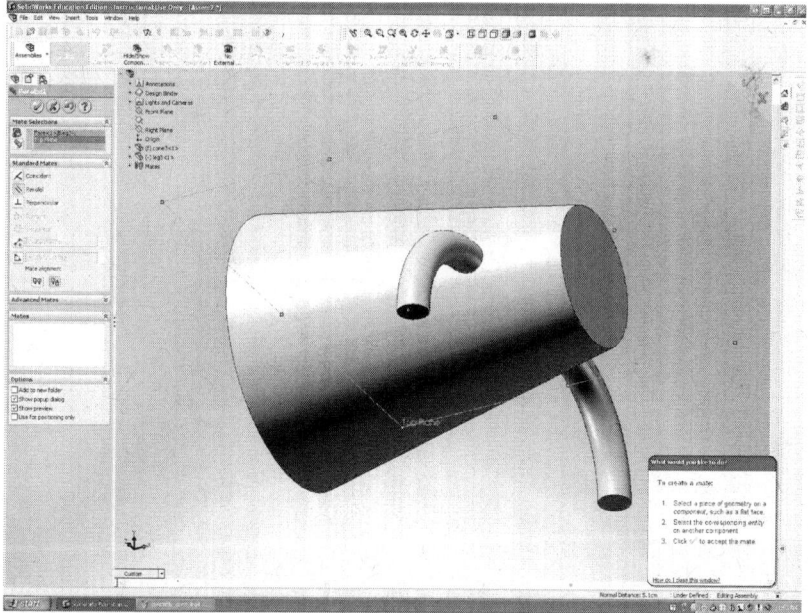

FIGURE 11.40
Mating two surfaces to align components.

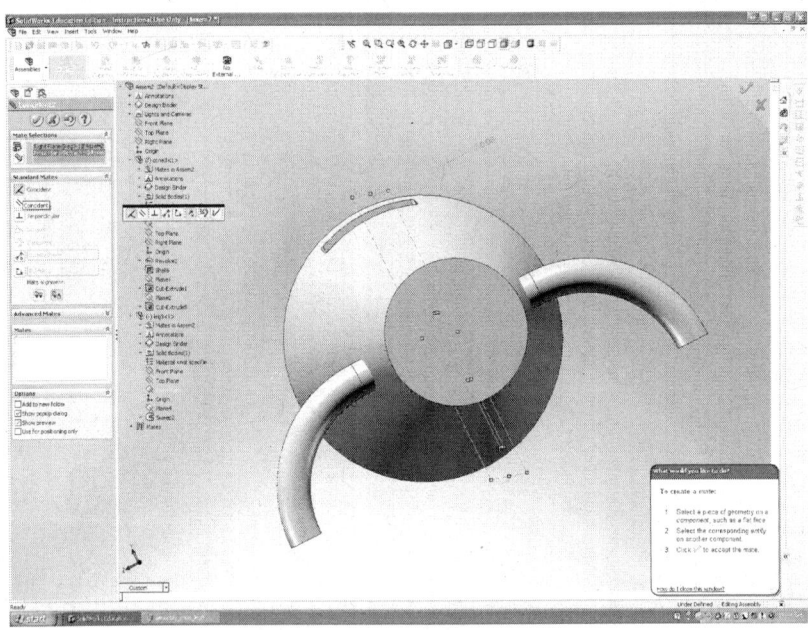

FIGURE 11.41
Fixing the lateral position.

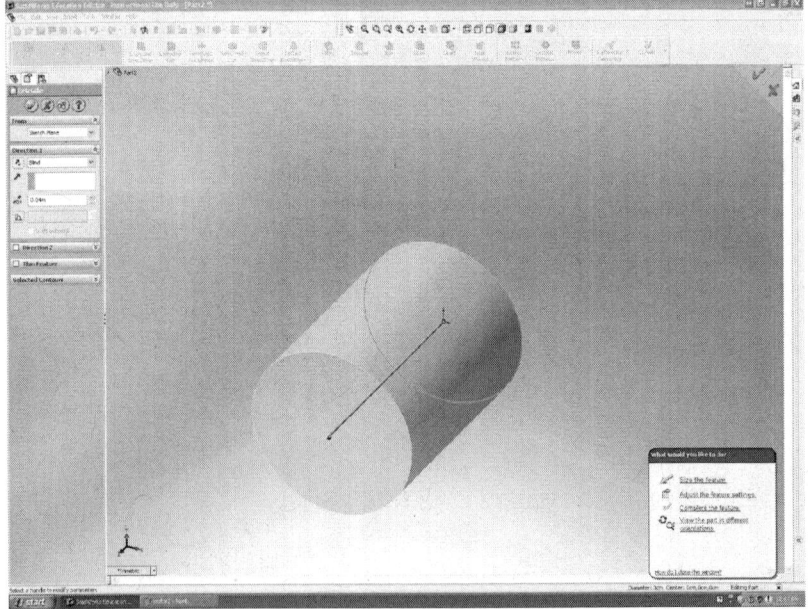

FIGURE 11.42
Extruded circle.

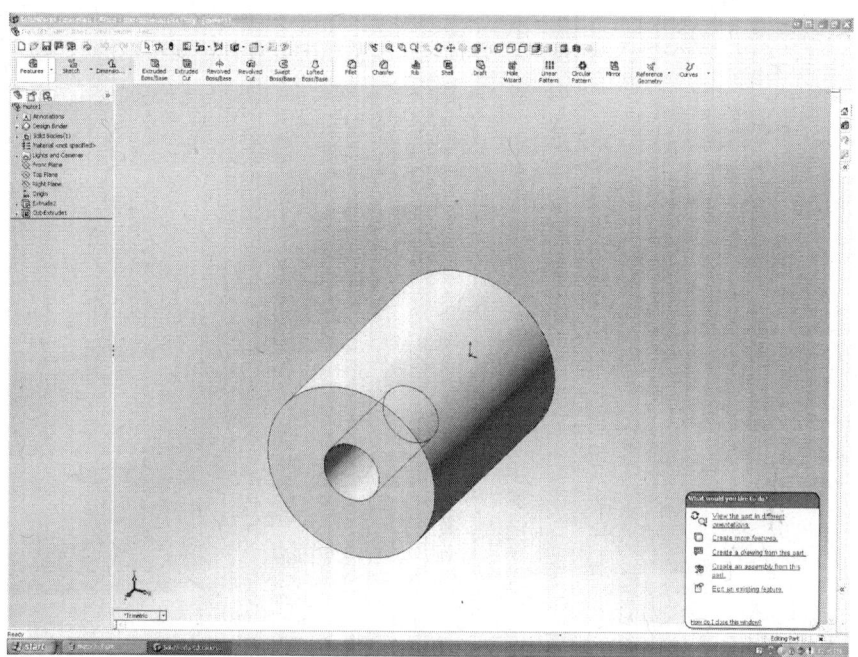

FIGURE 11.43
Extruded cut of a circle.

hollow from the front planer surface of the solid cylinder. Figure 11.43 shows the result of using the extruded cut feature to achieve it.

Figure 11.44 shows the components we have imported to assemble a motor and a base. The result is shown in Figure 11.45. We have skipped the details. Please plan how you would assemble the components using primitive mating features.

Once we have assembled the components, we have to click on the rotary objects on the design tree on the left side and select "float" before simulating. Then, in the "simulation" tab, select the type of motion you need (rotary, linear, etc.) and optionally select the strength and direction of gravity, and then select the object that is subjected to such motion properties. Figure 11.46 shows how we have selected the motor shaft to be a rotary motor.

Figures 11.47 through 11.49 show three sample states of the rotary shaft after we run the simulation by selecting "calculate simulation" in the simulation menu.

We wish to break the Solidworks guide to design the MURALI robot here because the above discussion should give you enough background to continue the rest of the design using your own creativity. The rest of the work will include adding the other half of the robot, as shown on the upper left-hand side of Figure 11.52. What is not shown in Figure 11.52 is an L-shaped

330   *Intelligent Control Systems*

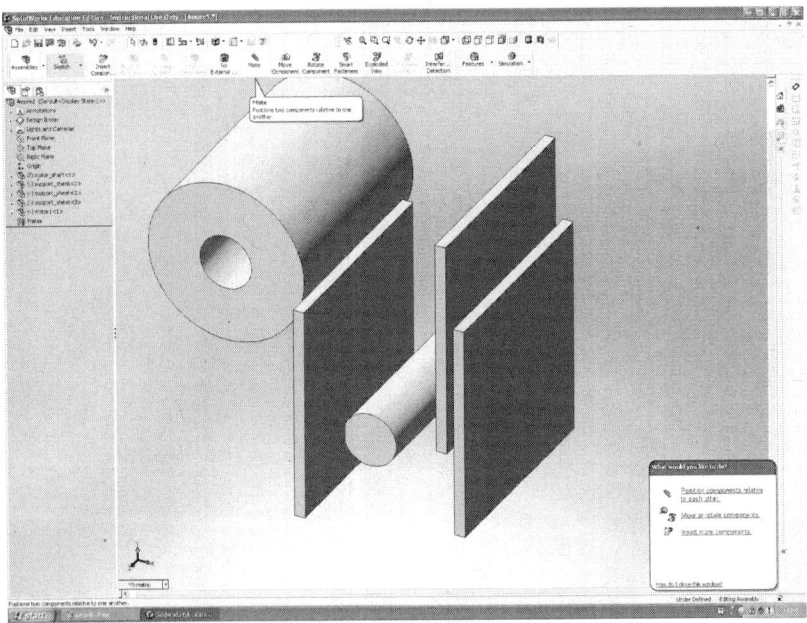

FIGURE 11.44
Components for a motor assembly.

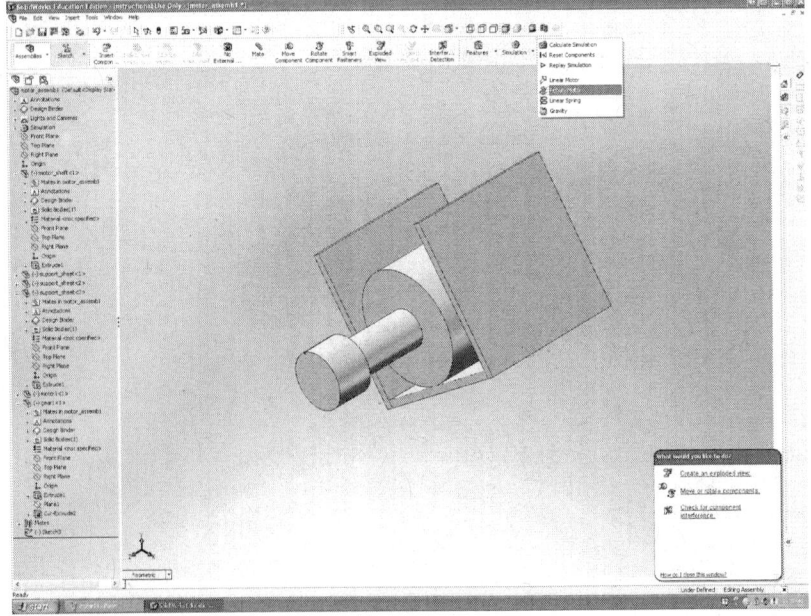

FIGURE 11.45
Starting a rotary motor simulation.

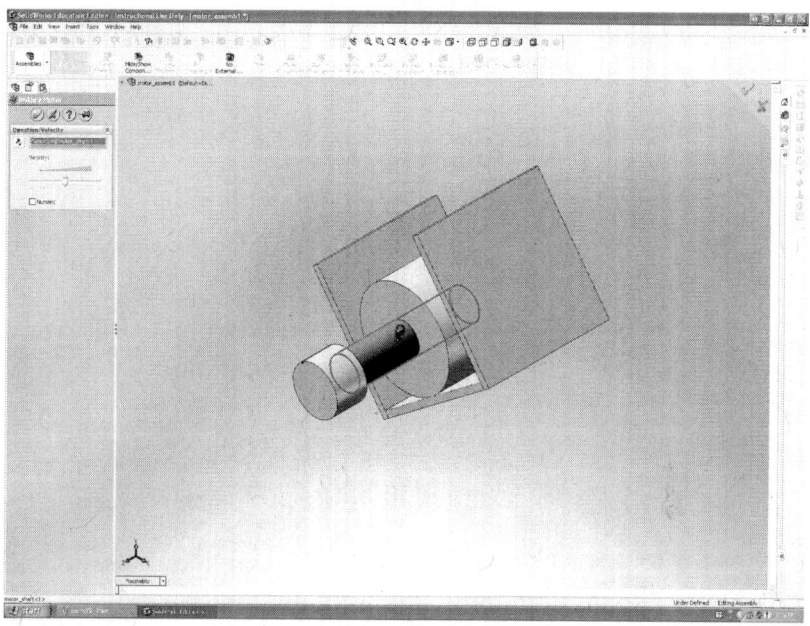

FIGURE 11.46
Selecting the speed and direction of the shaft of the motor.

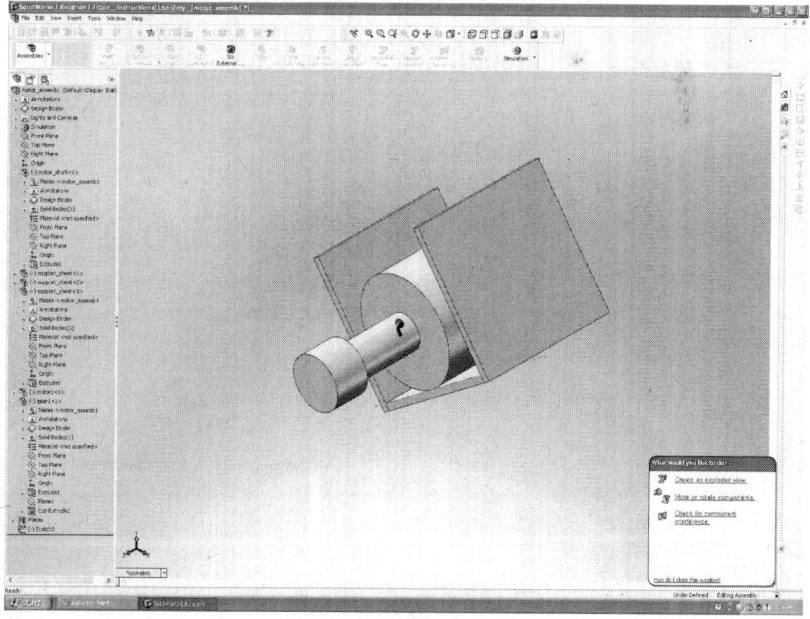

FIGURE 11.47
State 1.

332   *Intelligent Control Systems*

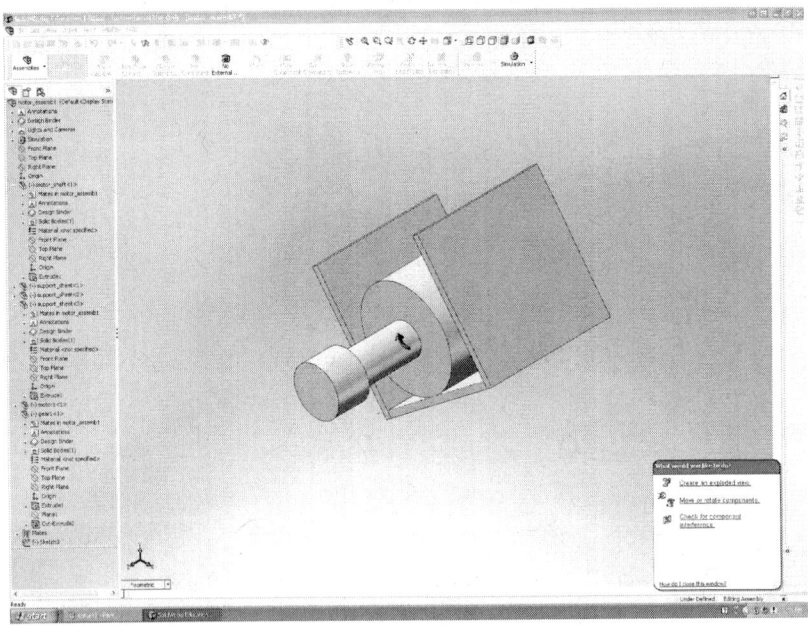

FIGURE 11.48
State 2.

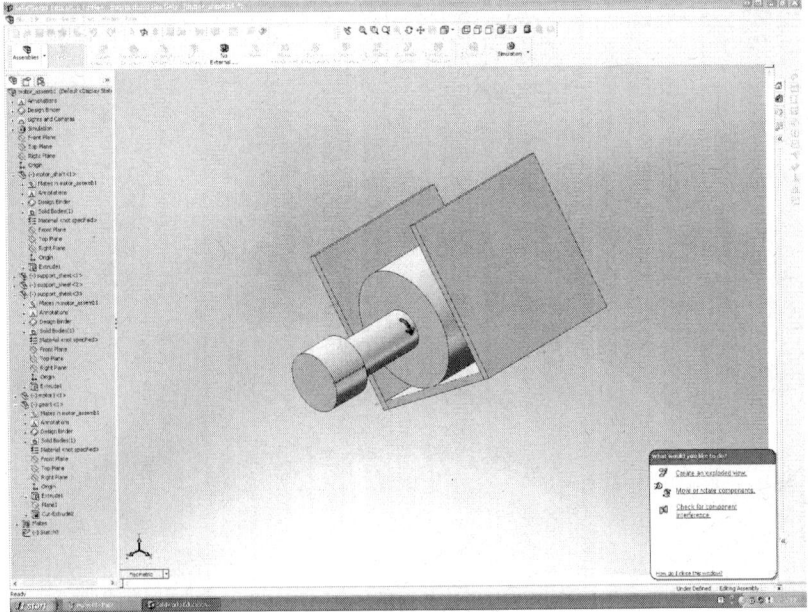

FIGURE 11.49
State 3.

A System of Intelligent Robots-Trained Animals-Humans 333

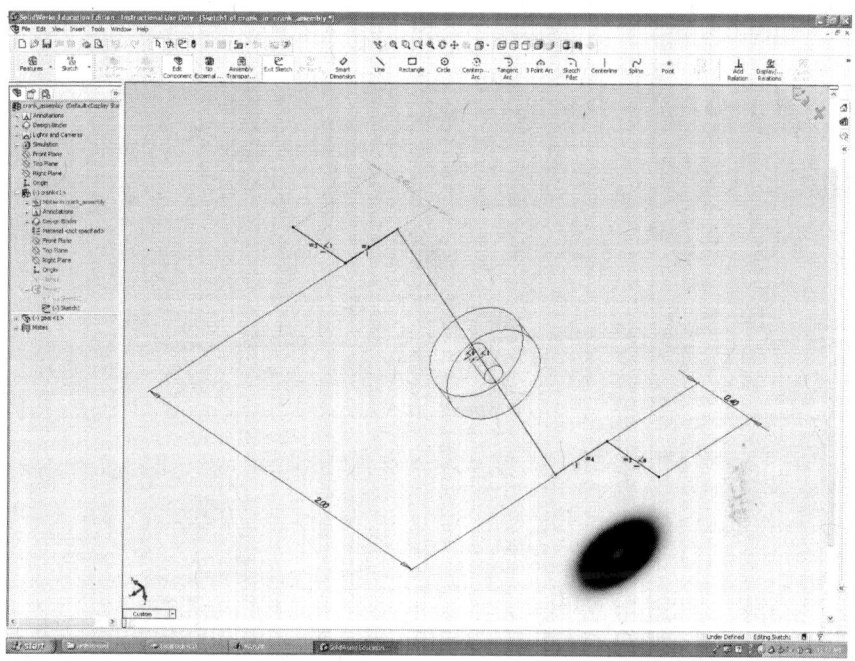

FIGURE 11.50
Skeleton of the crank shaft.

assembly that hooks onto the rectangular slot on the cone near the crank shaft. This is a vital mechanism to restrict the movements of the front cone to give the desired movements of the legs. This part requires a lot of imagination and careful design. We hope you will enjoy doing the rest of the simulations with Solidworks (Figures 11.50 through 11.52).

11.2.4 Motherboard to Control the Robot

In this subsection, we shall discuss some of the important things one should keep in mind when designing a controller for a robot. Figure 11.53 shows the complete circuit diagram of an early version of the robot's motherboard. In the following figures, we will discuss some important submodules. Figure 11.54 shows some essential electronic components used to make the motherboard, such as a microcontroller, crystal oscillator, voltage regulator, etc.

The heart of the motherboard is the microprocessor. In this type of field application, we have to use a rugged microprocessor that can withstand high tropical ambient temperatures, humidity, and dust. Due to the space limitation on board a microrobot, we also need a processor that does not require many interfacing boards to do computations such as analog-to-digital conversion, pulse-width modulation to drive motors, interface analog and digital

334 Intelligent Control Systems

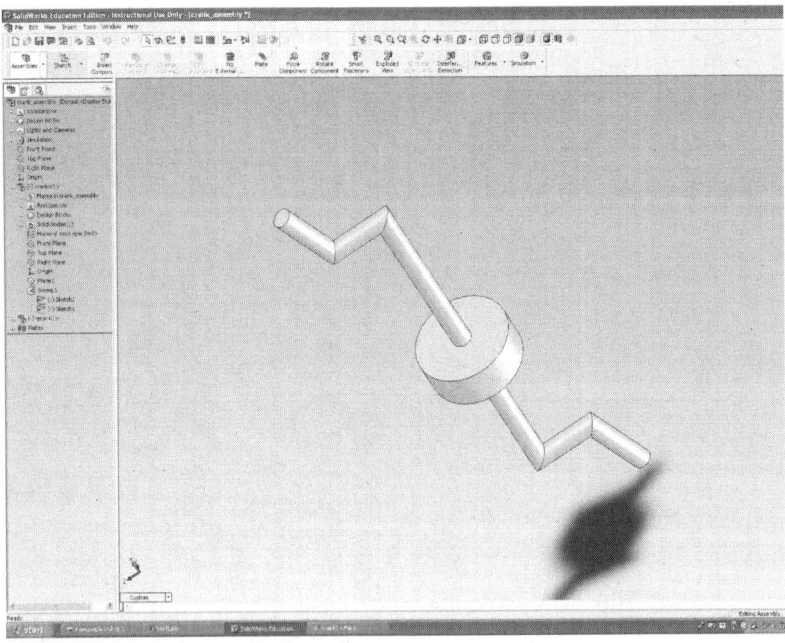

FIGURE 11.51
Crank shaft after swept boss/base.

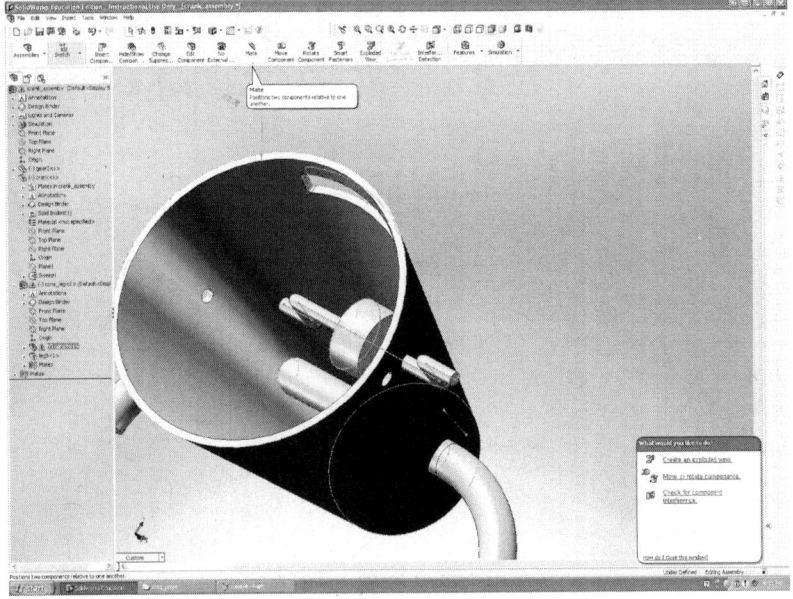

FIGURE 11.52
Crank-body assembly for one half of a robot.

A System of Intelligent Robots-Trained Animals-Humans 335

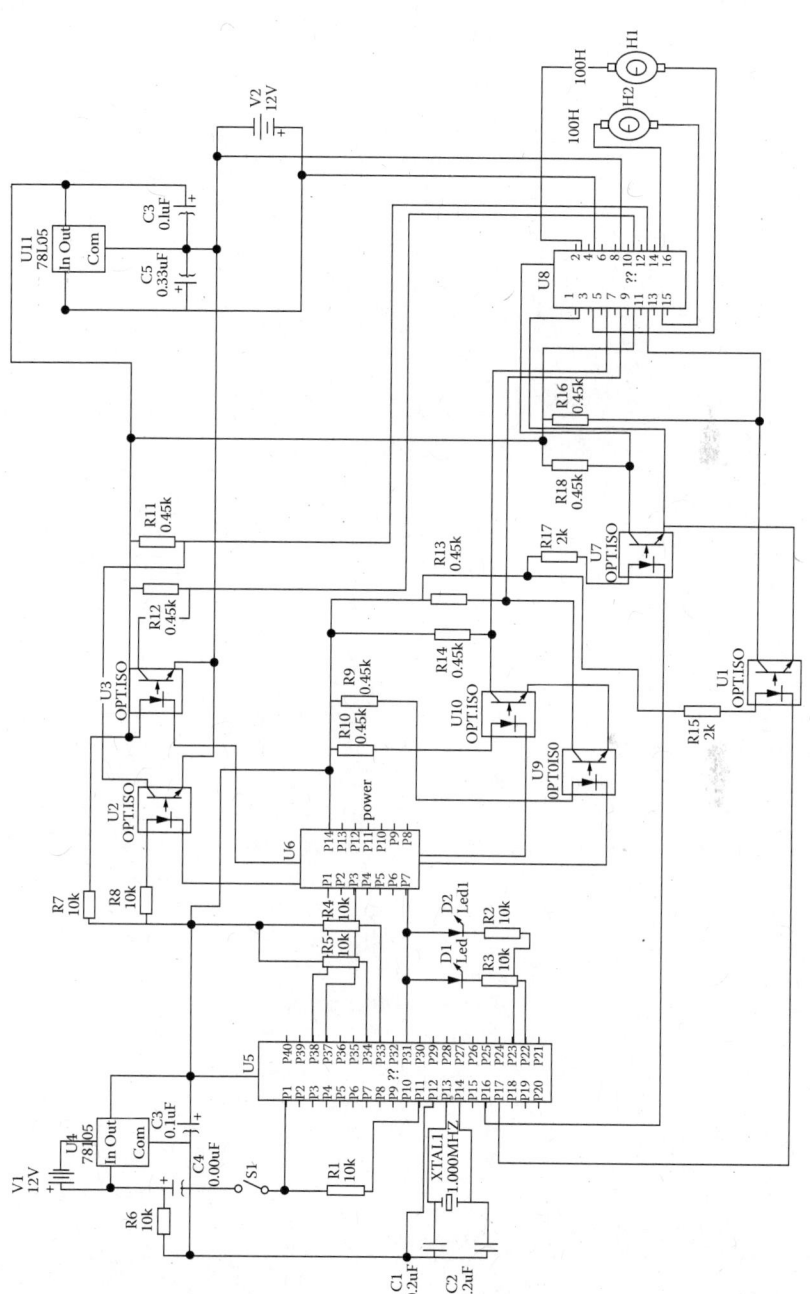

FIGURE 11.53
Motherboard of the robot.

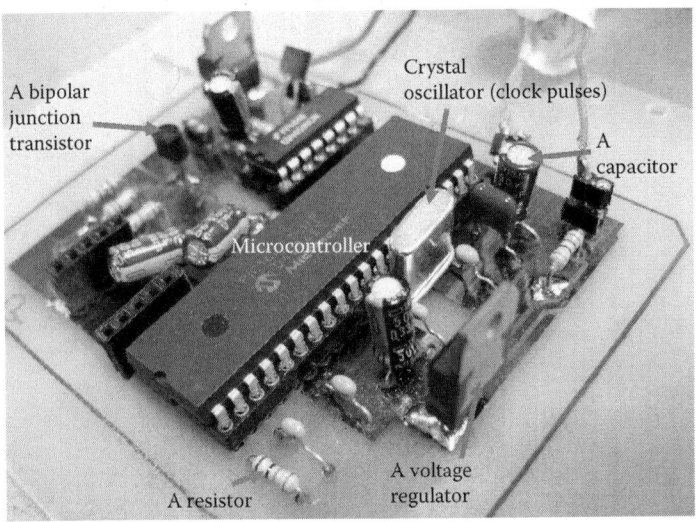

FIGURE 11.54
Few essential circuit components.

sensors, interrupt management, data storage, etc. The special class of microprocessors that can handle all these functions in one chip is called microcontrollers. However, microcontrollers come with very limited processing capacity and onboard memory. There are a few things we have to do before getting a microcontroller to work. First, we have to design a stable power supply. Microcontrollers can restart or malfunction if the power source is noisy. In addition, we should provide protection to the microcontroller against abrupt surges in the voltage signal. As shown in Figure 11.55, the alternating current power supply is first full-wave rectified by a diode bridge. A diode is a semiconductor device that passes current in only one direction.

In Figure 11.55, the current passes only in the direction of the arrow head of the diode. The capacitor next to the rectifier bridge poses low-impedance to high-frequency current components and high-impedance to low-frequency currents. Therefore, all high-frequency ripples will drain through the capacitor, leaving the DC component. The LM7805 is a regulator integrated circuit that takes voltages exceeding 5 V and sends out a stable 5-V signal. Again, a capacitor filters out any high-frequency voltage components before applying the voltage on the microcontroller at the output terminal. This motherboard interfaces sensors such as bumper switches, potentiometers, sonar sensors, and a metal detector, as shown in Figure 11.56.

11.2.5 Basics of Sensing and Perception

Before discussing how we can write software to interface a sensor to a microcontroller, let us first discuss the function of a sensor. A sensor converts some

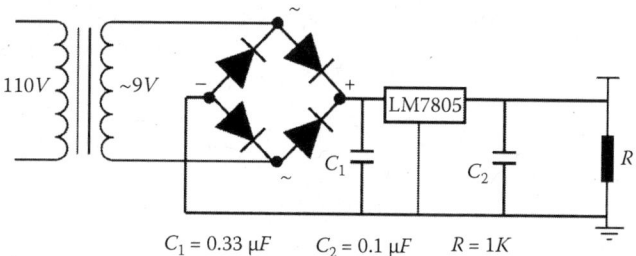

FIGURE 11.55
A circuit of a power supply—values were taken from LM7805 data sheet of Fairchild Semiconductors.

measurand such as acceleration, distance, pressure, chemical concentration, etc., to a transmittable variable as shown in Figure 11.57. It is typically an electric signal. There are several reasons as to why we use an electric signal as the output.

1. Electric signals can be transmitted from one place to another, enabling us to detect the behavior of a measurand from a remote location.
2. Electric signals can be converted to magnetic patterns and stored for later usage. This enables us to analyze the behavior of the measurand offline.
3. Electric signals can be processed. This includes filtering and amplification. This is particularly important for processing very weak signals.

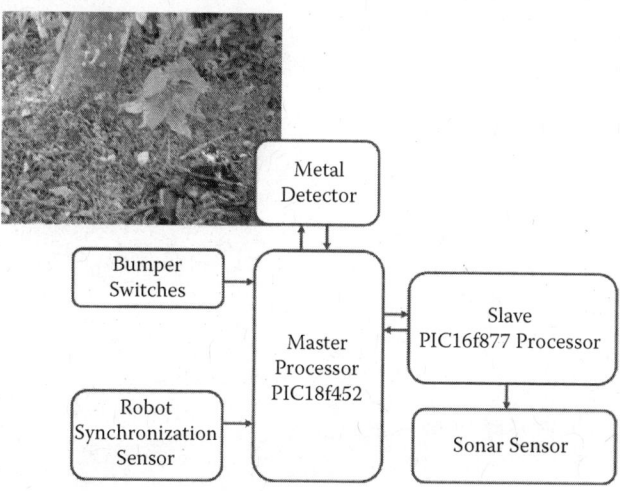

FIGURE 11.56
Sensor interfacing using microcontrollers.

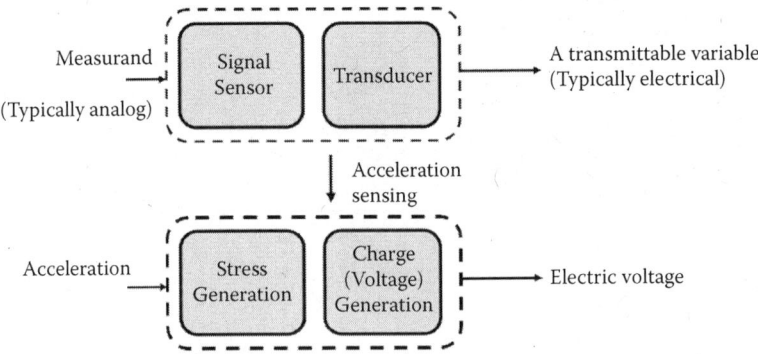

FIGURE 11.57
Function of a sensor.

4. Electric signals can be mixed together. This enables us to perform algebra on different measurands to derive higher-order information. This is similar to how we combine vision, taste, smell, touch, and audition to derive complex perceptions.
5. Electric signals can be digitized. This enables us to interface sensors to microprocessors.

Some sensors give an analog voltage output. These are called analog sensors. Other sensors give out a digital signal. They are called digital sensors. Depending on the output format, we have to use an analog port or a digital port of the microcontroller. For instance, Figure 11.58 shows the basic

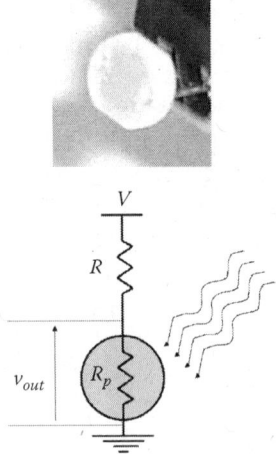

FIGURE 11.58
A photo resistor.

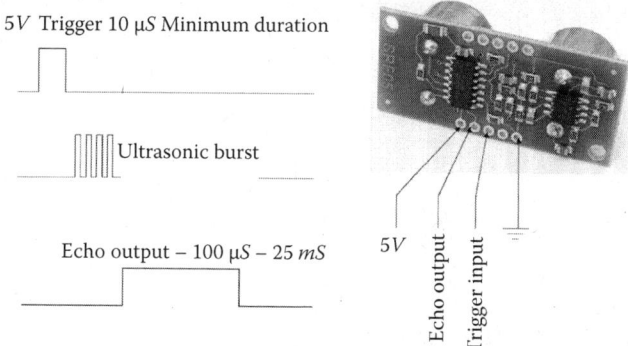

FIGURE 11.59
Devantech SRF05 sonar sensor.

function of a photo resistor. This is the normal resistor except for the fact that it changes resistance in response to background light. Therefore, the voltage across the resistor given by $v_{out} = R_p V / (R+R_p)$ can be used as an analog signal that reflects the background light level. We can use an analog port of a mircrocontroller to interface this type of a sensor.

The Devantech SRF05 sonar proximity sensor we have used in this application is a digital sensor because it gives out a digital signal as shown in Figure 11.59. Whenever we want to measure the proximity to an obstacle, we give a 5-V trigger pulse of more than 10 μS duration. Then, the sensor gives out a sonar wave burst and sets the voltage of the echo output pin to 5 V. We connected this pin to an interrupt pin of our microcontroller and set the interrupt to be sensitive to a rising edge. Therefore, we can keep a record of when the sonar burst was sent out. If there is an obstacle, the reflected wave will be detected by the sensor and set the echo output pin to 0 V. We use the same interrupt pin to detect this falling edge by setting it to be sensitive to a falling edge immediately after it detected the rising edge at the time the sonar burst was sent out. The total time between the rising edge and the falling edge is the time of flight of the sonar burst. Since we know the velocity of sound, we can calculate the distance to the obstacle.

11.2.6 Perception of Environment and Intelligent Path Planning

Perception involves interpretation of sensory signals. In this case, the robot tries to perceive two things. One is the statistical features of the distribution of plants in front of the robot, and the other is the mechanical impedance of obstacles that come in contact with the robot. The former perception is useful for proactive planning and the latter is useful for reactive navigation. We do these using very simple sensors, such as one sonar proximity sensor and a bumper switch, Nanayakkara et. al., 2005).

FIGURE 11.60
The sonar sensor system (left) and the robot (right).

Let us discuss how the robot perceives the statistical properties of the distribution of trees in front of it. First, let us discuss why a robot is better off perceiving the statistical properties of the environment than individual distances to each obstacle. Remember that the robot is in a forest environment. Therefore, the robot is always surrounded by diverse types of plants. Therefore, the conventional way of looking at every little blade of grass will be a futile task. Therefore, a more sensible thing to do would be to get a rough understanding of the sparsity of the distribution of obstacles, such as trees. Even humans tend to adopt this strategy when they move in cluttered environments. For instance, if we enter a room with various objects, we do not measure the distance to every single obstacle. Instead, we make a rough guess of the spatial distribution of objects and plan how to move; it could be changing the direction toward more sparse areas, slowing down when we go through heavily cluttered areas, etc.

The robot does this by swinging the sonar sensor back and forth to collect an array of proximity data as shown in Figures 11.60 and 11.61. Then, it calculates the mean distance to obstacles, the direction of the closest obstacle, and the variance in the distances. Then, we interpret these summary statistics in different classes of environments inside a forest. Figures 11.62 through 11.66 show five such classes.

Table 11.1 summarizes the interpretation given to the above classes of environments. Table 11.2 summarizes the statistical features in the respective classes of environments.

Then, we generalize these statistics using fuzzy membership functions as shown in Figure 11.67. Generalization of discrete observations helps the robot to classify any other environment to the closest class of known environments, which is done using a fuzzy inference engine shown in Figure 11.66.

A System of Intelligent Robots-Trained Animals-Humans

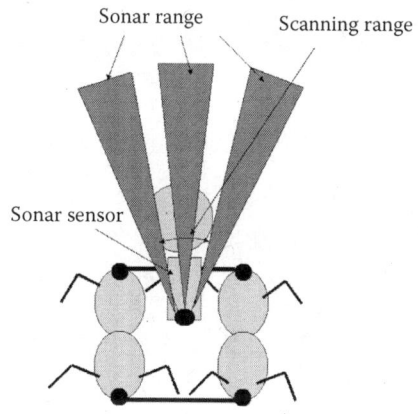

FIGURE 11.61
Scanning action of the sonar sensor.

The fuzzy sets are given by

$$\mu_S = \begin{cases} 1 & : \mu \leq 1 \\ -0.67\mu + 1.67 & : 1 \leq \mu \leq 2.5 \end{cases}$$

$$\mu_M = \begin{cases} 0.67\mu + 0.67 & : 1 \leq \mu \leq 2.5 \\ 0.67\mu + 2.68 & : 2.5 \leq \mu \leq 4 \end{cases}$$

$$\mu_L = \begin{cases} 0.67\mu - 1.67 & : 2.5 \leq \mu \leq 4 \\ 1 & : 4 \leq \mu \end{cases}$$

FIGURE 11.62
Situation 1 characterized by a large tree surrounded by few small plants spread with a mean distance of 2.3 m, and a variance/mean ratio of 4.3.

FIGURE 11.63
Situation 2 characterized by an open area between the robot and a cluster of densely located trees spread with a mean distance of 3.4 m and a variance/mean ratio of 2.8.

FIGURE 11.64
Situation 3 characterized by a large tree compared to the size of the robot with a mean distance of 0.53 m and a variance/mean ratio of 4.3.

FIGURE 11.65
Situation 4 characterized by an open space with trees located sufficiently far away from the robot (>4 m). In this case the variance was reset to zero.

A System of Intelligent Robots-Trained Animals-Humans 343

FIGURE 11.66
Situation 5 characterized by a cluster of sparsely distributed trees located with a mean distance of 1.9 m and a variance/mean ratio of 5.5.

TABLE 11.1
Interpretations of Environments

Situation	Interpretation and the Behavior Recommendation
Situation 1	There is a large tree at a medium distance from the robot. The rest of the environment is fairly clear except for few plants over which the robot can go. Therefore, it is recommended to navigate with a reduced speed directly toward the tree with the hope of skirting around it.
Situation 2	There is a fairly comfortable space before the robot will encounter a densely located cluster of trees. Therefore, it is recommended to navigate with a reduced speed while scanning the environment to locate a gap between the trees.
Situation 3	The robot has come very close to a large obstacle. It is recommended to start a skirting behavior with the help of tactile sensors.
Situation 4	There is an open field. Navigate with normal speed.
Situation 5	There is a sparsely distributed cluster of trees. Navigate with the normal speed unless it comes closer to a tree.

TABLE 11.2
Statistical Properties of the Distance from the Robot to the Trees in the Respective Environments When Scanned Using a Sonar Sensor

Situation	Statistics	
	Mean	Variance/Mean
Situation 1	2.3	4.3
Situation 2	3.4	2.8
Situation 3	0.53	4.3
Situation 4	>4	0
Situation 5	1.9	5.5

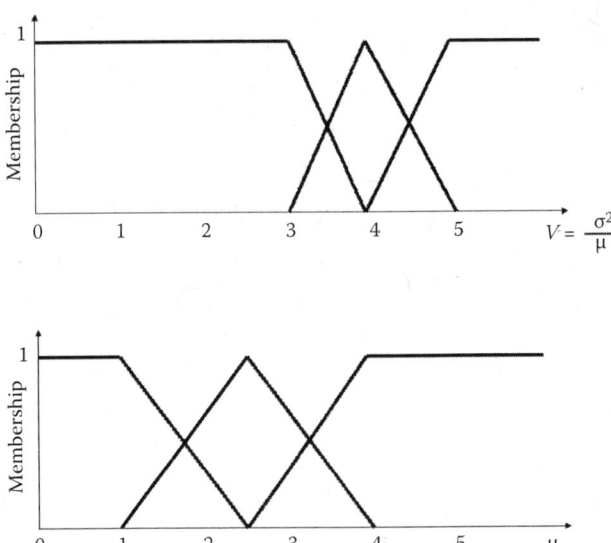

FIGURE 11.67
Linguistic labels (fuzzy sets) for the mean distance and ratio between the variance and the mean. The shape and location of the labels were derived from the experimental data given in Table 11.2.

Similarly, for the ratio between the variance and the mean, the fuzzy sets are given by

$$V_S = \begin{cases} 1 & : \mu \leq 2 \\ -\mu+3 & : 3 \leq \mu \leq 4 \end{cases}$$

$$V_M = \begin{cases} \mu-2 & : 3 \leq \mu \leq 4 \\ \mu+4 & : 4 \leq \mu \leq 5 \end{cases}$$

$$V_L = \begin{cases} \mu-3 & : 4 \leq \mu \leq 5 \\ 1 & : 5 \leq \mu \end{cases}$$

In addition to the information shown in Figure 11.68, the algorithm uses the angle to the closest obstacle in the process of mapping a new environment to one of the five (behavior, confidence) pairs. However, it should be noted here that this behavior planning is providing only a higher-level guidance to the robot. Detailed maneuvering is done using the tactile sensors mounted on the robot. Since the robot encounters obstacles from all directions in a forest environment, raw tactile information does not give any useful information. On the other hand, we wanted to minimize onboard signal processing because that would increase the complexity and cost of the robot.

A System of Intelligent Robots-Trained Animals-Humans 345

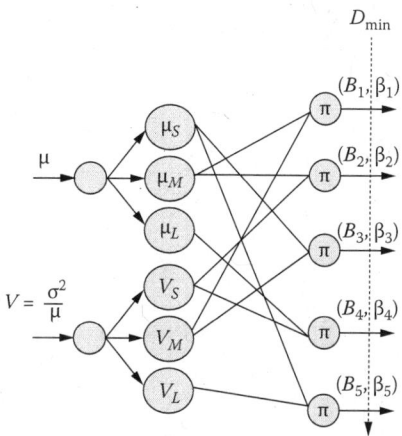

FIGURE 11.68
A simple fuzzy neural network to associate the statistical properties of the distance from the robot to the trees in the respective environments when scanned using a sonar sensor, with a set of behaviors.

Therefore, we designed a mechanical filter that would filter out obstacles with low mechanical impedance, such as grass, tree leaves, etc., and give a signal only for those high-impedance obstacles, such as tree trunks. This can be manifested simply by mounting a set of bumper switches behind a spring-loaded buffer as shown in Figure 11.69.

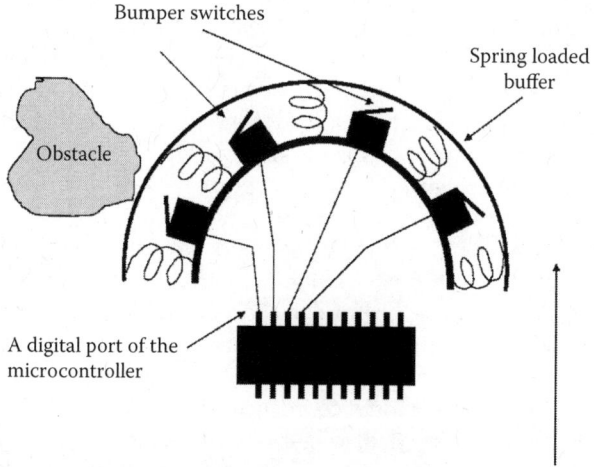

FIGURE 11.69
Mechanical impedance filter, manifested by mounting bumper switches beneath a spring loaded buffer.

11.3 Combining a Trained Animal with the Robot

Now, let us discuss how we could combine the strengths of the robot described above with a trained mongoose, which has very powerful olfactory capacity, so that a remote human operator can control the rodent through the robot. This reduces the risk to the humans who would otherwise be employed in the minefield. It also couses no damage to the sensative vegetation because both the robot and the mongoose can burrow through the vegetation without having to cut trees. Furthermore, neither the robot nor the rodent will set off a mine because their weight is too low to do so.

The rodent. None of the existing chemical sensors are sensitive enough to find smell gradients leading to a source. However, the primary survival strength of rodents such as mongooses or rats is their powerful olfactory system, which can find fine smell gradients leading to chemical sources. By carefully selecting the right rodent to suit a given environment, we expect to exploit the extremely powerful olfaction capabilities of these animals to search for target chemicals along smell gradients. The rodent is attached to a semiautonomous field robot through an elastic cord with an angle sensor at the robot's end. The angle sensor will provide information about the rodent's wandering behaviors. Subjected to the final recommendation of the remote human operator, some of the rodent's behaviors will help the robot to learn how to move in a forest environment in an elegant manner. This will sharpen the behavior of the robot and improve the efficiency of using inexpensive sensors for the robot to move in a cluttered environment.

The architecture and the role of the semiautonomous field robot. The primary role of the robot is to guide the rodent along a desired path. There will be a mechanism to keep the rodent informed about the orientation and dimensions of the robot. For example, the rodent should not try to creep through gaps through which the robot cannot move. This will improve the capabilities of the rodent to cooperate with the robot. The main design goal will be to enable the robot to be efficient in soft terrain conditions such as mud, grass, and sand.

The role of the remote human operator. The remote human operator can use the visual feedback sent by the robot and return control commands to augment the autonomous decisions taken by the robot and the rodent. For instance, if the rodent tries to move through a gap that is too narrow for the robot to move through, the human operator can send a command to the robot to stop and pull the rodent back and guide it to go through another gap. In this case, the human control command suppresses the behavior recommended by the rodent, and it will help the rodent to improve its capability to think as a system. These augmentations will also help the robot and the rodent to improve their capabilities to navigate efficiently. Consider another situation, where the rodent consistently pulls toward a particular spot where a target chemical is suspected to be located; the human operator can send commands to the robot to be compliant with the rodent. Depending on the success rate, the human can improve his/her capability to judge the rodent's behaviors when it comes into contact with various chemical agents of interest to it.

11.3.1 Basic Background on Animal–Robot Interaction

Animal-robot interaction. Experiments have been conducted to observe how a chicken behaves when a robot elicits different behaviors in the same cage (Bohlen, 1999). It has been found that the chickens react to the relative velocities and accelerations of the robot, the sounds it generates, and how close it comes to them. Sometimes, animals attach value to colors. Experiments done on male sticklebacks have shown that any odd-shaped thing with a red bottom will make them think that it is a threat to them because stickleback males show a red belly during courtship season to differentiate them from females (De Schutter et al., 2001). Experiments conducted on miniature robots working in societies of cockroaches show that, once the insects accept the robot to be part of their society, the robot can influence some of the collective behaviors of the animals, such as foraging under a shade, moving as a group, etc. (Caprari et al., 2005). An algorithm to accelerate reward-based training of rats is proposed by Ishii et al. (2005), in which a rat learns to press the levers of a robot to obtain food and water. It is observed that the training process should be phased out in order to accelerate training. In this case, the first phase was to remove the anxiety of the rat to face the robot, the next phase was to reward the rat for coming closer the robot, and the last phase was to motivate the rat to press the appropriate levers on the robot to obtain food and water. Based on the work done so far on understanding how different animals react to the presence of robots, and how an animal can be trained to interact with a robot, we expect to investigate the features of the robot's appearance and behaviors in order to make the best use of the animal's natural sensory system. A major novelty of the proposed study is that the robot does not duplicate any hardware needed to sense the environment wherever the animal can do it better. Hence, the robot becomes simpler and cheaper. Furthermore, the robot learns from the animal how to move in a cluttered environment while restricting the animal's movements to a given area of interest.

In the area of forcing an animal to behave in a given manner, some work has been done on invasive techniques, such as invoking behaviors through artificial stimulation of the animal's nervous system. A biorobotic system has been explained where a cockroach is electrically stimulated to turn left and right to keep it on a black strip (Holzer and Shimoyama, 1997). A similar attempt to guide a rat along a desired path is described by Talwar et al. (2002), where electrodes were planted in the rat's somatosensory cortical whisker representations to give sensory cues and the medial forebrain bundles to give the rewards. The rat was guided along a given path using a wireless data communication system between a backpack mounted on the rat and a remote supervisor. However, it is our belief that invasive techniques should be avoided wherever possible in order to make the best use of the animal's natural sensory system.

Characterizing the environment. An image-based 3-D map-building approach for forest environments has been proposed by Forsman and Halme (2005). Yet, just how far this can be applied in a minefield is questionable because the robot has to recognize the environment in the first trial. The sonar and laser range finder-based map building and robot localization have been extensively studied in recent history, especially in indoor environments (Thrun et al., 1998; Burgard, 1998; Ayache and Faugeras, 1989; Bozma and Kuc, 1991; Chong and Kleeman, 1996; Dudek et al., 1996; Gonzalez, 1992). In most cases, a sonar sensor is used to alert the laser range finder to look for finer details. In a forest or a densely wooded environment or in a highly congested area, the sonar will always give alerts. Therefore, these methods may not be as good as having an animal to signal the robot to change its path to avoid an obstacle.

11.3.2 Background on Reward-Based Learning

The plasticity of an animal's brain can be effectively used to develop a motivation to carry out new tasks. Basically, the animal is repeatedly given a reward whenever it carries out a desired task. This type of learning is known as reward-based learning or reinforcement-based learning (Ayache and Faugeras, 1989; Bozma and Kuc, 1991; Chong and Kleeman, 1996; Dudek et al., 1996; Gonzalez, 1992). Experimental results on how the activity of the dopamine neurons in the brain is related to reward-based learning suggests that the brain tries to give priority to actions that maximize the total expected rewards given by Equation (11.1) where $V(t)$ is the total expected reward, $E(\bullet)$ is the expectation function, $r(t)$ is the reward at time t, and γ is the discounting rate.

$$V(t) = E(r(t) + \gamma r(t+1) + \gamma^2 r(t+2) + \cdots) \tag{11.1}$$

In machine learning, a control policy π is updated using a critic that estimates $V(t)$ given a situation s, and a policy π. The policy is defined as $\pi: s \to a$, where s is the situation and a is the action. Therefore, it is clear that the speed and accuracy of the process of improving the control policy depends on the accuracy of the critic that estimates $V(t)$ given in Equation (11.1).

At present, the mongooses are trained manually. The training involves bringing a stick closer to the cage where a mongoose is kept. The stick is presented 40 times a day. The number of times a stick with explosives is presented varied across training sessions. If the mongoose comes to sniff the stick with explosives, a slice of cheese is presented, accompanied by a tone. The process of presenting sticks will be automated so that different training protocols can be programmed with minimum labor involvement. The preliminary designs were proposed in the final year project report (Watkins, 1989).

11.3.3 Experience with Training a Mongoose

11.3.3.1 *Phase 1: Conditioning Smell, Reward, and Sound*

The objective of phase 1 was to study the general response of the mongoose to the exposure of explosives, a food reward, and a sound. A small amount

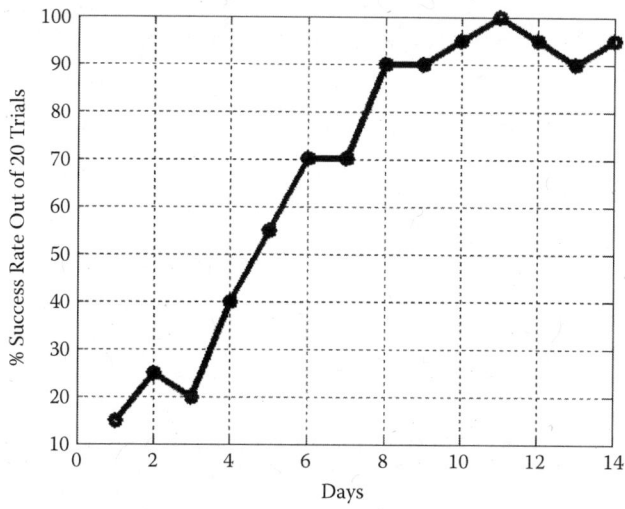

FIGURE 11.70
Learning curve of the first phase. The graph shows the percentage of success rate out of 20 trials a day.

of C4 explosives (<1 g) wound to one end of a stick was brought close to the mongoose. If the mongoose came closer to the stick, a "beep" sound was given with a slice of cheese as the reward. This was repeated 20 times a day. Figure 11.70 shows how the mongoose improved during the experiment.

11.3.3.2 Phase 2: Learning in a Paradigm Where the Degree of Difficulty of Correct Classification Was Progressively Increased

In the second phase, 40 trials were tested each day. The number of times a stick with explosives was presented decreased from 26 to 9 out of 40 trials. The cage was covered with a black polyester cloth to remove any biases due to visual cues. Therefore, in this paradigm, the difficulty of making a correct detection increased over time. Despite this progressively growing difficulty, the mongoose was able to improve the detection capability to near perfect levels within 2 weeks, as shown in Figure 11.71. To the best of our knowledge, this is the first time that such quantified data characterizing the learning curve of the mongoose has been presented.

Figure 11.72 shows the odor-guided target localization behavior of a trained mongoose. The fact that each path has different curvatures suggests that the odor gradient-seeking behavior may depend on factors other than distance to the target.

Images shown in Figure 11.73 show how the mongoose was connected to the robot. The link could swivel around the axis of the potentiometer as shown in Figure 11.74. This arrangement enabled the robot to sense the

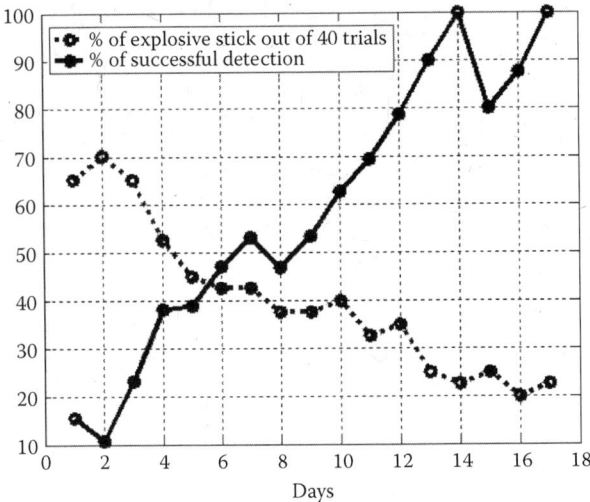

FIGURE 11.71
Learning curve of the second phase. The graph shows the percentage of success rate out of 40 trials a day.

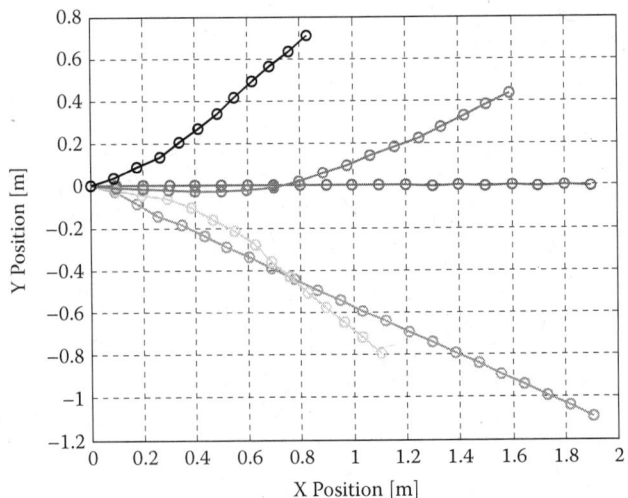

FIGURE 11.72
Path taken by the animal to locate the buried explosives.

A System of Intelligent Robots-Trained Animals-Humans 351

FIGURE 11.73
Mongoose-robot combined system.

FIGURE 11.74
Angle sensor mounted on the robot's side to sense the error between the direction of movement of the robot and the mongoose.

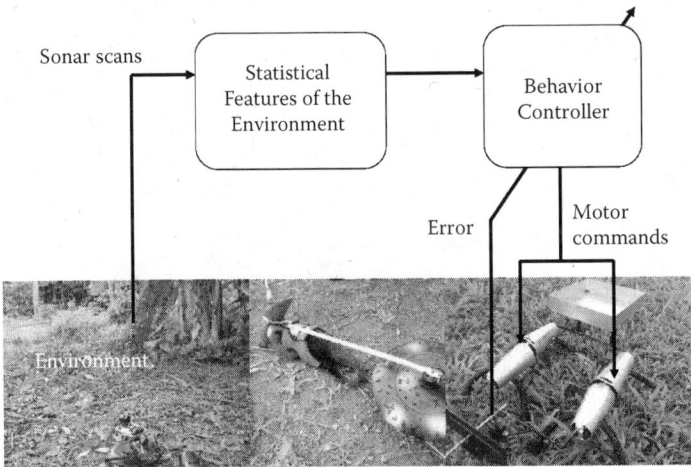

FIGURE 11.75
Error feedback based behavior learning paradigm.

"error" between the two directions taken by the mongoose and the robot. This error signal could be used for many purposes:

1. *To learn how to navigate in an unstructured environment.* We know that rodents like mongooses are very elegant navigators in forest environments. On the other hand, it is very difficult to build a set of rules in the brain of a robot to move in a forest as efficiently as a mongoose. Therefore, it would be best for the robot to compare its own behaviors with those of the rodent and improve its own internal representations of locomotion in a forest. Figure 11.75 shows how this can be done.

2. *To detect a consistent movement toward an odor source.* The biggest advantage of using a rodent with a powerful olfactory capacity is that it can sense odor gradients leading to the source. In the case of landmines, this happens due to leakage of explosives into the surrounding soil. Due to moisture, these explosive traces move in the soil, making an odor profile with a peak concentration around the landmine. We found out that mongooses can start from minute concentrations of explosive molecules and trace their way toward to the mine. Therefore, the angle sensor mounted on the robot can give vital information as to the mongoose's behavior when it has found such an odor trace.

3. *To tame the mongoose.* In the early stages of training, it is hard to tame and discipline the mongoose to move along set tracks. Usually, demining is done by dividing a given block of land into several straight

strips about 1 m wide because we know by experience that mines are buried roughly 1 m apart. Therefore, it is very important that the robot-mongoose duo travel along these tracks to make sure the best land coverage is done. The robot could sense distractions of the mongoose and, with the help of a remote human operator, who would give a second opinion, request the mongoose to come back on track. This can be done by actuating the same rod that connects the mongoose to the robot.

Figure 11.26 shows the proposed error feedback-based behavior learning paradigm. It works by comparing the behaviors recommended by the internal models of the robot with the actual decisions taken by the rodent. It proved to be a very simple but powerful mechanism. One of the most crucial problems faced by many learning algorithms for outdoor robot navigation is the difficulty to construct a cost function to be minimized, due to the unstructured nature of the natural environments. For instance, even a very simple task such as obstacle avoidance expands to a daunting coding problem if the robot is required to move through trees of a forest because every plant is virtually an obstacle.

11.4 Simulations on Multirobot Approaches to Landmine Detection

Swarm robotics is a related area of research, where a complex collective behavior is emerged in a group of relatively simple robots through interrobot and robot-environment interactions (Goldberg and Matarić, 1997; Hayes, 2002; Jones, and Matarić, 2003; Krieger and Billeter, 2000). A given robot is generally very simple and inexpensive, capable of eliciting a limited array of primitive behaviors, and equipped with a limited number of sensors and actuators. Given a task to be accomplished, such as walking over a gap that none of the individual robots can accomplish, the simple robots may share their diverse sensor information, actively support each other by joining hands, and coordinate movements to achieve the common goal. This concept can be useful in a task to find a hazardous object in a cluttered environment where one robot cannot carry all the required sensors. Yet, in that case, there is nothing that improves the behavior of an individual robot. The focus is to emerge a collective behavior based on the rules that apply to each robot to interact with other robots and the environment, and not to improve individual robotic behaviors through peer interaction. To the best of our knowledge, there has been no work done to study a scenario where each robot is directly influenced by a real animal as a peer who can sniff for the target objects and, at the same time, support the robot to learn how to navigate in a cluttered environment.

FIGURE 11.76
A typical minefield in the affected northern regions of Sri Lanka (courtesy of the United Nations Development Program).

Antipersonnel landmines are buried to protect a boundary of a military establishment to maim intruders. Inspections in the minefields in the north and the east of Sri Lanka helped us to understand that landmines are buried according to certain patterns. One such commonly used pattern is shown in Figure 11.76. Three lines of mines are buried parallel to an FDL. In a given line, mines are buried in a T-shaped pattern. This pattern can be modeled by three parameters and their variances given by Equation (11.2).

$$P_i^m = R \begin{bmatrix} N((C_1/2)i, \sigma_{C_1}^2) \\ N((C_2/2)(1-(-1)^i), \sigma_{C_2}^2) \end{bmatrix}$$

$$R = \begin{pmatrix} \cos\alpha & -\sin\alpha \\ \sin\alpha & \cos\alpha \end{pmatrix}, \alpha = N(\alpha^*, \sigma_\alpha^2)$$

(11.2)

where P_i^m is the coordinates of the ith mine, α and σ_α are the angle between the FDL and the line of mines and its variance, α^* is the expected value of that angle, C_1 and σ_{C_1} are the horizontal distance between two mines and its variance, and C_2 and σ_{C_2} are the distance between the top row and the bottom

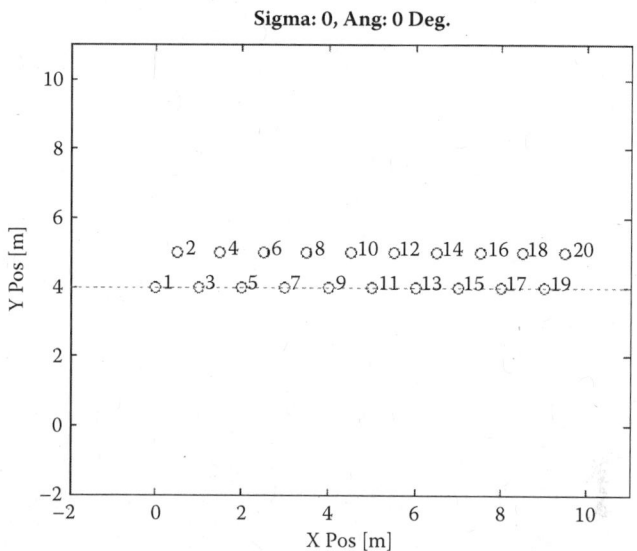

FIGURE 11.77
Distribution of landmines in the case of $\alpha = 0$, $C_1 = C_2 = 1$, $\sigma_{C_1} = \sigma_{C_2} = \sigma_\alpha = 0$.

row and its variance. The uncertainty of the locations of mines parameterized by σ_{C_1} and σ_{C_2} stems from natural disturbances such as rain, tree roots, mud slides, etc. The uncertainty in the angle of the mine front quantified by σ_α comes from the loss of landmarks due to aging. Figures 11.77 and 11.78 show a simulated minefield for two different cases. Figure 11.77 shows the case where $\alpha = 0, C_1 = C_2 = 1, \sigma_{C_1} = \sigma_{C_2} = \sigma_\alpha = 0$. Figure 11.78 shows the case where $\alpha = \pi/6, C_1 = C_2 = 1, \sigma_{C_1} = \sigma_{C_2} = \sqrt{0.5}, \sigma_\alpha = 0$.

Therefore, the detection of landmines can be accelerated if the above parametric model can be identified as fast as possible. This can be done by a distributed sensing mechanism supported by a swarm of lightweight field robots that can traverse the minefield without detonating mines (<7 kg per contact point).

We propose the following Bayesian parameter learning algorithm given in Equation (11.3), where the parameter vector W has to be learned. In Bayesian inferencing, we start with prior knowledge of a distribution for W and update the probability distribution $P(W)$ when actual data D are obtained by traversing the minefield.

$$P(W/D) = P(D/W)P(W)$$

$$W = [\alpha \quad C_1 \quad C_2 \quad \sigma_\alpha \quad \sigma_{C_1} \quad \sigma_{C_2}]^T \tag{11.3}$$

Simulations were done on a 10 × 10 m minefield with 20 landmines. The parameter values were set to as $W : \alpha = \pi/6, C_1 = C_2 = 1, \sigma_{C_1} = \sigma_{C_2} = 1, \sigma_\alpha = 0$.

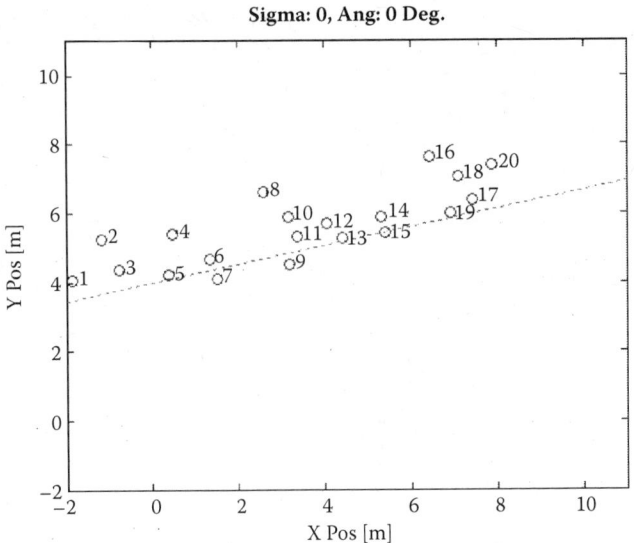

FIGURE 11.78
Distribution of landmines in the case of $\alpha = \pi/6$, $C_1 = C_2 = 1$, $\sigma_{C_1} = \sigma_{C_2} = \sqrt{0.5}$, $\sigma_\alpha = 0$.

The state-space equation of the robot is given by Equation (11.4)

$$x = [v \quad \varphi \quad \dot\varphi]^T$$

$$\dot x = \begin{bmatrix} \dfrac{-2c}{(Mr^2 + 2I_w)} & 0 & 0 \\ 0 & 0 & 1 \\ 0 & 0 & \dfrac{-2cl^2}{(I_v r^2 + 2I_w l^2)} \end{bmatrix} x + \begin{bmatrix} \dfrac{kr}{(Mr^2 + 2I_w)} & \dfrac{kr}{(Mr^2 + 2I_w)} \\ 0 & 0 \\ \dfrac{-krl}{(I_v r^2 + 2I_w l^2)} & \dfrac{krl}{(I_v r^2 + 2I_w l^2)} \end{bmatrix} u$$

$$u = [\tau_l \quad \tau_r]^T$$

(11.4)

where v is the velocity of the robot, φ is the azimuth, M is the mass, l is the distance between the two wheels, r is the radius of a wheel, I_v is the moment of inertia of the robot around the center of gravity, I_w is the moment of inertia of a wheel, k is a gain constant, c is the friction coefficient, and τ_r and τ_l are the torques of the right- and left-hand side wheels, respectively.

Since the efficiency of any search algorithm largely depends on the estimate of the parameter α, simulations were performed using an algorithm that adaptively changes the behaviors of a colony of three robots based on a recursively updated estimate of α. This method can be extended to estimate the other parameters of W.

The following steps describe the algorithm:

Step 1: Initialize the parameters W: $\alpha = 0$, $C_1 = C_2 = 1$, $\sigma_{C_1} = \sigma_{C_2} = \sqrt{0.1}$, $\sigma_\alpha = 1$.

Step 2: Set desired values for state variables $v_i^d = 0.1 m/\sec$, $\varphi_i^d = \pi/2$, $i = 1, 2, 3$ and the initial position of the three robots given by $p_1^r = [1,0]^T, p_2^r = [3,0]^T, p_3^r = [5,0]^T$.

Step 3: Calculate feedback control commands given by $F_i^v = k_1(v_i^d - v_i)$, $F_i^\varphi = k_2(\varphi_i^d - \varphi_i)$,

$$\begin{bmatrix} \tau_l \\ \tau_r \end{bmatrix} = \begin{bmatrix} \frac{r}{2} & \frac{r}{2} \\ \frac{r}{2l} & \frac{-r}{2l} \end{bmatrix}^T \begin{bmatrix} F_i^v \\ F_i^\varphi \end{bmatrix}$$

where v_i^d and φ_i^d are desired velocity and the azimuth of the ith robot. Consequently, F_i^v and F_i^φ are the force command to change the linear speed and that to change the azimuth of the ith robot, respectively.

Step 4: Detect mines and update the model parameters. It is assumed that if the robot comes within a 0.3-m radius of a landmine, the sensory system can detect the mine. In the case of mongooses, this radius can be as much as 2 m if the mine is old.

Calculate the best linear fit to the cluster of detected mines

$$y = \tan^{-1}(\hat{\alpha})x + \hat{c}$$

$$P(\hat{\alpha}/D) = P(D/W)P(\hat{\alpha})$$

$$P(\hat{\alpha}) = N(\hat{\alpha}, \sigma_{\hat{\alpha}}^2)$$

$$P(D/W) = N(\alpha^*, \sigma_{\alpha^*}^2)$$

where $P(\hat{\alpha})$ is the prior belief of the distribution of $\hat{\alpha}$, $P(D/W)$ is the likelihood of the location of a detected mine given the parameter vector W, and $P(\hat{\alpha}/D)$ is the posterior estimate of the probability distribution of $\hat{\alpha}$ after the data has been obtained. Then, the parameters of the posterior distribution are calculated by

$$\sigma_{\hat{\alpha}}^2 \leftarrow \frac{\sigma_{\alpha^*}^2 \cdot \sigma_{\hat{\alpha}}^2}{\sigma_{\alpha^*}^2 + \sigma_{\hat{\alpha}}^2}, \quad \hat{\alpha} \leftarrow \frac{\sigma_{\alpha^*}^2 \cdot \hat{\alpha} + \sigma_{\hat{\alpha}}^2 \alpha^*}{\sigma_{\alpha^*}^2 + \sigma_{\hat{\alpha}}^2}$$

Step 5: Generate a new desired azimuth for each robot in 40-s intervals given by

$$\varphi_i^d = \text{atan2}\left(\left(T_y - p_y^{r_i}\right), \left(T_x - p_x^{r_i}\right)\right)$$

$$T_y = \tan^{-1}(\hat{\alpha})T_x + \hat{c}$$

$$T_x = U(-L, L)$$

where (T_x, T_y) is the coordinates of the target location, $(p_x^{r_i}, p_y^{r_i})$ is the current coordinates of the ith robot r_i, $i = 1,2,3$, $U(-L, L)$ is a uniform random number between $(-L, L)$, $L = 10$, and parameter $\hat{c}$ is updated by $\hat{c} \leftarrow \hat{c} + \eta$, $\eta = 0.1m$ in 40-s intervals so that the robot traverses along isoclines parallel to the estimated gradient $\hat{\alpha}$.

Step 6: Go to step 2.

Figure 11.79 shows the simulation results of how a team of three robots self-organize their behaviors to estimate the parameter α. Figure 11.80 shows the history of parameter estimation. It is clear from Figures 11.79 and 11.80 that 100% coverage of mines laid in a 10 × 10 m area can be done even with an approximate estimate of α.

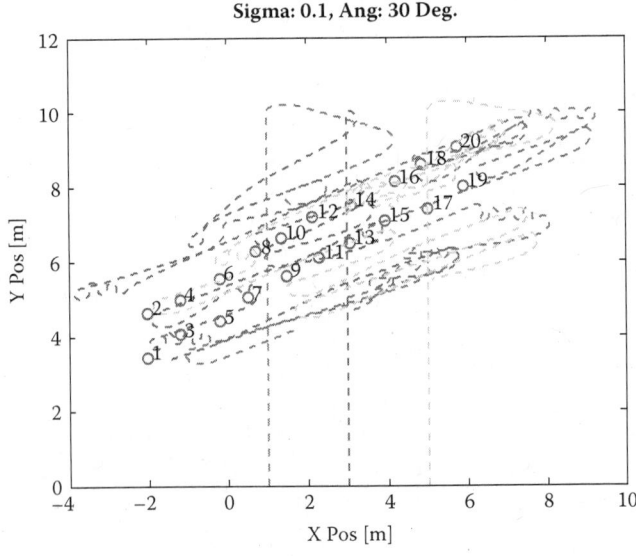

FIGURE 11.79
A team of three robots self-organize their behaviors to sweep accross the mine distribution.

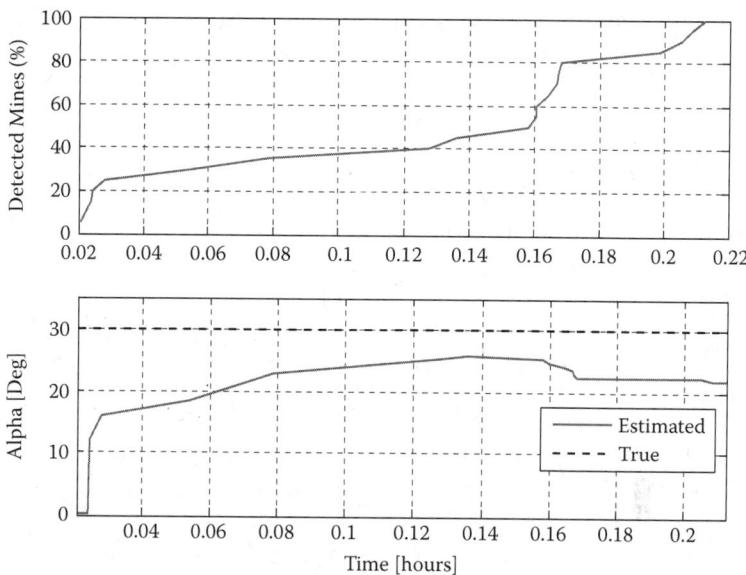

FIGURE 11.80
Percentage of detected landmines (top), the history of estimated α.

References

Arkin, R. C., *Behavior-Based Robotics*, MIT Press, Cambridge, MA, 1998.
Ayache, N., and Faugeras, O. D., "Maintaining representation of the environment of a mobile robot," *IEEE Transactions on Robotics and Automation* 5(6), 804–819, 1989.
BIOSENS Consortium (2004) BIOSENS Final Report, www.eudem.info.
Bohlen, M., "A robot in a cage," in *Proceedings of IEEE International Symposium on Computational Intelligence in Robotics and Automation*, pp. 214–219, 1999.
Bozma, O., and Kuc, R., "Building a sonar map in a specular environment using a single mobile sensor," *IEEE Transactions on Pattern Analysis and Machine Intelligence* 13(12), 1260–1269, 1991.
Brooks, R. A., "A robust layered control system for a mobile robot," *IEEE Journal of Robotics and Automation* RA-2(1), 14–23, 1986.
Bunyan, M., and Barratt, J., *AMS Guidance on Technology Readiness Levels (TRLs)*, GBG/36/10, UK MOD, 4 February 2002.
Burgard, W., Fox, D., Hennig, D., and Schimidt, T., "Estimating the absolute position of a mobile robot using position probability grids," in *Proceedings of the Thirteen national Conference on Artificial Intelligence*, AAAI Press/MIT Press, Menlo Park, CA, 1998.

Caprari, G., Colot, A., Seigwart, R., Halloy, J., and Deneubourg, J. L., "Animal and robot mixed societies-building cooperation between micro robots and cockroaches," *IEEE Robotics and Automation Magazine*, pp. 58–65, 2005.

Chong, K. S., and Kleeman, L., "Mobile robot map building from an advanced sonar array and accurate odometry," Technical Report MECSE-1996-10, Department of Electrical and Computer Systems Engineering, Monash University, Melbourne, 1996.

Collins, L., Gao, P., Schofield, D., Moulton, J. P., Makowsky, L. C., Deidy, D. M., and Weaver, R. C., "A statistical approach to landmine detection using broadband electromagnetic induction data," *IEEE Transactions on Geoscience and Remote Sensing* 40(4), 2002.

Crabbe, S., Eng, L., Gårdhagen, P., and Berg, A., "Detection of explosive as an indicator of landmines—BIOSENS project methodology and field tests in South East Europe," in *Proceedings of SPIE Conference on Detection and Remediation Technologies for Mines and Mine-like Targets X*, Vol. 5794, Orlando, pp. 762–773, 2005.

Crabbe, S., Sachs, J., Alli, G., Peyerl, P., Eng, L., Khalili, M., Busto, J., and Berg, A., "Results of field testing with the multi-sensor DEMAND and BIOSENS technology in Croatia and Bosnia developed in the European Union's 5th Framework Program," in *Proceedings of SPIE Conference on Detection and Remediation Technologies for Mines and Mine-like Targets IX*, Vol. 5415, 2004.

Cumming, C., Aker, C., Fisher, M., Fox, M., laGrone, M., Reust, D., Rockley, M., Swager, T., Towers, E., and Williams, V., "Using novel fluorescent polymers as sensory materials for above-ground sensing of chemical signature compounds emanating from buried landmines," *IEEE Transactions on Geoscience and Remote Sensing* 39(6), 1119–1128, 2001.

Daniels, D. J., Curtis, P., Amin, R., and Hunt, N., "MINEHOUND production development," in *Proceedings of SPIE Conference on Detection and Remediation Technologies for Mines and Mine-like Targets X*, Vol. 5794, Orlando, FL, pp. 488–494, 2005.

De Schutter, G., Theraulaz, G., and Deneubourg, J. L., "Animal-robot collective intelligence," *Annals of Mathematics and Artificial Intelligence* 31, 223–238, 2001.

Dissanayake, D. T. A., Sanjaya, K. A. G., and Mahipala, M. M. P. P., "Development of a mechanism to train mongooses to detect explosives," Final year thesis, Department of Mechanical Engineering, University of Moratuwa, Sri Lanka, 2007, http://people.seas.harvard.edu/~thrish/pubs.htm.

Doheny, R. C., Burke, S., Cresci, R., Ngan, P., and Walls, R., "Handheld Standoff Mine Detection System (HSTAMIDS) field evaluation in Thailand," in *Proceedings of SPIE Conference on Detection and Remediation Technologies for Mines and Mine-like Targets X*, Vol. 5794, Orlando, FL, pp. 889–900, 2005.

Donskoy, D. M., "Non-linear seismo-acoustic technique for landmine detection and discrimination," in *2nd International Conference on Detection of Abandoned Land Mines, IEEE Conference Publication No. 458*, Edinburgh, UK, pp. 244–248, 1998.

Dudek, G., Freedman, R., and Rekleitis, I. M., "Just-in-time sensing: Efficiently combining sonar and laser range data for exploring unknown worlds," in *Proceedings of the IEEE International Conference on Robotics and Automation*, Washington, pp. 667–672, 1996.

Ferrari, S., and Vaghi, A., "Demining sensor modeling and feature-level fusion by Bayesian networks," *IEEE Sensors Journal* 6(2), 471–483, 2006.

Fisher, M. E., and Sikes, J., "Minefield edge detection using a novel chemical vapor sensing technique," in *Proceedings of SPIE Conference on Detection and Remediation Technologies for Mines and Mine-like Targets VIII*, Vol. 5089, Orlando, FL, pp. 1078–1087, 2003.

Forsman, P., and Halme, A., "3-D mapping of natural environments with trees by means of mobile perception," *IEEE Transactions on Robotics* 21(3), 482–490, 2005.

George, V., Jenkins, T. F., Leggett, D. C., Cragin, J. H., Phelan, J., Oxley, J., and Pennington, J., "Progress on determining the vapor signature of a buried landmine," *Proceedings of SPIE, Detection and Remediation Technologies for Mines and Minelike Targets IV*, Vol. 3710, part 2, pp. 258–269, 1999.

Goldberg, G., and Matari'c, M., "Interference as a tool for designing and evaluating multirobot controllers," in *Proceedings of the 14th National Conference on Artificial Intelligence (AAAI-97)*, MIT Press, Cambridge, MA, USA, pp. 637–642, 1997.

Gonzalez, J., "An iconic position estimator for a 2D laser rangefinder," in *Proceedings of the IEEE International Conference on Robotics and Automation*, Los Alamitos, CA, pp. 2646–2651, 1992.

Hayes, A., "How many robots? Group size and efficiency in collective search tasks," in *Procs. of the 6th International Symposium on Distributed Autonomous Robotic Systems (DARS-02)* (H. Asama, T. Arai, T. Fukuda, and T. Hasegawa, Eds.). Springer Verlag, Heidelberg, Germany, pp. 289–298, 2002.

Holzer, R., and Shimoyama, I., "Locomotion control of a bio-robotic system via electric stimulation," in *Proceedings of IEEE/RSJ IROS International Conference on Intelligent Robots and Systems*, vol. 3, Grenoble, France, pp. 1514–1519, 1997.

Ishii, H., Nakasuji, M., Ogura, M., Miwa, H., and Takanashi, A., "Experimental study on automatic learning speed acceleration for a rat using a robot," in *Proceedings of the IEEE International Conference on Robotics and Automation*, Barcelona, Spain, pp. 3078–3083, 2005.

Jones, C., and Matari'c, M., "Adaptive division of labor in large-scale minimalist multi-robot systems," in *Proceedings of the IEEE/RSJ International Conference on Intelligent Robots and Systems*, Vol. 2, IEEE Press, New York, NY, USA, pp. 1969–1974, 2003.

Krieger, M., and Billeter, J. B., "The call of duty: Self-organized task allocation in a population of up to twelve mobile robots," *Robotics and Autonomous Systems* 30(1–2), 65–84, 2000.

Lloyd, C. P., Piyathilake, J. M. L. C., Manatunga, M. M. A. S., and Masinghe, W. D., "Automated robot for landmine detection," Final year thesis, Department of Electrical Engineering, University of Moratuwa, Sri Lanka, 2002, http://people.seas.harvard.edu/~thrish/pubs.htm.

Mataric, M. J., "Reward functions for accelerated learning," in *Proceedings of the Eleventh International Conference on Machine Learning*, Morgan Kaufmann, San Francisco, CA, pp. 181–189, 1994.

Nanayakkara, T., Piyathilaka, L., and Subasingha, A., "A simplified statistical approach to classify vegetated environments for robotic navigation in tropical minefields," in *Proceedings of the Internatonal Conference on Information and Automation*, Colombo, Sri Lanka, pp. 337–342, December 15–18, 2005.

Ng, A. Y., Harada, D., and Russell, S., "Policy invariance under reward transformations: Theory and applications to reward shaping," in *Proceedings of the Sixteenth International Conference on Machine Learning*, Morgan Kaufmann, San Francisco, CA, pp. 278–287, 1999.

Perrin, S., Duflos, E., Vanheeghe, P., and Bibaut, A., "Multisensor fusion in the frame of evidence theory for landmine detection," *IEEE Transactions on Systems, Man, and Cybernetics-Part C: Applications and Reviews* 34(4), 485–498, 2004.

Sato, M., Fujiwara, J., Feng, X., and Kobayashi, T. "Dual sensor ALIS evaluation in Afghanistan," *IEEE Geoscience and Remote Sensing Society Newsletter*, 22–27, 2005.

Savelyev, T. G., Kempen, L. V., Sahli, H., Sachs, J., and Sato, M., "Investigation of time-frequency features for GPR landmine discrimination," *IEEE Transactions on Geoscience and Remote Sensing* 45(1), 118–129, 2007.

Scott, W. R., Martin, J. S., and Larson, D., "Experimental model for a seismic landmine detection system," *IEEE Transactions on Geosciences and Remote Sensing* 39(6), 1155–1164, 2001.

Sun, Y., and Li, J., "Time-frequency analysis for plastic landmine detection via forward-looking ground penetrating radar," *IEEE Proceedings on Radar Sonar Navigation* 150(4), 253, 2003.

Talwar, S. K., Xu, S., Hawley, E. S., Weiss, S. A., Moxon, K. A., and Chapin, J. K., "Rat navigation guided by remote control," *Nature* 417, 37–38, 2002.

Tan, Y., Tantum, S. L., and Collins, L. M., "Kalman filtering for enhanced landmine detection using Quadrupole resonance," *IEEE Transactions on Geoscience and Remote Sensing* 43(7), 1507–1516, 2005.

Thrun, S., Burgard, W., and Fox, D., "A probabilistic approach to concurrent mapping and localization for mobile robots," *Journal of Autonomous Robots* 5(3–4), 253–271, 1998.

Watkins, C. J. C. H., *Learning from Delayed Rewards*, PhD dissertation, Cambridge University, 1989.

Xiang, N., and Sabatier, J., "Laser Doppler vibrometer-based acoustic landmine detection using the fast M-sequence transform," *IEEE Transactions on Geosciences and Remote Sensing Letters* 1(4), 292–294, 2004.

Xu, X., Miller, E. L., Rappaport, C. M., and Sower, G.D., "Statistical method to detect subsurface objects using array ground-penetrating radar data," *IEEE Transactions on Geoscience and Remote Sensing* 40(4), 963–976, 2002.

12

Robotic Swarms for Mine Detection System of Systems Approach

12.1 Introduction

In this chapter, we present an ant colony optimization (ACO)-based mine detection algorithm and its implementation with a group of robotic swarms. We evaluate robotic swarms in the context of the system of systems (SoS) concepts because of their similar characteristics, such as interoperability, integration, and adaptive communications. Thus, robotic swarms can be evaluated and studied as SoS based on these common characteristics. This chapter also explores the hardware and software designs of the modular microrobots used in the application and testing of the ACO-based mine detection algorithm. In addition, this chapter presents the details of SoS characteristics of robotic swarms. Next, the characteristics of both swarms and SoS are evaluated.

12.1.1 Swarm Intelligence

Swarms in nature are found as large groups of small insects. In these swarms, each member performs a simple task, but the actions they produce present complex behaviors as a whole [1]. High-order animals, such as ant colonies, bird flocks, and packs of wolves, also exhibit this emergent behavior. These groups present swarm behaviors in many ways. Many areas in computer science and engineering have explored the complex problem-solving capability of swarms [1].

The swarm intelligence definition was first generated by Beni et al. [2–4] in the context of cellular robotics. Intelligent swarms have members that can exhibit independent intelligence. Intelligent swarm members can be homogenous or heterogeneous based on their environmental interactions [1]. A more general definition of swarm intelligence is provided Bonabeau, Dorigo, and Theraulaz as "any attempt to design algorithms or distributed problem solving devices inspired by the collective behavior of social insects and other animal societies" [5]. The behavior of the social insects is determined by the use of pheromones. For example, the ants are capable of foraging in unknown environments by secreting and following pheromone trails. Ant colony-based algorithms are generally focused on the short-range recruitment (SRR) and long-range recruitment

(LRR) behaviors of the ants [6–8]. Dorigo and Di Caro [9] provided a common framework for the algorithms related to ant colonies, ACO.

On the other hand, swarm intelligence techniques are different from intelligent swarms in the sense that they are population-based stochastic methods and are mainly used in combinatorial optimization problems [10–13]. In swarm intelligence methods, the collective behavior is created by the members' local interactions with their environment to come up with functional global patterns [1]. A recent example of such a swarm intelligence technique is called particle swarm optimization (PSO) [13].

The PSO algorithm involves the structure and behavior of a population of cooperative agents, called particles, randomly in the multidimensional search space. Each particle has an associated fitness value, which is evaluated by the fitness function to be optimized, and a velocity that determines its motion. Each particle can keep track of its solution that resulted in the best fitness (personal best), as well as the solutions of the best performing agents in its neighborhood (neighborhood best). The trajectory of each particle is dynamically governed by its own and its companions' historical behavior. Kennedy and Eberhart [13] view this adjustment as conceptually similar to the crossover operation used by genetic algorithms. Such an adjustment maximizes the probability that the particles are moving toward a region of space that will result in better fitness. At each step of the optimization, the particle is allowed to update its personal best position by evaluating its own fitness and the neighborhood best solution. The PSO algorithm is terminated when the specified maximum number of generations is reached or when the best particle position of the entire population cannot be improved further after a sufficiently large number of generations.

In recent years, PSO [10–14] and ACO [15,16] have been studied and applied to various problems. Swarm intelligence has been applied to important problems and domains such as telecommunications [17–20], business [21], robotics [6–8,22–26], and optimization [27,28]. There are other emerging swarm intelligence techniques, such as artificial immune systems [29,30]. This chapter explores the ACO-based mine detection problem and its implementation on micromodular swarm robots, namely, GroundScouts.

12.1.2 Robotic Swarms

Swarm robotics is the application of swarm intelligence techniques to the analysis of activities in which swarm members are physical robotic devices that can change their environments by intelligent decision-making based on various inputs [1]. These robots are mobile robots with various locomotion architectures, such as legged or wheeled ground robots [25,31–34], underwater robots [35,36], and flying robots [37]. In addition, there are efforts in designing microrobots and microelectromechanical system-based swarm robots [38,39].

Control strategies developed for swarming robots are generally inspired by biological systems [37,40–43] as well as hierarchical [44] and cooperative control schemes [45,46]. In most control schemes, the swarm individuals have a common communication medium and very similar hardware and software components. In addition, modular robots are mainly used in swarm robotics applications because they are easy to assemble and, by definition, share the same design architecture, which makes them highly compatible [31–33].

Modular robotics is a rapidly growing research field [31–33,47,48]. Modular robotic systems are inherently robust and flexible. These properties are becoming increasingly important in real-world robotics applications. In recent years, special attention has been given to *self-reconfigurable* robots [33], that is, modular robots whose components can autonomously organize into different connected configurations. In most current modular robotics implementations, modular robots are initially manually assembled and, once assembled, they are incapable of adding additional modules without external help. In this chapter, we present and use a modular robot design with hot swappable modules, called GroundScouts [31,32]. Next, we study and evaluate the characteristics of SoS since the swarming robots exhibit similar characteristics with SoS characteristics.

12.1.3 System of Systems

Recently, there has been a strongly growing interest in SoS concepts and strategies, especially in aerospace and defense systems. Performance optimization among a group of heterogeneous systems in order to achieve a common task is the major focus of a diverse range of applications, including military, security, aerospace, and disaster management [49,50]. In the literature, the issue of coordination and interoperability in an SoS has been addressed by several researchers [51–53].The concept of SoS is essential to more effectively implement and analyze large, complex, independent, and *heterogeneous* systems working (or made to work) cooperatively [34]. The main thrust behind the desire to view the systems as an SoS is to obtain higher capabilities and performance than would be possible with a traditional system view. The SoS concept presents a high-level viewpoint and explains the interactions among the independent systems. However, the SoS concept is still in its developing stages [54,55]. The literature has revealed that much of the recent work introduces new concepts toward an SoS approach [56,57]. However, there are only a few applications of SoS in real-world scenarios [52,56,57].

SoS are super systems consisting of systems that are independent complex operational systems interacting among themselves to achieve a common goal. Each element of an SoS achieves well-substantiated goals even if they are detached from the rest of the SoS [58,59]. For example, a Boeing 707 airplane, as an element of an SoS, is not an SoS, but an airport is an SoS, or a rover on Mars is not an SoS, but a *robotic colony* (or a *robotic swarm*) exploring the red planet is an SoS. Associated with SoS, there are numerous problems and open-ended issues that need a great deal of fundamental advances in theory and verifications [56,57,60].

Based on the literature survey on SoS, there are several definitions of SoS [60–66]. Detailed literature survey and discussions on these definitions are given by Kotov [62] and [56,57]. Our favorite definition is "Systems of systems are large-scale concurrent and distributed systems that are comprised of complex systems." This definition focuses on *information systems*, which emphasizes the *interoperability* and *integration* properties of an SoS [56,57,67,68].

Interoperability in complex systems (i.e., multiagent systems) is very important because the agents operate autonomously and interoperate with other agents (or nonagent entities) to accomplish better actions. Interoperability requires successful communication among agents (systems). Thus, the systems should carry out their tasks *autonomously* and communicate with other systems in the SoS in order to take actions for the overall benefit of the SoS, not just for themselves [56,57,67,68].

Integration implies that each system can communicate and interact (control) with the SoS components regardless of their hardware and software characteristics. This means that they need to have the ability to communicate with the SoS or a part of the SoS without compatibility issues, such as with operating systems (OS), communication hardware, communication protocol, and so on. For this reason, an SoS needs a common language that the SoS components can speak. Without having a common language, the SoS components cannot be fully functional and the SoS cannot be adaptive in the sense that new components cannot be integrated into the SoS without major effort. Integration should also consider meaningful control aspects of the SoS [67,68]. For example, a system within an SoS should understand commands and/or control signals from other SoS components.

In a robotic swarm, similar approaches to integration are used to achieve effective communication among the components of the swarm. The next section explores how SoS concepts can be used on robotic swarms.

12.2 SoS Approach to Robotic Swarms

As stated earlier, robotic swarms have common characteristics with SoS, such as interoperability and integration. Interoperability and integration require an effective adaptive communication scheme among the systems (swarm members). Adaptivity of communication is essential because it allows the SoS components to be able to communicate independently of the system dynamics. Next, we explore the interoperability and integration characteristics of the SoS and their reflections on robotic swarms [32].

12.2.1 Interoperability

In a swarm, members of the swarm can survive independently, although their individual performance is generally very limited compared to the performance of the swarm overall. However, in an SoS, the systems are often independently operable and their individual performance may be the same as the performance they would have in an SoS. This suggests that interoperability is stronger in an SoS than in a robotic swarm, that is, interoperability of an SoS is more general compared to the limited interoperability of a swarm member. Therefore, in the study of robotic swarms such as SoS, the interoperability of the swarm should be achieved, but should not be an integral part of the design. In a swarm, if a member leaves the swarm, it can survive by itself but it immediately starts to try to rejoin the swarm. The robotic swarm design presented in this chapter has modular microrobots with various sensors, which can autonomously operate in the environment. Thus, they definitely satisfy the interoperability characteristics of an SoS.

12.2.2   Integration

Integration is essential for both SoS and robotic swarms because both need to expand and shrink with ease during the operation. In an SoS, it is critical to have the ability to bring a system into the SoS without any major problems, such as redesigning, restarting, and/or reprogramming the system. Integration in an SoS has two major components: (1) adaptive communication methods that allow the inclusion of a new member; (2) a common language such that any new system can send its data to other systems in the SoS without any difficulty or significant effort. Thus, a common language such as Extensible Markup Language (XML) is important for an SoS. The architecture for such a language can be combined with a discrete event simulation tool so that the SoS concepts can be simulated.

Consequently, the integration characteristics of an SoS need effective and adaptive communication among the systems. When the systems in an SoS are heterogeneous in hardware and software, the XML-like common language becomes very critical. However, since all members in a swarm have either identical or similar hardware and software components, the necessity for a common language is not so critical in general. Therefore, in a robotic swarm, the design does not have to have an XML-like language. Instead, a common protocol or a common format will be sufficient.

Another component of the integration characteristics is the need for an adaptive communication schemes. This characteristic is more important for a swarm robot since members of a swarm move in and out of the swarm very often. Thus, the communication scheme needs to have the ability to adapt itself for population changes in the swarm. To achieve this goal, processes for accepting new members to the swarm and removing a member from a swarm need to be implemented as part of the communication scheme. Details of the communication scheme used in the robotic swarm are discussed in this chapter.

12.3 Designing System of Swarm Robots: GroundScouts

In recent years, there has been a significant amount of research in reconfigurable and modular robotics [31–33,46,47,69–71]. The majority of the research has been focused on multiple identical modules that can construct a single robot. In this chapter, we present a modular design that has a vertical modularity. In this approach, the modules are not identical to each another [31–33]. This modular design slices a robot into functional abstract modules such as locomotion, control, sensors, communication, and actuation. Any mobile robot can be constructed by combining the above abstract modules for a specific application. A submodule is a piece of hardware that accomplishes the functionality of an abstract module. For example, a wireless communication submodule is hardware for a communication module. A similar modular robotics approach is explored by Fryer and McKee [72]. Fryer and McKee [72] focus on sensor-oriented modular robotics systems (MRS), where modules contain a mobile base, a camera mount, and a camera [72]. In our architecture [31–33], a similar robot can be constructed by combining a locomotion module, a sensory module, and an actuator module. Fryer and McKee [72] did not explore software modularity intensively. Our MRS model [31–33], shown in Figure 12.1, has both hardware and software modularity.

In our MRS model, each abstract module can be implemented by the corresponding hardware. For example, a sensor module may be an ultrasonic sensor board or a proximity sensor board, or perhaps both. The submodules are designed such that they have a unique signature and standard pin connections. Submodules can be added at any level and the position does not affect its operation. An application-specific robot can be easily constructed because

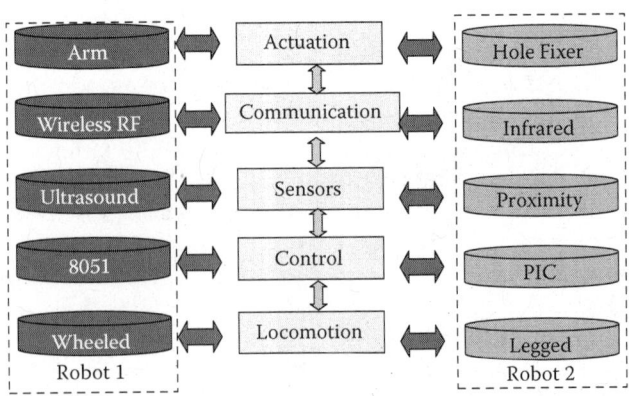

FIGURE 12.1
Modular hardware architecture [31,32].

modules can be combined in any order. For example, if the problem domain requires legs rather than wheels, the wheeled submodule can be instantly swapped with the legged submodule. This is essential for swarm intelligence applications since agents might be equipped with complementary abilities instead of the same abilities. For example, in ant colonies, different tasks are handled by different types of ants using special abilities [15,16].

In addition, there is no standard software approach for programming robots. The main reason is that each robot is composed of very special hardware designed for a specific goal or domain. Thus, software components also become specific for each robot. Current available systems are not capable of providing modules that are easily maintained and reusable [73]. An attempt was made by Denneberg and Fromm [73] to develop an open software for autonomous mobile robots. Their approach, known as Open Software Concept for Autonomous Robots (OSCAR), is built on a layered model with four software levels: command layer, execution layer, image layer, and hardware layer. Each layer works as an interface between the upper layer and the lower layer. Therefore, the upper layer does not need the knowledge of the lower layer [63].

Our MSR approach constitutes software modularity that matches the hardware modularity mentioned above. In our approach, each hardware module corresponds to a software module as well, instead of each hardware layer dealing with only the hardware components, as in OSCAR. An object-oriented paradigm was used to design our modular software methodologies. The approach has abstract classes and their concrete subclasses. The abstract classes are *robot*, *locomotion*, *control*, *communication*, *sensor*, and *actuators*. The concrete classes are derived from these abstract classes. For example, a *wireless* layer may be instantiated from the *communication* class. The *wireless* object represents the robot's communication module. The concrete classes inherit functionality from the abstract classes, in addition to their hardware-specific functionalities. Even though the modules contain their own software modules, the central controller (or OS) identifies newly inserted modules and creates the corresponding objects and interfaces to the hardware module. Figure 12.2 shows the class structure for the modular software architecture.

12.3.1 Hardware Modularity

Based on the approach described in the previous subsection, modular microrobots, called GroundScouts, are designed by taking advantage of a layered-design approach. Even though there are five levels in the hardware and software architectures, the implementations (submodules) of the levels (modules) can be more than one. In addition, each level may also involve multiple closely related functionalities. The following sections present hardware (sub)modules in the GroundScouts: locomotion, control, sensor, communication, and actuation. More detailed explanations of the modules are given by Sahin [31,32].

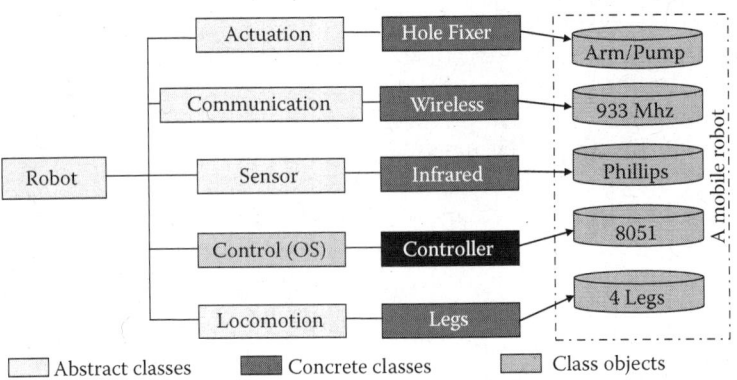

FIGURE 12.2
Modular software architecture [31,32].

12.3.1.1 Locomotion

The locomotion module houses a mechanical base and locomotion module hardware (submodule). The mechanical base is composed of an aluminum frame (designed by a CNC), two DC motors, some gearing, four wheels and associated ball bearings, and the batteries. Figure 12.3 (left) shows the base with wheels, gears, and motors. A legged version of the base (right) is also designed as an alternative locomotion to be used in different applications.

The locomotion module hardware, the first electronic level shown in Figure 12.4, is the most important level in the operation of the robot. It has the circuitry for an H bridge, charging circuitry, and a power system. The power system generates +5 V for the entire robot, and will accept 1.5 to 15 V for its input. This is great flexibility since the robot can operate even if the battery voltage level is very low. Finally, this layer contains three proximity sensors, one centered at the front and two facing 45° off center on each side

FIGURE 12.3
Robot base with four wheels and legged [31,32].

FIGURE 12.4
Locomotion submodule top (left) and bottom (right) view [31,32].

of the front sensor. These sensors detect small objects missed by the sensor layers. They operate like bumper sensors since they will stop the robot when the robot gets very close to a small object.

12.3.1.2 Control

This layer, shown in Figure 12.5, houses the main controller, an 8051-based microcontroller running at 11.0592 MHz, a 64k × 8 flash memory integrated circuit (IC), a 32k × 8 SRAM IC, and an 8-bit latch. The flash memory can be removed to be reprogrammed with ease. With improved memory architecture, GroundScouts are able to run a modular embedded OS, which is explained later. The current microcontroller is a Philips 80C552-5, which has no internal ROM and vectors directly to the external flash for its program. In addition, it has five 8-bit registers, two of which are used to access external program and data memory. One of the registers has an 8-bit analog-to-digital (A-to-D) converter. The microcontroller also has two pulse-width modulation lines and some program and data memory access lines.

FIGURE 12.5
Microcontroller board top (left) and bottom (right) view [31,32].

FIGURE 12.6
Infrared submodule [31,32].

12.3.1.3 Sensor

GroundScouts have two sensor submodules. The infrared (IR) submodule, shown in Figure 12.6, has five IR sensor pairs. Each pair consists of an emitter and a detector for detecting objects in the area of sight. The IR emitters of all five pairs of sensors are enabled simultaneously via a logic level field-effect transistor, and they are turned on by the microcontroller. The detectors require that the emitters be driven at 38 kHz modulated on 833 Hz, which is done by the microcontroller. The detectors have built-in filters and provide the logic 1 if there is an object reflecting the transmitter IR signal.

The ultrasonic sensor submodule, shown in Figure 12.7, consists of three ultrasonic receiver and transmitter pairs. These sensors measure the distance to an object in the range of 2 cm to 300 cm. In order for the distance to be measured, the object must have at least one section that is normal to the sensors, and it must be of a material that will reflect a majority of the ultrasonic

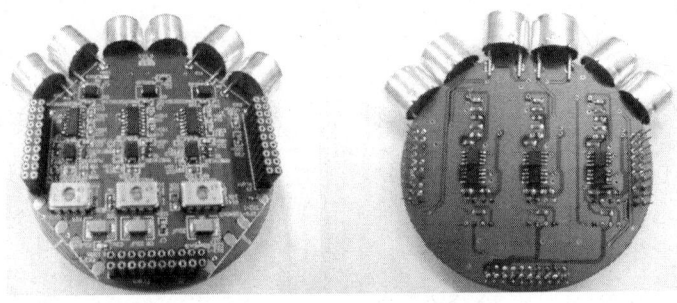

FIGURE 12.7
Ultrasonic submodule top (left) and bottom (right) view [31,32].

FIGURE 12.8
Communication submodule for GroundScouts [31,32].

energy. There are two possible configurations for the ultrasonic sensor layer: one front and centered and two on either side at 45° from center, and one front and centered and two facing rearward at 45° from center.

12.3.1.4 Communication

The communication submodule, shown in Figure 12.8, has a PIC microcontroller that uses a wireless transceiver by LINX Inc. The transceiver uses frequency shift keying (FSK) centered at 913 MHz with a 20-MHz bandwidth. The PIC microcontroller handles the forming of the packets and does address decoding. The PIC communicates with the main controller (80552) on an RS-232 bus. When multiple users are present, a medium access control (MAC) protocol must be implemented in order for data to be successfully transmitted. Time division multiple access (TDMA) is designed as the main protocol because it is easier to implement and is fairly well known.

Since the number of individuals in a swarm changes based on the swarm characteristics, conventional TDMA presents some inefficiencies when all users (swarm members) are not present. Thus, unused time slots could be used to increase the transfer rate of users who are present. An adaptive TDMA (ATDMA) protocol was developed to utilize these unused time slots [26,31,32,74]. As swarm members get in and out of the network, the number of time slots will change accordingly, creating a TDMA network with no unused time slots. The cost of this protocol is one time slot per frame, where everybody is in receiving mode. This time slot is designated for users to enter the network.

12.3.1.5 Actuation

As an actuation layer, a global positioning system (GPS) submodule, shown in Figure 12.9, is designed for GroundScouts. The chosen GPS receiver was the Lassen SQ GPS receiver. It operates on the standard L1 carrier frequency of

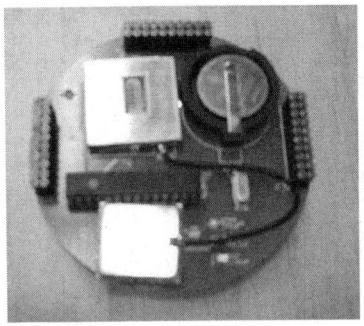 

FIGURE 12.9
GPS submodule top (left) and bottom (right) view [31,32].

1575.42 MHz and has the capability to automatically get GPS satellites and track up to eight of them. It can also compute location, speed, heading, and time.

The GPS receiver sends GPS data through serial bus. The communication protocol is dependent on the standard chosen, either Trimble Standard Interface Protocol (TSIP) or National Marine Electronics Association. TSIP was chosen for our robotic swarm applications, having an input and output baud rate of 9600, 8 data bits, odd parity, 1 stop bit, and no flow control. TSIP automatically sends information on GPS time, position, velocity, receiver health/status, and satellites in view. GPS time and receiver health/status are sent every 5 s, and all others are sent every 1 s. Position and velocity measurements can be sent as singles or doubles, depending on the desired precision. GPS information is received and stored by a PIC16F73 using its onboard UART configured at the aforementioned settings. The I2C bus is used for the communication between the main microcontroller (80552) and the PIC when GPS information is requested by the 80552.

This completes the modular hardware building blocks of GroundScouts. Figure 12.10 shows pictures of a completed GroundScout. The next section presents the software modularity of the GroundScouts.

12.3.2 Software Modularity

Controlling many cooperative robots as a swarm is not an easy task. The required software must have the multitasking abilities. To ease the complexity, many developers have begun using real-time operating systems (RTOSs). In addition to providing the ability to precisely synchronize multiple events, RTOSs give the application programmer predefined system services and varying degrees of hardware abstraction, both of which are aimed at making software development easier and more organized [31,32]. Thus, in addition to the software architecture described previously, we have developed a micro-RTOS with the GroundScouts' software architecture. This was essential

FIGURE 12.10
A GroundScout with all submodules [31,32].

especially for dynamic task uploading, which allows the transfer of executable software components through the communication medium.

12.3.2.1 Operating System

Based on the software architecture described previously, a significantly small-sized basic OS is designed with dynamic task loading over a serial port (via wireless transceiver). Each robotic submodule has an associated task (driver). Each task can be seen as an object of the corresponding concrete class defined earlier. These drivers are kept separate from the OS in order to limit the OS size. This separation makes the dynamic task uploading and execution an essential component of the OSs.

The current OS is an RTOS and can accommodate the modular nature of the robots [75,76]. The main properties of the OS are as follows:

1. It is generic and promotes portability.
2. It was programmed in C, except for processor-specific assembly.
3. It provides modular source code.
4. It is a multitasking OS that supports dynamic loading and task execution.

12.3.2.2 Dynamic Task Uploading

When implementing dynamic task uploading (execution), the main challenge comes from the fact that there are fundamental OS limitations. An RTOS such as the μC/OS-II is preemptive, which means that the highest priority

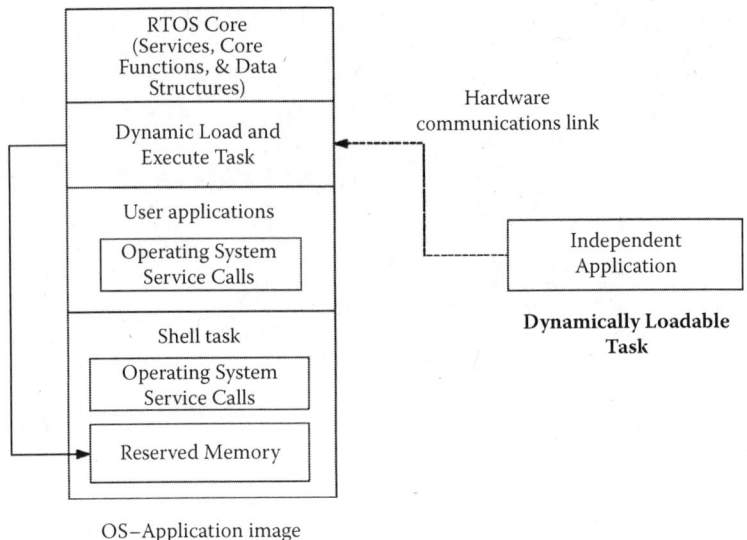

FIGURE 12.11
The high-level view of µC/OS-II with dynamic load and execute functionality [31,32].

task waiting for execution will be executed unless it is suspended, delayed, or waiting for an event. The µC/OS-II OS, written by Jean J. Labrosse [77], was specifically designed for embedded systems. The µC/OS-II runs on most 8-, 16-, 32-, and even 64-bit microprocessors, including the 8051. The most practical solution is to use µC/OS-II to create a shell task as part of the OS-application image so that the dynamically loaded task can be loaded into this location at a later time. This shell task remains dormant (not executed) until a dynamic task is loaded into the shell. Figure 12.11 presents the dynamic load and execute feature. The human supervisor via the software interface and wireless communication link sends a task to the GroundScouts. The OS receives the task and passes it to a reserved memory space. Finally, the task is executed when it is needed. The human supervisor can also be substituted by a master robot, which would have a library of tasks (drivers) to be dynamically uploaded. The dynamic task uploading has not been used in our robotic swarm experiments since the swarm intelligence techniques we explored had identical swarm members and tasks.

12.3.3 Communication Protocol: Adaptive and Robust

In a swarm, the number of swarm members is dynamic most of the time. In addition, communication among swarm members is not a global one where everybody can talk to everybody. In general, there is a small group of swarm elements that need to communicate in function well. Thus, the

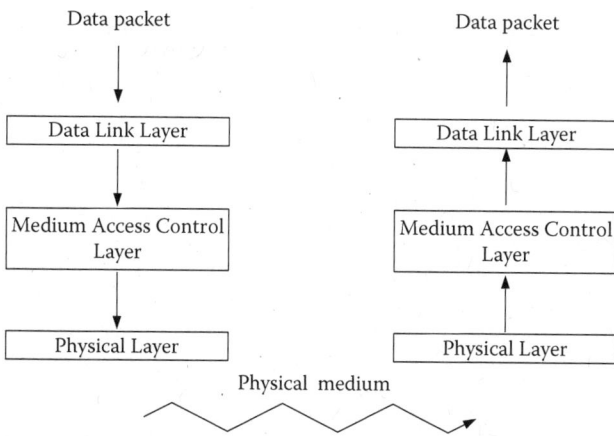

FIGURE 12.12
Layered network structure [26].

communication protocol for a swarm of robots should be adaptive and/or dynamic. As noted earlier, one of the suitable protocols for adaptive communication is ATDMA. To fully understand the development of this adaptive protocol, the different layers will be defined as well as how they interact with one another. A picture of these layers is shown in Figure 12.12.

The data link layer links the data together with the addition of different information, depending on the protocol. The new packet then goes into the MAC layer, where the protocol controls the data being placed on the physical layer in order to prevent collisions that could occur on the physical layer. The physical layer consists of the actual physical medium that the signal is going to travel across, in this case, the air. The specific layers are described in detail in the following subsections [32].

12.3.3.1 Physical Layer

The communication module on the GroundScouts has a PIC microcontroller that controls a transceiver (LINX Inc.) that uses FSK to transmit the data. FSK changes the frequency corresponding to whether the datum is 0 or 1. The biggest advantage of this technique is that the reception of the signal is not as dependent on signal power in contrast to other techniques. The data can still be successfully received down to a certain power threshold.

12.3.3.2 MAC Layer

Since there will be multiple users on the network, a MAC protocol must be designed to ensure that data are successfully transmitted through the physical medium. There are three suitable MAC protocols: frequency division multiple access (FDMA), code division multiple access (CDMA), and TDMA.

FDMA separates the users in frequency by placing the messages on different carriers such that their bandwidths do not overlap. The difficulty with this protocol is that the receiver must know what carrier the signal that is being transmitted is on or else the data will not be received. This protocol is not feasible because the carrier frequency is set in hardware by the transceiver and cannot be altered.

CDMA incorporates frequency separation as well as time separation, creating a signal that is only coherent to the receiver that has the valid "code" word. The signal is transmitted in a spread form that spreads the power spectrum in the frequency band. This results in the signal power being below the noise level. The receiver then uses a correlator that is based on the code word to acquire and decode the spread signal. The spectrum cannot be spread with the current hardware [26]. A correlator is hard to implement. It can be done with either a DSP (digital signal processor) chip or an FPGA (field-programmable gate array). Also, an A-to-D converter must be present. The current hardware does not have an A-to-D converter.

The last MAC protocol that is considered is TDMA. This protocol separates the users in time, giving each user a time slot to transmit while all other users listen to the channel. This is feasible to implement using the current hardware and characteristics of swarming robots. One of the microcontroller's timers must be used to keep track of the time so that each node knows when to transmit. Thus, this protocol is the best candidate for the required communication protocol.

The basis of TDMA [78] is to divide the users by giving each node (or link) a "slot" to transmit data. The time it takes for all users on the network to have a chance to transmit is referred to as a "frame." As noted before, the ATDMA scheme is used as the communication protocol among the GroundScouts. Next, we present the details of the ATDMA implementation.

12.3.3.2.1 ATDMA Development

Under bursty traffic conditions, TDMA-based protocols perform badly since a large number of slots can remain idle [79]. This observation raises the question of whether a simple adaptive strategy can be developed that can utilize unused slots to increase the performance of the nodes that need more bandwidth. The following presents a few distinct protocols that attempt to create the ideal adaptive network.

A dynamic protocol for *ad hoc* networks was developed by Kanzaki et al. [80], who introduced a few different techniques for slot allocation. Their protocol is based on the work of Young et al. [81] in the *unifying slot assignment protocol* (USAP) and *USAP-multiple access* (USAP-MA). The protocol starts with the first slot being reserved for new nodes to transmit control packets to request slot assignments. There are four different control packets

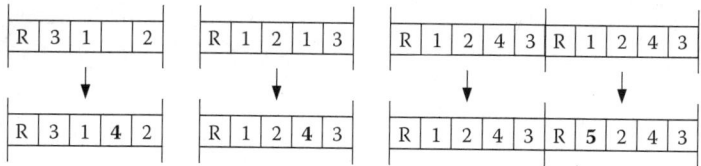

FIGURE 12.13
Slot allocation: (a) node taking unused slot, (b) node taking multiple slot, node doubling frame to split slot [26].

associated with this protocol: the request (REQ), information (INF), suggestion (SUG), and reply (REP) packets. When a REQ packet is received, the network switches into a control mode in which all of the other nodes send INF packets. The new node will accordingly set its frame length. It then needs to get a time slot to transmit. This can be done in one of three ways. Either it takes an unused slot (if there are no unused slots, it asks a node that has multiple slots to release one) or, if there are no nodes with multiple slots, it doubles the frame length. These are illustrated in Figure 12.13.

Conflicting slots are similarly treated where, if they both have single slots, the frame is doubled and they share the slot in opposite frames. If one has multiple slots, it hands the conflicting slot to the other node. This strategy is proven to be affective, but it is fairly complicated to implement on the GroundScouts.

Burr et al. [82] propose an adaptive strategy that is based on interference in the channel. The forward error correction (FEC) coding technique used was modified according to the amount of error correction needed to successfully transmit the packet. In this case, the tradeoff is bandwidth. When low interference is detected by real-time channel evaluation, the FEC code is lowered, which will improve the transmission rate, therefore exploiting the channel characteristics. When the bandwidth is changed, either slots will become vacant or more slots will be necessary. The protocol will continuously monitor the slots and allocate them accordingly. When there is interference, packets will be lost. Therefore, this protocol has a one packet buffer that holds a packet until it is successfully transmitted.

Ali et al. [83] propose a protocol that has three major advantages: the information that is necessary for the protocol is collected locally, global coordination is not necessary, and changes to the state of a specific terminal do not need to be passed to all of the other terminals in the network. To create these advantages, two algorithms were developed [83]. The first algorithm varies the slot assignments if an added link is detected. The second algorithm updates slot assignments to increase the efficiency when two terminals are no longer in range of each another. The main constraint in this approach is that the two scenarios (a node leaving and a collision occurring) must be detectable. A flag header is appended to the beginning of the each packet in order to send control-type signals for new users to enter the network. The details of the algorithms are given by Ali et al. [83]. It is critical to note that the flag field is a waste of bandwidth when it is not used and that only one change at a time can be handled by the network.

These protocols were studied and used to develop a protocol that attempts to exploit all positive features of each network to produce a more efficient and reliable adaptive network. In this protocol, we designated the first slot of each frame for new users to enter the network. This idea was adopted from Kanzaki et al. [80]. The one-packet buffer described by Burr et al. [82] allows packets to be transmitted without any loss. The constraints given by Ali et al. [83] are used in this protocol, allowing the network to adjust to users entering and leaving the network. Thus, our adaptive protocol only has contention when a new user is attempting to enter. The resulting network is at its highest efficiency at all times, with the only cost being one time slot per frame designated for new users to enter the network.

12.3.3.2.2 ATDMA Protocol for Swarm Robots

Following is the sequence of activities that constitute our ATDMA protocol:

1. At startup, there is no master in the network. Each node waits for a sink signal. When the waiting period times out, the node assumes master responsibilities. Figure 12.14 illustrates the initial frame structure for the network. Assuming the new node is node A, it takes the first slot as the transmit window.
2. If there is another node, node B, it receives the sink signal and locks on to the master. The node then asks for a transmit (TX) window. The master grants the TX window if there is no contention. Thus, node B is placed into the network in the slot right after the request slot, that is, node A's current slot. The master now does not need to know who is on the network. Instead, it needs to know how many people are on the network. The master knows that the sink signal needs to be sent right after the master's TX window. Figure 12.15 illustrates the new frame structure.
3. If there is contention in the new node slot, the nodes select a random number between 0 and 7, which represents how many more frames the node will wait before reattempting to request a TX window.
4. As more nodes enter the network, the same procedure is followed.

New Node	Node A TX	Sink

FIGURE 12.14
Frame structure at startup [26].

New Node	Node B TX	Node A TX	Sink

FIGURE 12.15
Frame structure after second node has joined [26].

The procedure just described focuses on nodes that are entering the network. The protocol should also be adaptive when the nodes leave the network. When a node does not transmit a packet, the network concludes that the node left the network. When a window is not used, all of the nodes that are in later time slots decrement their time slot for the next frame. Thus, the slot that is leaving the network is closed. However, this may cause some contention. Since all nodes must hear one another, contention can happen if one of the nodes does not receive the signal regarding their ID number change. If this situation occurs, the two nodes that are colliding can sense a collision when an acknowledge packet is not received, meaning the packet was trashed. Both of the nodes will then withdraw their transmit windows and reapply to enter the network. No packets are lost during this transition because one packet buffer is used.

12.3.3.3 Implementation Results

Both TDMA and ATDMA protocols are implemented in the communication module of the GroundScouts. In the swarm intelligence applications, we tested both protocols with success. Both protocols can be used for swarm intelligence, but ATDMA can handle larger swarms by allowing multiple swarms.

The TDMA protocol has the following specifications:

- Static master—the master has different code from the slaves
- Error control—done with packet numbering and checksum
- Fixed frame length—the number of robots in the network is fixed (1 to 15).

The ATDMA has the following specifications:

- Dynamic master
- Ability to create subnetworks
- Error control—done with packet numbering and checksum
- Dynamic frame length—a maximum of 255 users can be on the network
- Minimization of unused slots—maximizes data transfer rates

The only disadvantage of the ATDMA system is that robots are required to send a NULL packet when they have no data to transmit. This wastes power, but the communication module on the robot does not consume very much power compared to other modules. Thus, this waste of power is considered negligible.

We have presented modular microrobots, called GroundScouts, designed for swarm intelligence research and applications. Their hardware and software modularity present great opportunities for testing and experimenting with swarm intelligence-based algorithms and applications. Next, we present a widely know swarm intelligence algorithm, ACO, and a widely known humanitarian application, mine detection. The next section presents an ACO-based mine detection algorithm and experimental results of the algorithm using GroundScouts.

12.4 Mine Detection with Ant Colony-Based Swarm Intelligence

The first ant colony-based algorithm was developed by Colomi et al. [84] in 1991. They carried out an experiment that showed how ants lay *pheromone* trails as they move, which creates a high probabilistic path for other ants to follow. As more ants follow the same path, the pheromone strength increases. Colomi et al. [84] states that this is a form of an *autocatalytic* behavior—where the more the ants follow the path, the higher the probability that other ants will follow the path. This behavior results in a group of agents having low-level interactions/communication that create a cooperative network.

To test the results of the observation, Dorigo et al. [16,85] applied the algorithm to the traveling salesman problem. This inspired other researchers to apply this technique to other optimization problems [15]. In the field of communications, researchers have been studying different means of solving the ever-growing network-routing problem. Some researchers have shown great success in applying this technique to mapping out various optimal routes for networks [18,86,87]. Other studies have been conducted with ACO-based algorithms in data mining [88] and scheduling [89] with good results. These developments suggest strongly that the mine detection problem could benefit from the ACO-based algorithm as well because of excellent optimization characteristics.

We (Multi-Agent Bio-Robotics Laboratory Group at the Rochester Institute of Technology) have done simulation studies of the application of the ACO to the mine detection problem [6–8]. Results showed that the algorithm was well suited to the problem. Then, we implemented the ACO-based mine detection algorithm on GroundScouts and tested their performance. Next, we give some details about the ACO algorithm implemented on GroundScouts and the results of their performance.

12.4.1 Simulation of Mine Detection with ACO

In this section, we first give an overview of ACO. Then, we present our ACO-based mine detection algorithm. Finally, we present some simulation results regarding the mine detection algorithm with the ACO approach.

12.4.1.1 Ant Colony Optimization

Ant colonies are well-coordinated and well-organized entities exhibiting a bottom-up approach with no rigid hierarchy. In an ant colony, individual manifestations of stimuli and responses are in response to local problems. These local solutions then grow collaterally to solve the complex global problem. The major characteristic of the ant colony system is the absence of central control. All of the ants are assumed to have absolutely no or very few means of direct communication among themselves, yet they have the responsibility of carrying out the routine tasks of the colony system. Deneubourg's model emphasizes positive feedback obtained by individuals that reinforces the ability of the ants to interact at the colony's hierarchical levels [90]. Positive feedback is the imposition of popular actions performed by agents, which is caused by the performance of the action itself. Positive feedback often results in chaos, which is then countered by negative feedbacks. A dynamic system becomes successful when there is a balance of forces resulting from positive and negative feedbacks. Another important and interesting feature of the ant colony is the amplification of fluctuations. Positive feedback acts as an important component in amplifying the noise-like random fluctuations inherently present in the system (mostly produced without a source, as in the case of oscillators) to produce varied and newer properties for exploration that could enhance the overall performance of the system.

12.4.1.1.1 Navigation in Ants

Navigation in ants is a task involving the coordination of nestmates. The basic means of communication during navigation among the individual ants in an ant colony is done by creating pheromone trails. Pheromones are special chemical substances secreted by an ant during motion to convey the information that it has followed a particular route. Ants tend to follow routes rich in pheromone concentration.

The use of *cognitive* or *visual cues* (terrain and celestial cues) by ants during navigation has been reported in *Polyrhachis laboriosa* or tree-dwelling ants [91,92]. This type of navigation is generally known as vision-based navigation. In certain ant colonies, the visual cues act as a means for individuals to evaluate their position with respect to certain known coordinates (usually the nests). This can be done in two ways: path integration and the ability of ants to remember their positions during motion. In certain species, a correlated random walk is also seen [91,92].

12.4.1.1.2 Collective Transportation in Ants

Collective transport is a complex phenomenon in insect colonies that involves detection and physical transportation of an object (mainly food particles or brood) from one location to another. Collective transportation involves features such as the coordination in collective movement and stagnation recovery. One of the main aspects of collective transportation is the study of cooperative prey retrieval in ants. Cooperative prey retrieval is the art of finding the prey, deciding on how it can be transported to the destination (nest), and the physical transportation of it to the nest [90].

When prey is detected, a decision has to be made as to whether the object can be lifted or pulled by the detector alone or whether more of the nestmates have to assist. In most cases, the main factor that leads to group transportation is the weight of the prey. A solitary ant initially tries to realign itself and the prey to test for solitary transportation. In cases when this turns out to be futile, group transportation is preferred. Other factors that can lead to group transportation are the prey's resistance to motion (*Pheidole pallidula*) and the preference of the prey [90]. We can thus see that it is the manifestation or accumulation of the individual's inability in execution of the task that is responsible for group's responsive patterns toward the task change.

12.4.1.1.3 Recruitment of Nestmates

Recruitment of nestmates is the process of attracting nestmates when an ant succeeds in detecting prey and, after certain attempts, realizes that solitary transportation is not possible. Holldobler showed that, in the species of *Novomessor* (*Novomessor albisetosis* and *Novomessor cockerelli*), two distinctive processes were involved in the process of recruiting [93]. In the first process, the ant that detects the prey spreads a scent around the prey, which attracts the ants present in the local vicinity of the prey. This is called SRR [90,93]. In most cases, the concentration of the secretion decreases spatially. It was also observed that, as time progresses, the number of ants following the scent decreases, which may be due to the scent evaporating or to allow for some other effective techniques to be followed by the ants at the prey so that collective transportation is possible. SRR is followed by LRR, a process wherein the ant lays a pheromone trail from the point of the scent to the nest so that other ants may follow to assist in collective transportation [90]. It was shown by Holldobler [93] that LRR is chosen only when the inability of the group of ants in moving the prey cumulates over time.

The behavior as to how an ant resorts to attracting other ants around it or in the ant colony gives us important clues on how coordination in a collective system can be achieved. The scent acts as the component of stigmergy in the case of the SRR, whereas the pheromone trail does this part in the case of the LRR [90,93]. This type of behavior is also noted in the case of defense strategies in an ant colony system. The basic aim of the coordination mechanism is to bring the optimal number of ants to the point of action. We may see the scent concentration in the region around the prey decreasing as more and more ants are recruited for the action. The number of ants required is decided based mainly on the size of the prey to be retrieved. As the concentration of the scent decreases, one can see that the tendency of the ants to react toward the SRR is reduced.

12.4.1.2 The Mine Detection Problem

The standard set for humanitarian demining is 99.6% guaranteed clearance by the Humanitarian Demining Development Programme [94–97]. The use of mines both in war and defense is a growing concern for international peace and stability. Traditional or classical techniques of control engineering were primarily used for the technologies involved in this area, but recently, due to the boom in distributed control, a trend advocating the use of independent intelligent agents (usually multiagents) in demining has emerged. The adaptation of a particular trend in the field of demining has not been easy. Much scrutiny would be done in deciding to adopt a particular technology because of the sensitivity of the whole issue.

The number of people killed by land mines is ever increasing. More people die or get maimed by land mines than by intercontinental (ballistic) or nuclear weapons [94]. In fact, there seem to be more concrete international conventions and laws on ballistic and nuclear arms than on land mines. One of the most lucrative weapons for war or terrorism is land mines because of their low cost and easy deployment. Most often, a single land mine costs less than $10, which is a very low price in the international market. The deployment also does not take a lot of effort or manpower. Figures state that on average, as many as 110 million remain planted in the ground on every continent, with more than 80% of them found in Asia and Africa alone [94–96]. It is estimated that, at the present pace of clearing mines, another thousand years would be consumed in de-mining them completely if no more are planted in this period. Every day, about an average of 70 people are directly killed or maimed due to land mines [94]. The placement of land mines is normally in difficult terrains. Mainly hilly regions, mountain slopes, river beds, and forests are chosen for their placement, which makes them all the more difficult to remove. Eighty percent of the mine removal process directly involves humans, which results in greater causalities of life [94]. Because land mines are normally found in Third World countries, availability of funds for such a cause has become the reason for involving humans in mine detection and removal.

Problems posed by sensors in the mine detection process are getting narrower due to advances in technology [95–97]. When sensors with a considerable degree of accuracy are available, using robots in the process can be more feasible both in true detection of mines and in minimizing false alarms. Some of the main aspects to be looked at when robots are (re)deployed for the detection process are their ability to distinguish between mines and nonmines (reduce lost time due to false alarms), their ability to operate under varied types of terrain and climatic conditions, their ability to detect a variety of mines, and their cost-effectiveness in construction. In general, robots are used in minefields where locations of mines are unknown. Under such a scenario, a proper time-effective foraging strategy has to be devised for the robot(s). A multitude of heuristic and random strategies are used for this purpose. Most of these strategies depend on the minefield and the construction of the robots themselves. These strategies should de designed to include the following issues: low cost and simple operation for the built robots, durability of the robots, and quicker detection of the mines. Some examples of mobile robots used in mine detection are lightweight robots such as Pemex-BE, legged robots COMET-II, vehicle mounted mine detectors, and suspended robots [98].

Before using any robot for the demining process, the terrain should be prepared artificially according to the needs of the mechanical aspects of the robots. This may require removing vegetation and other obstacles in the field, either by mechanical flails or human hands. When robots are used for mine detection, two questions arise: "How are the robots detecting the mines?" and "How are the robots removing the detected mines?" Robots can be built with separate detection and removal strategies. Many robots are programmed to detonate mines when they detect them. However, this could be dangerous in the sense that it can lead to physical damage to the robots used. In other strategies, mechanisms can be devised to carry the detected mine to a monitoring place where it can be safely detonated, but this would demand special abilities of the robots for the removal process [94].

In the literature, most research focuses on the utilization of heuristic techniques for robots in the mine detection process [98]. Research has experimentally found that using a number of robots could be more effective in the detection process as against a single robot in terms of cost effectiveness and time consumed [96,98,99]. Intelligence can be bestowed to the robots as centralized or distributed. In centralized intelligence, a robot could be partially or totally dependent on a central agency for advice and feedback. This is a type of master-slave relationship between the controlling agency and the robot entity. The central monitor could be humans and/or GPS, which can communicate directly with the robot(s). Traditional centralized control techniques are usually computationally intensive and lack robustness and flexibility. However, a distributed control environment does not have any centralized control monitoring the individual robots. All of the robots are independent in decision-making and execution, constrained by certain control laws. It is basically in this regard that multiagent studies or distributed artificial intelligence comes into the mine detection domain [98].

The most inspiring form of distributed intelligence is biological intelligence. Nature bestows intelligence in various forms. Swarm robotics is division in multiagent theories that has been vastly explored in this regard. In swarm robotics, the individual robots (which are equivalent to swarm members in nature) do not need to have high coordination and communication complexity to execute their common goal. All of the robots manifest information on the environment and react accordingly to the environment. The main problem that robots in a real mine detection situation could face is that they can get disoriented or lost on their path. However, since swarm members do not share anything in common (no direct communication or reference), the system, as a whole, would not be affected much. In fact, these fluctuations from normal behavior can be used for more explorations, which could expedite better results. A good example for swarm robots is the Growbots of the Idaho National Engineering and Environmental Laboratory [95]. A good method of adaptive learning can be implemented in the robots both about the environment (terrain) and foraging knowledge. Algorithms should be robust enough to accommodate for false alarms, robot damages (which can result in the loss of robots in service), and misses in finding mines.

In this chapter, by the mine detection problem, we mean the problem of detecting and fusing randomly placed mines over a field. We assume that the agents are equipped with the ability to detect mines when they approach them physically. The fusing of the mines demands the collective action of a certain number of agents. The agents are considered to be individual ants performing the task of prey detection and retrieval. We assume that the mines are tantamount to the prey that the ants, which are all identical in behavior and movement, have to detect over a field area. We give the ants a foraging strategy in order to scout for the mines. When an ant reaches a mine, it spreads a scent around the mine in resemblance to the SRR found in certain ant colonies. This scent decreases exponentially in a region around it. The ants around the mine follow the scent's increasing gradient and finally land themselves at the mine. We should note that the ants would not lose their foraging behavior even when they are intimated by the scent about the presence of a mine in the locality. The following of the scent is deterministic, and no random movements are found inside the area covered with the scent. Thus, the mine would be eventually defused when the required number of ants arrives at the mine's location.

12.4.1.3 ACO Model Used for Mine Detection

The approach that we have used for the mine detection problem is a combination of both deterministic and stochastic methods. We bring in the stochastic component only in the foraging stage, whereas in the stage when an ant enters a scent area, it does not behave stochastically, but rather in a deterministic manner.

An important concept that has to be observed during the process is when all of the ants go into the state of waiting for the others to come to their respective mines. Factors leading to this can be the field size, the initial distribution of the mines, and the ratio of the number of ants deployed to the number of mines in the field. We call such a situation *freezing*, where there is no motion of the ants and all of the ants are waiting infinitely for the others [6–8].

TABLE 12.1

Condition for Behavioral Changes

Behavior Transition	Conditions
F to TF	When a foraging ant detects the scent
TF to W	When an ant follows a scent concentration to a mine
TF to F	When the maximum of the scent intensity in the trail becomes zero
F to W	When a foraging ant detects a mine
W to F	When an ant has waited at the mine for a specific amount of time
	When the mine at which it has been waiting is defused
W to TF	When an ant has waited at the mine for a specific amount of time (the threshold time) or when the mine is defused, but the ant finds a scent around it

In our ACO-based mine detection algorithm, the ants have three main behaviors: foraging, trail following, and waiting. The ants change their behaviors (states) based on the environmental conditions. Table 12.1 lists the conditions for each behavioral transition. In the table, F, TF, and W stand for foraging, trail following, and waiting, respectively. Similarly, the ants produce reactions with the behavioral transitions. Table 12.2 lists the reactions ants (swarm robots) generate based on the behavioral transitions.

Simulations were run over a field size of 100×100 and results were obtained as shown in Figure 12.16. The distribution of the mines was done uniformly over the field. We acknowledged that the time required for the ants to completely clear the field (defuse all of the mines in the field) is a good criterion for reporting the efficiency of the mine detection process.

When the mines are densely concentrated over the field, there may be instances where two or more scents can overlap. Under such instances, the algorithm should ensure that the peaks in scent intensity occur only at the mines and that there are no or very few local maxima in the overlap region, which may mislead the ants as the ants within the scent area move only toward

TABLE 12.2

Reaction Produced on Behavioral Changes

No	Behavior Transition	Reactions
1.	F to TF	Initiate the scent sensor (in addition to the mine sensor, which should be read first), which can read scent intensities at all of the points around it and pick the point with maximum intensity
2.	TF to W	Do not activate scent spreading
3.	F to W	Activate scent spreading
4.	W to F	Deactivate scent spreading if it is activated
5.	W to TF	Read scent sensor

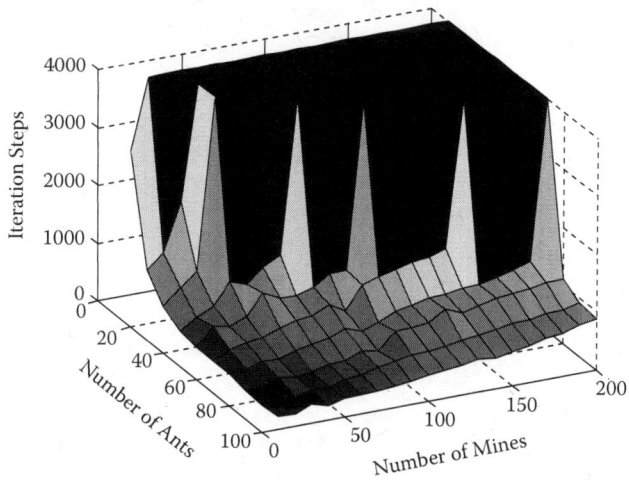

FIGURE 12.16
Results for runs on a 100 × 100 field.

the increase in scent concentration. Local maxima, which can normally occur in the overlap region, could be detrimental in restricting the movement of the ants within the scent-spread region. The behavior that the ants exhibit when they enter the scent-spread regions is following the increasing gradient of the scent concentration. They have the belief that the concentration of the scent peaks at the location of the mines. When local maxima occur within an overlapped region, ants could be trapped in regions where there are no mines. This causes a temporary stoppage of the ants' motion, which can be cleared only when the local maxima disappear. Thus, the ant is unavailable for the task for a short duration. It should be noted that this will not lead to a false detection of mines as the behavior of ants in the scent spread region is twofold: scent following and sensing for mines. At the local maxima, since there are no mines, the ant would not go into waiting mode. Therefore, although this problem is not drastically detrimental, it needs to be considered in the overall interest of the problem. We have devised a method to minimize the number of ants falling into the overlap region. The method assumes that, if we prevent ants that are normally outside the overlap region from coming into the overlap region, the probability of the ants being in the overlap region would be less during the time of the mine detection process.

For the spread of scent, we use a falling two-dimensional exponential curve. Thus, the exponent has to be less than 1 in the case when we use integers for raising the exponent. This may be good enough when there is no overlap in the scent regions, but when there are overlaps (shown in Figure 12.17) of the scent field (this can happen when two or more mines that are located relatively close together are detected by two ants simultaneously or nearly

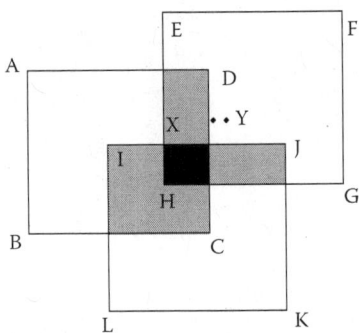

FIGURE 12.17
The overlap of three scent spread regions.

at the same time), care should be taken in fixing the exponent for the scent distribution. Let us explore a scene where two overlaps have occurred and how the exponent should be modified.

Overlap of scent regions on the simulation (rectangular overlaps of three scent regions) takes place in the following manner: The mines are located at the center of the rectangle. The scent concentration peaks at the center of the rectangle and exponentially falls to the boundaries.

Let the exponential fall go according to the function $f(x) = a^x$ for x values ranging from 0 (at the center of the mine) to m (the value it takes at the boundaries). Therefore, if a lies in the range from 0 to 1, m is the largest value that x can take. If we were to prevent an ant from entering an overlap region from outside (i.e., prevent it from entering the shaded region in Figure 12.17 from anywhere in the nonshaded region), the scent concentration along the boundaries (point X in Figure 12.17) should not be greater than that of the value just outside it (point Y in Figure 12.17).

Therefore,

$$a^{n+1} + a^m \leq a^n \tag{12.1}$$

where a^n is the scent contribution at point Y (point immediately outside the boundary) and a^m is the scent contribution at point X (point along the boundary). Scent contributions by independent rectangles are additive. From Equation (12.1), we can rewrite the equation as

$$a^n(1-a) \geq a^m \tag{12.2}$$

Then, taking the logarithm of both sides of the inequality, we get the following equations:

$$n \log(a) + \log(1-a) \geq m \log(a) \tag{12.3}$$

$$(m-n)\log(a) \leq \log(1-a) \tag{12.4}$$

The worst case [the case when the quantity $(m - n)\log(a)$ is closest to the constant $\log(1 - a)$] can occur when n takes the value $m - 1$. Therefore, when we substitute $n = m - 1$ in inequality Equation (12.4), we obtain the following:

$$\log(a) \leq \log(1 - a) \tag{12.5}$$

Therefore, we can write the following inequality:

$$a \leq 1 - a \tag{12.6}$$

This occurs only when a is in the range 0 to 0.5. Thus, we see that for Equation (12.1) to be valid, a (the exponent of the scent distribution) should be a number that is in the range 0 to 0.5. We see that for two or more overlaps, the range of values that the exponent can take becomes narrower. Thus, the concentration of the mines poses a constraint on the value that the exponent of the distribution can take. A feature to be observed is that the area of the scent spread has no effect on the exponent of the distribution, although it can affect the number of local maxima occurring in the overlap region.

As a measure of estimating the phenomenon of freezing, we define the curve of freezing. The curve of freezing is the curve showing the instances where freezing was seen 60% of the time on the simulation. Figure 12.18 shows the freezing curve for the number of mines versus the number of ants.

One solution to avoid the problem of freezing is to have a time bound for the ants to be in the waiting mode at a particular mine and, henceforth, resume their foraging. Research on natural ants suggests that the amount

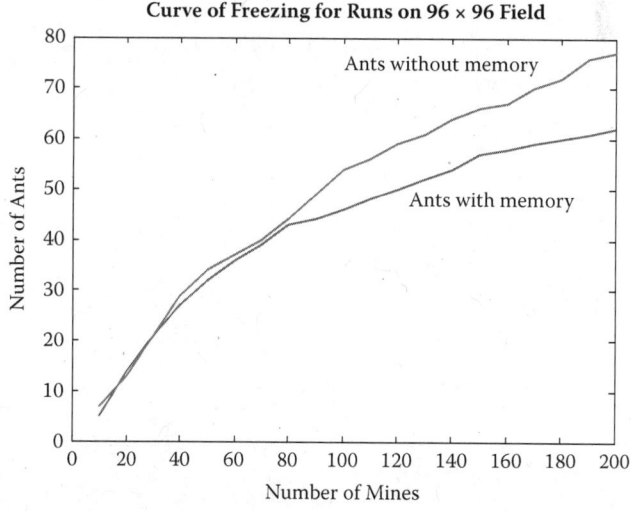

FIGURE 12.18
Curve of freezing for a 96 × 96 field [8].

of time spent by them in trying to pull a food particle (prey) is proportional to the amount of traction that the food particle offers to motion [2]. Thus, we designed the ants such that they change their behavior to foraging after they wait around the prey for a certain amount of time. The main problem in doing this would be that the released ant would be attracted back to the same mine (the mine it was waiting at when the time bound was crossed) by the scent concentration that would be around. To avoid this problem, we associate the scent spreading of an ant to be activated only when a foraging ant on its course finds a mine without the scent surrounding it for a fixed time, called *threshold time*. This implies that all ants, when they detect a mine for the first time, get their scent spreading activated, but the converse may not be true. All of the ants, when they reach a mine, set a counter that runs for a certain amount of time. When the timer reaches a specific value (the threshold time), the ant changes its behavior to foraging. In this way, the ant has a probabilistic chance of getting away from the mine or staying at the same mine, as the foraging behavior of the ants is random. All ants, which have their scent field activated at the mine, have the responsibility of deactivating the scent field when their time bound at the mine exceeds its limit. This approach can assure that the freezing would largely not occur. However, the time taken by the ants to defuse all of the mines would increase.

Figure 12.19 shows the performance of the ants with different threshold times. The results suggest that there could be an empirical optimal value determined by the minimal point on the curves shown below. The results

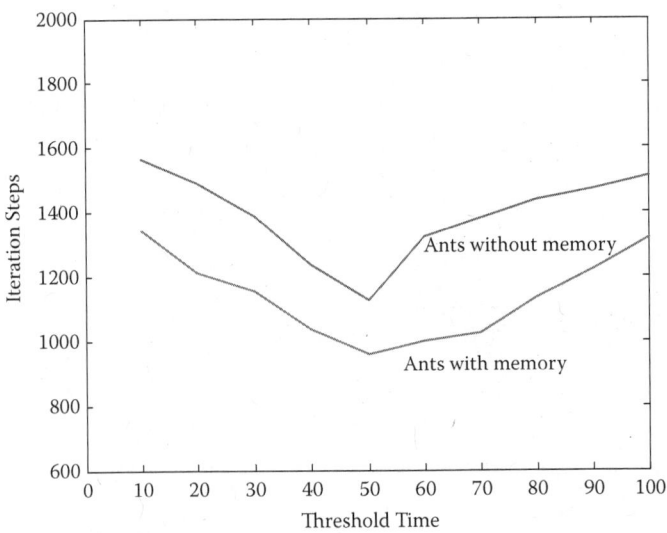

FIGURE 12.19
Iteration steps versus threshold time [8].

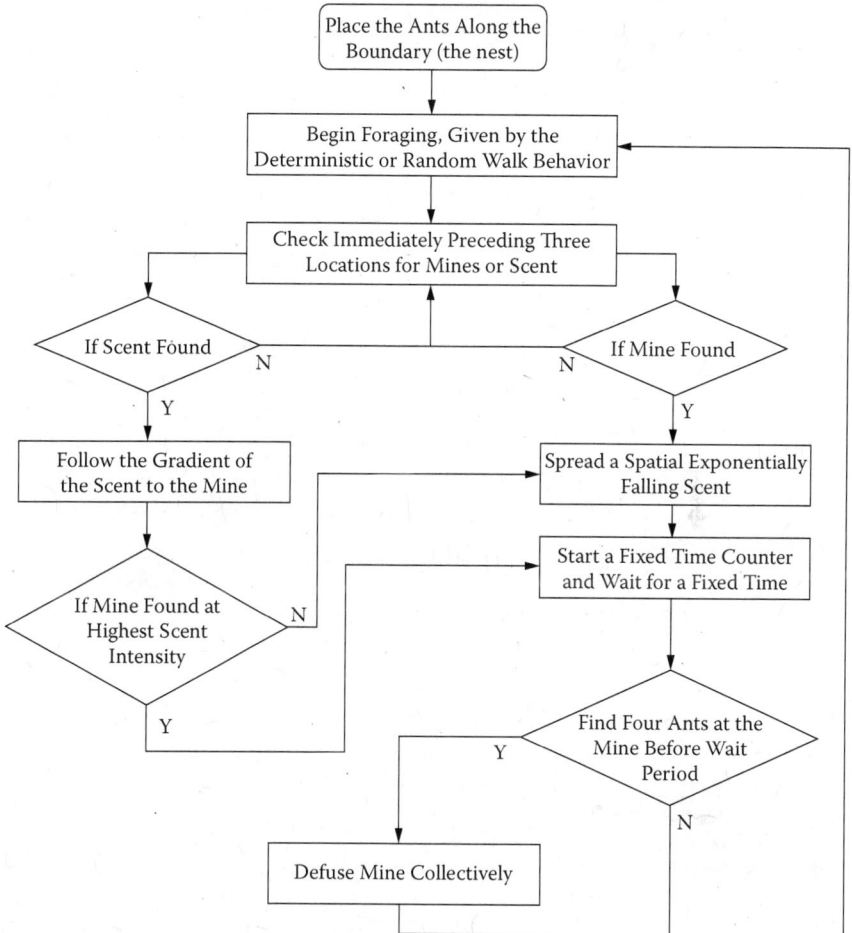

FIGURE 12.20
Flow chart description of the individual ant's tasks [8].

are given for runs over a field size of 96 × 96 with 100 ants and 50 mines. The entire simulation process as a flowchart is given in Figure 12.20.

12.4.1.3.1 Cognitive Maps

Another important issue to be considered apart from that of partner recruitment is that of having an effective foraging strategy. *Foraging* is the process by which the robots move over the field containing mines, trying to detect them. A foraging strategy can be defined as an action–reaction mechanism exhibited by the group of robots that aims at optimizing certain criteria such as the time involved in the demining process and assuring that the whole terrain is covered. The exercise of foraging is divided into subtasks called

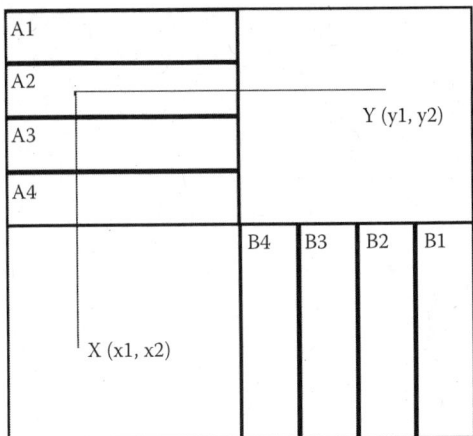

FIGURE 12.21
Motion of an ant over the field in accordance with its cognitive map [8].

navigations. Gallistel [100], in *The Organization of Learning*, defines navigation as follows:

> "Navigation is the process of determining and maintaining a course of trajectory from one place to another. Processes for estimating one's position with respect to the known world are fundamental to it. The known world is composed of surfaces whose locations relative to one another are represented on a *map*."

One such map can be a *cognitive map*. Cognitive maps are representations of a geographical, quantitative model of a decision maker's subjective belief, the ants in our case. Cognitive maps in ants are suggested for devising the foraging strategy [91,92]. For the ants with memory, local interactions result in *knowledge* (useful cognitive map information) being interchanged. Thus, the ant's foraging strategy is devised with enriched knowledge as time progresses. The foraging strategy decisions are made with some cost functions. Figure 12.21 shows the motion of an ant. The motion is from a point X to point Y cutting across regions A2, A3, and A4 in the cognitive map.

The cost function we have chosen for our simulation is a probability distribution function given as follows: Let $a1^d, a2^d, \ldots, an^d$ and $a1^u, a2^u, \ldots, an^u$ be the knowledge of the defused and undefused mines in regions A1, A2, ..., AN on the field. An ant would tend to select a new course according to the following probability distribution:

$$P(y2 \in Ak) = \frac{1}{(1+\beta)}\left(\frac{N - ak^d}{(n-1)N} + \beta\frac{ak^u}{M}\right) \qquad (12.7)$$

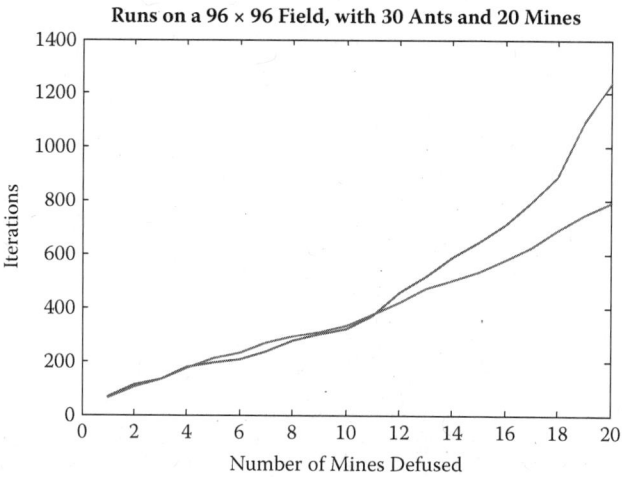

FIGURE 12.22
Comparison or performance of the ants with and without memory [8].

In the probability distribution equation, the notations M and N are given as follows:

$$N = \sum_{k=1}^{n} ak^d \qquad (12.8)$$

$$M = \sum_{k=1}^{n} ak^u \qquad (12.9)$$

The symbol β is a user-defined variable weighing the importance of the undefused and defused mines with respect to each other. The plot in Figure 12.22 is done based on a β value of 1.

As a comparison, to show the advantage of ants empowered with memory over memory-less ants, we define a quantity called the *reduction factor*, T_r. Variation in β is also found to affect the value of T_r. The reduction factor can be defined as follows:

$$T_r = \frac{T_f}{T_w} \qquad (12.10)$$

The variables T_f and T_w are the respective time iterations necessary to complete a specific mine detection task by the ants without and with memory, respectively. Figure 12.23 illustrates the trend in the reduction

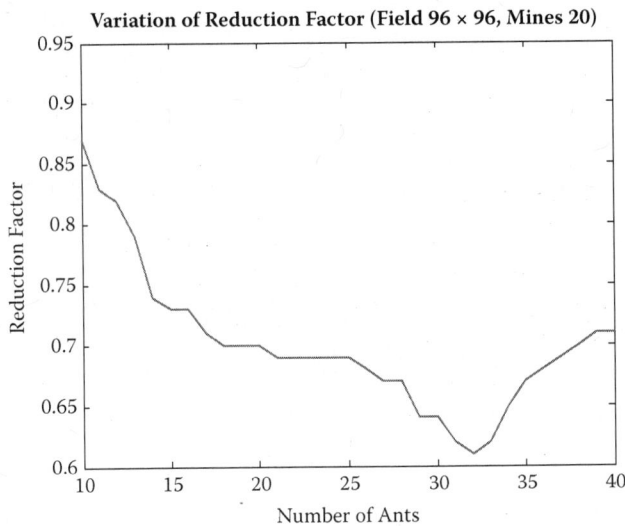

FIGURE 12.23
Variation of reduction factor with the number of ants used for foraging [8].

factor as the number of ants is increased for a specific combination of field size and mines.

Another measure of performance is the measure of the reduction factor, T_r, against changes in β. Figure 12.24 shows this effect when 30 ants are foraging a 96 × 96 minefield.

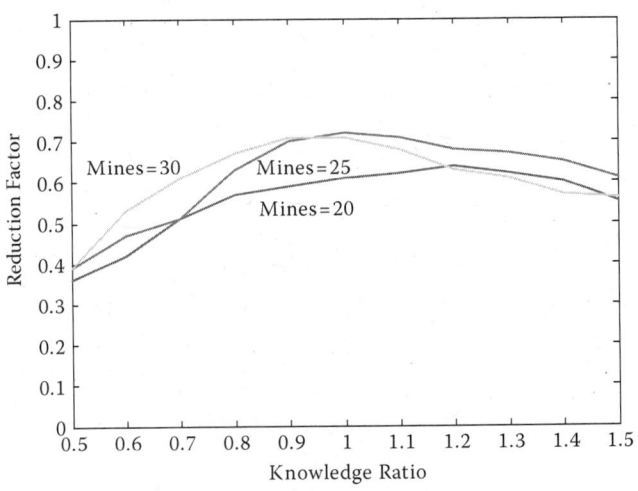

FIGURE 12.24
Variation of reduction factor with β [8].

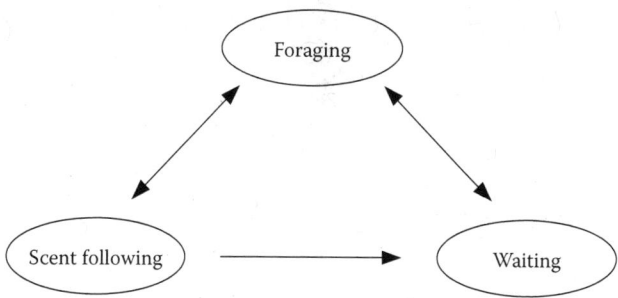

FIGURE 12.25
State diagram of the different ant behaviors [26].

12.4.2 Implementation of ACO-Based Mine Detection Algorithm on GroundScouts

In this section, we explore the implementation of the ACO-based algorithm to mine detection using modular swarm robots, GroundScouts. As explained in previous sections, in ACO-based swarm approach, swarm elements have three behaviors: foraging, trail following (scent following), and waiting [6,7]. A diagram of the behaviors (states) is shown in Figure 12.25.

Foraging in our approach is defined as the ants' random search for mines. The action is characterized by random movements throughout the search space. As mentioned before, the addition of cognitive maps to the state can increase the efficiency of the foraging state by giving the ant a memory mechanism so it does not continue to study the same location multiple times [7,8].

The scent following state is characterized by an ant following a scent (pheromone) left by another ant. For simulation purposes, the scent was represented as an exponential function that grew stronger as an ant got closer to the source. This provided a means for the ants to have a path to follow [7,8].

The waiting state is also referred to as the recruitment state. This involves the ants releasing the pheromone to attract other ants to come and assist using SRR and LRR approaches [7,8]. The mine detection problem is an optimization problem composed of a number of mines randomly dispersed throughout a space and swarm agents looking for mines to demine. The goal for the algorithm is to find and disarm all mines in as little time as possible using minimal effort from each agent [7,8]. As described earlier, our ACO approach to mine detection utilizes the foraging capability of the ants, which allows them to search a large space in a short amount of time and uses the recruiting capabilities to attract more ants to collectively disarm the mine.

There are three events causing transitions from state to state: finding a mine, finding a scent, and waiting timeout. During the finding-a-mine transition, an ant switches into the waiting state regardless of what state the ant is currently in if a mine is found, since the goal of the algorithm is to find and disarm all mines. In the finding-a-scent transition, if an ant (robot) is currently in the foraging state, it will switch to the scent-following state if a scent is detected. If the ant is in the waiting state, it will not follow the scent since it is already at a mine. Finally, in the waiting-timeout transition, the waiting timeout occurs when an ant has been waiting for a long time for other ants to come and assist in disarming a mine. In our approach, an ant will leave the mine and begin foraging again after a certain amount of time, called a threshold, has passed. This mainly prevents the freezing problem [7, 8].

As noted in the previous section, simulation results of Kumar and Sahin [8] illustrated that, by varying the number of ants, the size of the minefield, and the number of mines, the ants could indeed disarm all of the mines. Thus, the feasibility of a real-time implementation of the algorithm should be explored and tested. This section explores such an implementation with GroundScouts.

To realistically test the mine detection algorithm, we need to implement mines, scents, and the main program for the mine detection algorithm. In the main program, we explore the implementations of foraging, scent following, and waiting. Finally, the graphical user interface (GUI) developed for the experiments is presented.

12.4.2.1 Implementation of Mines

The mines consist of two main hardware components: a communication module and an IR beacon that constantly transmits an IR signal modulated at 38 kHz in all directions [25]. A picture of the beacon is shown in Figure 12.26. The GroundScouts IR sensor layer can detect this IR signal within a 7-ft radius. In the foraging stage, the robot is constantly searching for the IR

FIGURE 12.26
(a) Outside view and (b) inside view of the mine [25].

signal transmitted by the mines. As soon as the signal is detected, the robot concludes that it is close to the mine and tries to move toward the mine.

The direction of the IR signals is found by viewing the five IR sensors in the IR sensor layer of the GroundScouts. This has proven to be difficult since the sensors are somewhat omnidirectional, resulting in a number of sensors having good signals, making it difficult to determine the exact direction of the mine. Finding the direction could be simplified by looking for the two sensors that have the worst signals. The robot could then move in the opposite direction, which would be directly toward the mine.

The mines are disarmed by sending a defuse signal to the communication module (shown in Figure 12.26a) on the mines. When enough robots (four robots) are surrounding the mine to disarm it, a message is sent by the command center (GUI on a laptop) to the communication board telling it to disarm the mine so that the robot can start foraging to find other mines in the field. The PIC on the communication board will then toggle a pin that will turn the mine off. As soon as the mine is disabled, the robots will then shift back into the foraging state since the signal from the mine will not be available.

12.4.2.2 Implementation of the Scent

The scent around the mines is implemented by the IR signals of the mines. The scents that robots secrete are implemented by the communication module in the form a packet. These packets have the information about the scent the robot is secreting. To make sure that the scent is detected only within a certain distance, the robots can process scent packets coming from the robots within a certain distance [26]. This poses a problem: How can the robots be programmed to ignore messages that are outside of their listening, or smelling, radius?

A realistic solution is to create an internal coordinate system on the robots so they know where they are in reference to one another. The robots maintain an X and Y coordinate, as well as a direction. At the start of the experiments, all of the robots start at the same position pointing in the same direction. When the robots send a packet to other robots, the coordinates are embedded in the packets. Then, each robot calculates distances to the transmitting robot using these embedded coordinates. There is a limit to the distance a robot can sense another robot's scent. If the distance is larger than this threshold, the robots simply ignore the messages from those robots. This is also effectively emulating vaporizing pheromone in an ant colony.

There are three different messages that are communicated among the robots. The first message signals other robots that a mine was found. This mimics the scent. The second message is broadcasted when a robot is leaving a mine. This message is sent when a robot times out after the waiting threshold has been reached. This message is also used to allow robots to determine exactly how many other robots are surrounding the mine. The third message is broadcasted when a robot has disarmed the mine. This

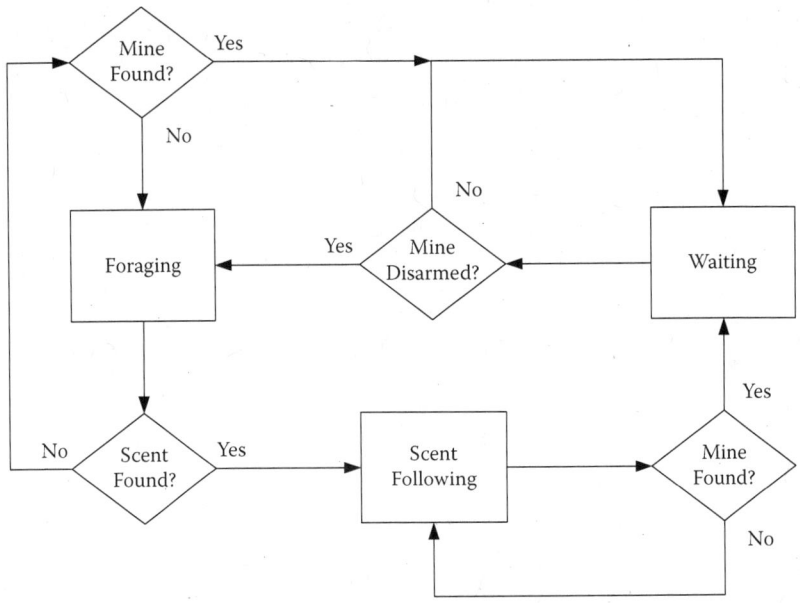

FIGURE 12.27
State machine for the overall program [26].

message also lets the robots waiting around the mine know that the mine was disarmed.

12.4.2.3 Main Program

The main program is implemented as a state machine with three possible states: foraging, scent following, and waiting, as shown in Figure 12.25. Figure 12.27 illustrates the flow chart of the main program state machine of the main program.

In the state machine of the main program, the *scent found* event occurs when a packet is received saying that the robot is at a mine when the robot is currently in the foraging state. The robot will then switch immediately to the scent following state, where it will go to the location of the mine. The *mine disarmed* event occurs when the front IR sensor no longer picks up a signal from the mine, meaning that the mine is off. Next, we describe the individual states in the state machine of the main program.

12.4.2.3.1 Foraging

The flow chart of the foraging scheme implemented on the GroundScouts is shown in Figure 12.28. In GroundScouts, every movement of the robot is set to be 6 in. Thus, setting $n = 6$ makes the robot move forward 3 ft, assuming

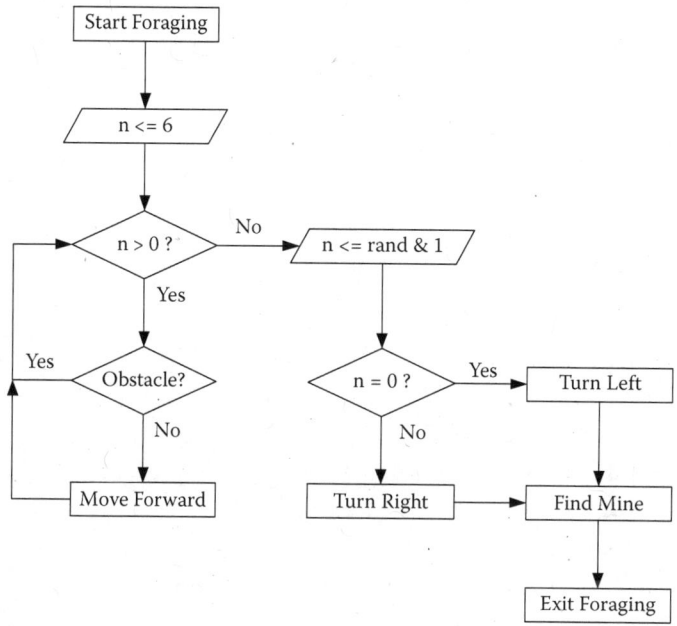

FIGURE 12.28
Flow chart for the foraging state [26].

there are no obstacles ahead. When the forward movement is finished, the robot will choose a random number in order to determine which direction it will turn. If the random number is even, then the robot will turn left. Otherwise, the robot will turn right. The robot stays in the foraging state unless it finds a mine.

12.4.2.3.2 Scent Following

Figure 12.29 illustrates the flow chart of the scent following algorithm (state).

12.4.2.3.3 Waiting

In the waiting state, the robot periodically transmits a packet that informs the other robots that it has found a mine and where the mine is. The flow chart of the waiting state is shown in Figure 12.30.

The front IR *if* statement in the chart checks to see if the mine has been disarmed. The following part of the algorithm checks whether the robot timed out. If the robot did indeed time-out, then it will turn around and travel 15 ft away from the mine. This ensures that the robot does not instantly find the same mine again since other robots around the mine may still send scent signals.

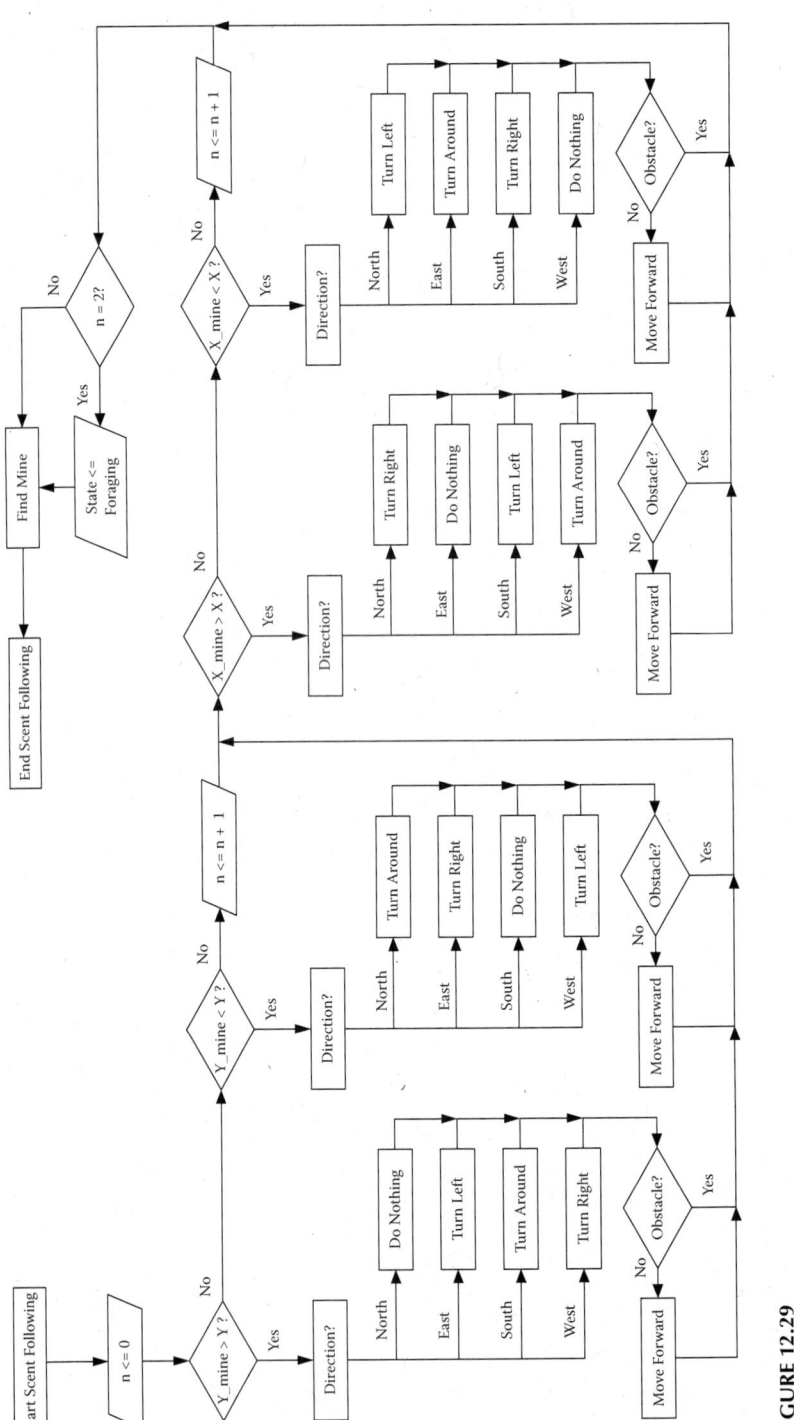

FIGURE 12.29
Flow chart for the scent following state [26]. The algorithm looks complex. However, all it does is compare the Y coordinate corresponding to the location of the mine to the current Y coordinate and decides the direction that the robot has to travel in to get closer to the mine. It does the same for the X coordinates. At the end, the find mine algorithm is run. If $n = 2$, then the robot is at the location of the mine and did not find the mine. The robot will then go back into the foraging state. This is a safeguard against the situation where the internal coordinates of the robot become off-center, causing the mine to be in a different place from where the robot thinks it is.

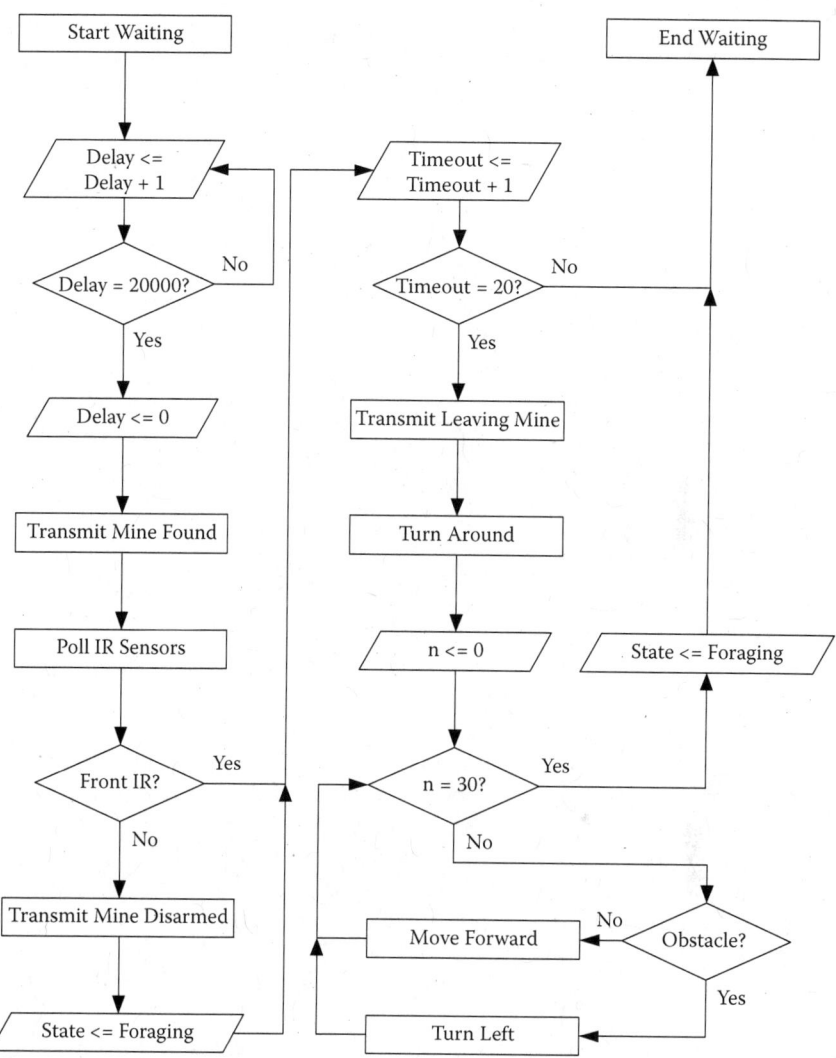

FIGURE 12.30
Flow chart for the waiting state [26].

All the three states are implemented on the 8051 controller in the controller module of the GroundScouts. As mentioned above, some functionalities of the mine detection algorithm are implemented by a laptop PC through a GUI.

12.4.2.4 Graphical User Interface

A GUI [74] is designed to function as the base station, human supervisor, or a master robot. It controls the deployment of each robot, shows the

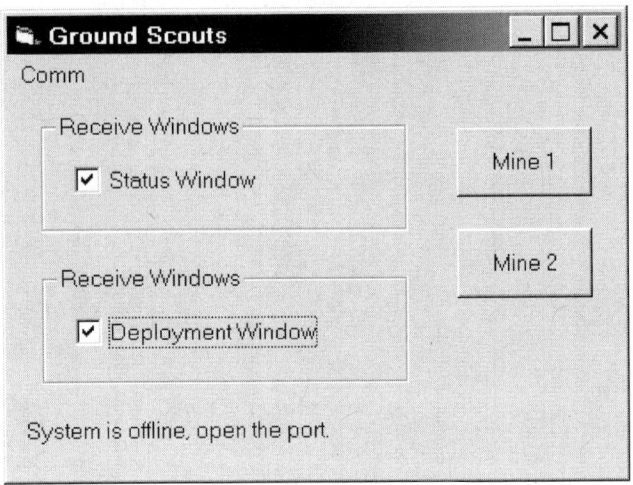

FIGURE 12.31
Initial GUI window.

current state of each robot in the system, and finally sends the "disarm" command to a surrounded mine. The GUI is able to monitor the crucial information about robots, including the status of each robot in the algorithm, sensor readings for each robot, and GPS readings. During system operation, many messages could be observed originating from any robot in the system detailing what each one was doing, be it searching for a mine, if it was locked onto a mine and tracking it, aiding another robot, or at a mine and signaling for aid. Each of these messages has a unique system code and could easily be seen by the user from a personal computer through the GUI. Figure 12.31 shows the initial window of the GUI from which the system status window, as well as the robot deployment window, could be turned on and off.

The *Comm* menu contains the option of enabling the wireless communications port that is needed in order to send and receive messages between the GUI and the system robots. In the TDMA scheme, the GUI is always assumed to be communication board "1" in the system as an identical communication module is connected to the computer (base station). Figure 12.32 shows the main screen of the GUI. The box in the lower panel is where status messages are printed.

It is up to the user monitoring this window to disarm a surrounded mine whenever it is seen that the necessary number of robots have surrounded a mine at the same position. This scenario is easily observed as each robot will be reporting that it has found a mine. The *Mine 1* or *Mine 2* buttons in the window from Figure 12.31 is used to disarm the first or second mine,

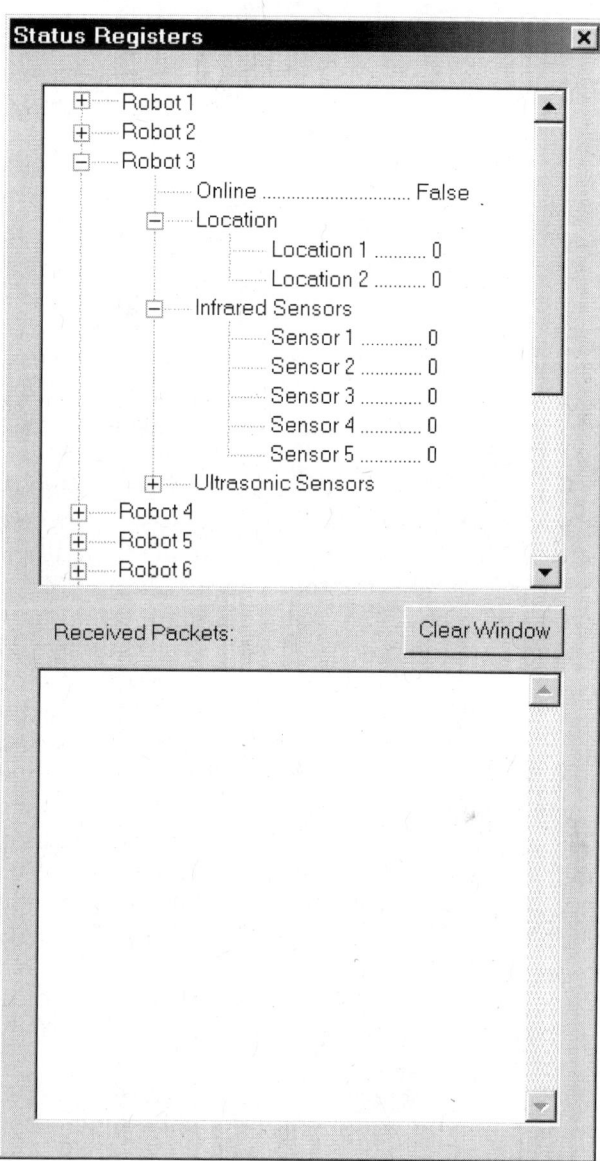

FIGURE 12.32
GUI robot status panel.

respectively. Disarmament is done by sending the disarm command to the communication board attached to the mine. The GUI is not only used as a monitor of the system but also as a command and control station. The GUI can send directives to the robots and mines if the experiment requires it.

12.4.3 Experimental Results

The experiments were performed in a basketball court at Rochester Institute of Technology so that the robots had plenty of room during the operation. Figure 12.33 illustrates a picture of the starting point of the experiments with five robots and two mines.

The robots were turned on one by one by allowing each robot to move about 3 ft before the next robot was turned on. As noted previously, the robots are in the foraging state at the start of the algorithm. Figure 12.34 clearly shows the robots randomly searching for mines, in the foraging state.

The robots near the top of Figure 12.34 are beginning to find the first mine. As robots detect mines, they signal other robots using scent packages in order to bring the other robots toward the mine. The robot continues to move until the back IR sensors have no signal and the ultrasonic sensors are picking up an object closer than 6 in. Figure 12.35 shows a robot at a mine.

As soon as the robot reaches a mine, it begins sending out the recruitment signals (scent packets) to other robots. Some problems were encountered with this approach. Since the implementation of the internal coordinate system neglects slippage, in time, the robots' internal coordinates accumulate errors and become off-center. As the robot at a mine sends out the recruitment signals, other robots within the physical distance may not hear (disregard) this signal since they are outside of the *listening* range according to the coordinate system. Another problem is that sometimes a robot would hear the signal but could not successfully reach the mine because of the inaccuracies of its internal coordinate system due to slippage.

As soon as four robots surround a mine, the mine can be turned off by the GUI. The robots determine whether a mine is turned off by checking their front IR sensor. If no IR signal is detected, then the robot assumes that the mine is disarmed. Then, they instantly switch into the foraging stage, which incorporates obstacle avoidance as shown in Figure 12.36.

Some problems were encountered during the disarming. First of all, the robots have difficulty seeing each other using the ultrasonic sensors because of the limited surface area of their physical structure. Once they decide that the mine is disarmed, they go back into the foraging state.

The implementation of the algorithm proved to be successful, but there were some problems worth mentioning. The following are the three main problems the experimental setup dealt with:

- As time progressed, the internal coordinate system became very inaccurate because of slippage. This is a common problem for a robot using dead reckoning for position. This is mainly caused scent-following problems.
- Robots that were right next to each other could not hear one another because of interference. This was because the transmission power level was full, causing the received signal to be saturated. Thus, it

FIGURE 12.33
Starting point of the experimental setup [26,32].

FIGURE 12.34
Robots in the foraging state [26,32].

FIGURE 12.35
Robot at the mine [26,32].

FIGURE 12.36
Robots disarm the mine and leave [26,32].

was impossible for a robot to know how many other robots were around the mine when they were very close to each other.
- The algorithm that brought the robots next to the mine using the IR sensors worked, but not with the desired efficiency. The robots would sometimes go directly next to the mine, miss it, and have to turn around and try again. However, this is a common behavior of a swarm member as well in many insect colonies.

As stated before, the implementation of a robotic swarm based on ACO for a real problem is accomplished successfully. The robotic swarms are designed with the characteristics of a swarm and/or an SoS. Mainly, their modular architecture and ATDMA for robust communication among the swarm members are inspired by these characteristics, such as integration. The GroundScouts are also used to test another swarm intelligence technique, artificial immune systems, for the same mine detection problem [74]. This also proved that GroundScouts' modular design and their communication scheme are suitable for swarm intelligence applications and concept tests.

12.5 Conclusion

In this chapter, we have presented a design of a modular robot for swarm intelligence applications, called GroundScouts, using some of the concepts from SoS. In addition, a mine detection problem is solved with an ACO-based algorithm and implemented with swarm robots, GroundScouts. The algorithm is tested both in simulation and in the field with real robots and mines. As mentioned before, a swarm can be considered an SoS since each member can operate independently, and they can accomplish a test collectively. These satisfy the integration and interoperability characteristics of an SoS. With these characteristics in mind, we have presented how the hardware and software architectures can be designed and tested on a real-world problem, mine detection.

As noted in the Introduction, the SoS field is new and there are no standards for the terminology and theory. We have presented existing definitions and their characteristics. Then, we evaluated similarities between robotic swarms and SoS concepts. As the concepts of SoS become clearer and a better theory of SoS is developed, one may design better swarm robots. In fact, robotic swarms could be a testing ground for the SoS concepts and theory. The MATLAB® files for the simulation of mine detection using the ACO-based algorithm can be found in the book's website.

Acknowledgment

The author appreciates the help of graduate students Eric Chapman, B. J. Opp, and Vignesh Kumar in preparing this chapter.

References

[1] Hinchey, M. G., Sterritt, R., and Rouff, C., "Swarms and swarm intelligence," *Computer Society Magazine* 40(4), 111–113, 2007.
[2] Beni, G., "The concept of cellular robotic systems," in *Proceedings of the IEEE Symposium on Intelligent Control*, Arlington, VA, pp. 57–62, August 1988.
[3] Beni, G., and Wang, J., "Swarm intelligence in cellular robotic systems," in *Proceedings of the NATO Advanced Workshop on Robots and Biological Systems*, I1 Ciocco, Tuscany, Italy, 1989.
[4] Hackwood, S., and Beni, G., "Self-organization of sensors for swarm intelligence," in *Proceedings of the IEEE International Conference on Robotics and Automation*, vol. 1, pp. 819–829, 12–14 May 1992.
[5] Kennedy, J., and Eberhart, R., "Particle swarm optimization," *Proceedings of the IEEE International Conference on Neural Networks*, 4, 1942–1948, 1995.
[6] Kumar, V., and Sahin, F., "A swarm intelligence based approach to the mine detection problem," *IEEE International Conference on Systems, Man and Cybernetics*, 3, 2002.
[7] Kumar, V., and Sahin, F., "Foraging in ant colonies applied to the mine detection problem," *IEEE International Workshop on Soft Computing in Industrial Applications*, 61–66, 2003.
[8] Kumar, V., and Sahin, F., "Cognitive maps in swarm robots for the mine detection problem," *IEEE International Conference on Systems, Man and Cybernetics*, 4, 3364–3369, 2003.
[9] Dorigo, M., and Di Caro, G., "Ant colony optimization: A new meta-heuristic," *Proceedings of the 1999 Congress on Evolutionary Computation*, vol. 2, 1999.
[10] Sahin, F., and Devasia, A., "Distributed particle swarm optimization for structural Bayesian network learning," *Swarm Intelligence: Focus on Ant and Particle Swarm Optimization*, Advanced Robotics Systems International, 2007.
[11] Sahin, F., Yavuz, C. M., Arnavut, Z., and Uluyol, O. "Fault diagnosis for airplane engines using bayesian networks and distributed particle swarm optimization," *Parallel Computing* 33(2), 124–143, 2007.
[12] Yavuz C. M., Sahin F., Arnavut Z., and Uluyol O. "Generating and Exploiting Bayesian Networks for Fault Diagnosis in Airplane Engines," in *Proceedings of IEEE International Conference on Granular Computing, GrC 2006*, pp 250–255, May 2006.
[13] Kennedy, J., and Eberhart, R. C., *Swarm Intelligence*, Morgan Kaufmann Publishers, San Francisco, CA, 2001.
[14] Taşgetiren, M. F., and Liang, Y.-C., "A binary particle swarm optimization algorithm for lot sizing problem," *Journal of Economic and Social Research* 5(2), 1–20, 2003.

[15] Sim, K. M., and Sun, W. H., "Ant colony optimization for routing and load-balancing: Survey and new directions," *IEEE Transactions on System, Man, and Cybernetics—Part A* 33(5), 560–572, 2003.
[16] Dorigo, M., Birattari, M., and Stutzle, T., "Ant colony optimization: Artificial ants as a computational intelligence technique," *IEEE Computational Intelligence Magazine* 1(4), 28–39, 2006.
[17] Di Caro, G., and Dorigo, M., "AntNet: Distributed stigmergetic control for communications networks," *Journal of Artificial Intelligence Research* 9, 317–365, 1998.
[18] Günes, M., Sorges, U., and Bouazizi, I., "ARA—The ant-colony based routing algorithm for MANETs," *Proceedings of IEEE ICPPW*, pp. 79–85, 2002.
[19] Kassabalidis, I., El-Sharkawi, M. A., Marks, R. J., Asabshahi, P., and Gray, A. A., "Swarm intelligence for routing in communication networks," in *Proceedings of IEEE Globecom*, San Antonio, TX, pp. 3613–3617, November 2001.
[20] Montresor, A., Meling, H., and Babaoglu, Ö., "Messor: Load-balancing through a swarm of autonomous agents," Dept. Comput. Sci., Univ. Bologna, Bologna, Italy, Tech. Rep. UBLCS-2002-11, 2002.
[21] Bonabeau, E., and Meyer, C., "Swarm intelligence, a whole new way to think about business," *Harvard Business Review* 79(5), 106–114, 2001.
[22] Beckers, R., Holland, O. E., and Deneubourg, J. L., "From local actions to global tasks: Stigmergy and collective robotics," in *Prog. Artificial Life IV*. MIT Press, Cambridge, MA, pp. 181–189, 1994.
[23] Hayes, A. T., Martinoli, A., and Goodman, R. M., "Swarm robotic odor localization," in *Proceedings of IEEE/RSJ Int. Conf. IROS*, pp. 1073–1087, October 2001.
[24] Leung, H., Kothari, R., and Minai, A., "Phase transition in a swarm algorithm for self-organizing construction," *Physical Review E, Statistical Physics, Plasmas, Fluids, and Related Interdisciplinary Topics* 68(4), 046 111.1–046 111.5, 2003.
[25] Chapman, E., and Sahin, F., "Application of swarm intelligence to the mine detection problem," *Proceedings of 2004 IEEE International Conference on Systems, Man and Cybernetics*, 6, 5429–5434, 2004.
[26] Chapman, E., "GroundScouts: Application of swarm intelligence to the mine detection problem," Master Research Paper, Rochester Institute of Technology, 2004.
[27] Dorigo, M., and Stützle, T., *Ant Colony Optimization*, MIT Press, Cambridge, MA, 2004.
[28] Dorigo, M., Maniezzo, V., and Colorni, A., "Ant system: Optimization by a colony of cooperating agents," *IEEE Transactions on Systems, Man, and Cybernetics, Part B* 26(1), 29–41, 1996.
[29] Sathyanath, S., and Sahin, F., "Application of artificial immune system based intelligent multi agent model to a mine detection problem, " in *the Proceedings of the SMC 2002, IEEE International Conference on Systems, Man, and Cybernetics*, vol. 3, Tunisia, October 2002.
[30] Sathyanath, S., and Sahin, F., "AISIMAM—An AIS based intelligent multi agent model and its application to mine detection problem," in *the Proceedings of the ICARIS 2002 1st International Conference on Artificial Immune Systems*, Canterbury, UK, September 9–11, 2002.
[31] Sahin, F., "GroundScouts: Architecture for a modular micro robotic platform for swarm intelligence and cooperative robotics," *Proceedings of 2004 IEEE International Conference on Systems, Man and Cybernetics* 1, 929–934, 2004.

[32] Sahin, F., "Robotic swarms as system of systems," *System of Systems–Innovations for the 21st Center* (M. Jamshidi, Ed.). Willey & Sons, New York, 2008.

[33] Groß, R., Bonani, M., Mondada, F., and Dorigo, M., "Autonomous self-assembly in swarm-bots," *IEEE Transactions on Robotics* 22(6), 1115–1130, 2006.

[34] Azarnoush, H., Horan, B., Sridhar, P., Madni, A. M., and Jamshidi, M., "Towards optimization of a real-world robotic-sensor system of systems," in *the Proceedings of World Automation Congress (WAC) 2006*, Budapest, Hungary, July 24–26, 2006.

[35] Gracias, N. R., Zwaan, S. v. d., Bernardino, A., and Santos-Victor, J., "Mosaic-based navigation for autonomous underwater vehicles," *IEEE Journal of Oceanic Engineering* 28(4), 609–624, 2003.

[36] Yu, J., Wang, L., and Tan, M., "Geometric optimization of relative link lengths for biomimetic robotic fish," *IEEE Transactions on Robotics* 23(2), 382–386, 2007.

[37] Hart, D. M., and Craig-Hart, P. A., "Reducing swarming theory to practice for UAV control," in *Proceedings of 2004 IEEE Aerospace Conference Proceedings*, pp. 3050–3063, 2004.

[38] Kim, S., Knoll, T., and Scholz, O., "Feasibility of inductive communication between millimeter-sized wireless robots," *IEEE Transactions on Robotics* 23(3), 605–609, 2007.

[39] Donald, B. R., Levey, C. G., McGray, C. D., Paprotny, I., and Rus, D., "An untethered, electrostatic, globally controllable MEMS micro-robot," *Journal of Microelectromechanical Systems* 15(1), 1–15, 2006.

[40] Pack, J. P., and Mullins, B. E., "Toward finding a universal search algorithm for swarm robots," in *Proceedings of the 2003 IEEE/RSL International Conference on Intelligent Robots and Systems*, pp. 1945–1950, 2003.

[41] Stewart, R. L., and Russell, R. A, "Building a loose wall structure with a robotic swarm using a spatio-temporal varying template," in *Proceedings of 2004 IEEE/RSL International Conference on Intelligent Robots and Systems*, pp. 712–716, 2004.

[42] Fritsch, D., Wegener, K., and Schraft, R. D., "Control of a robotic swarm for the elimination of marine oil pollutions," in *Proceedings of the 2007 IEEE Swarm Intelligence Symposium (SIS 2007)*, 2007.

[43] Meng, Y., Kazeem, O., and Muller, J. C., "A hybrid ACO/PSO control algorithm for distributed swarm robots," in *Proceedings of the 2007 IEEE Swarm Intelligence Symposium (SIS 2007)*, 2007.

[44] Kloetzer, M., and Belta, C., "Temporal logic planning and control of robotic swarms by hierarchical abstractions," *IEEE Transactions on Robotics* (23)(2), 320–330, 2007.

[45] Clark, J., and Fierro, R., "Cooperative hybrid control of robotic sensors for perimeter detection and tracking," in *Proceedings of 2005 American Control Conference*, pp. 3500–3505, 2005.

[46] Bishop, B. E., "Dynamics-based control of robotic swarms," in *Proceedings of the 2006 IEEE International Conference on Robotics and Automation*, pp. 2763–2768, 2006.

[47] Yim, M., Zhang, Y., and Duff, D., "Modular robots," *IEEE Spectrum* 39(2), 30–34, 2002.

[48] Rus, D., Butler, Z., Kotay, K., and Vona, M., "Self-reconfiguring robots," *Communications of the ACM* 45(3), 39–45, 2002.

[49] Crossley, W. A., "System of systems: An introduction of Purdue University Schools of Engineering's Signature Area," *Engineering Systems Symposium*, Tang Center—Wong Auditorium, MIT, March 29–31, 2004.

[50] Lopez, D., "Lessons learned from the front lines of the aerospace," in *Proceedings of the IEEE International Conference on System of Systems Engineering*, Los Angeles, April 2006.
[51] Wojcik, L. A., and Hoffman, K. C., "Systems of systems engineering in the enterprise context: A unifying framework for dynamics," in *Proceedings of the IEEE International Conference on System of Systems Engineering*, Los Angeles, April 2006.
[52] Abel, A., and Sukkarieh, S., "The coordination of multiple autonomous systems using information theoretic political science voting models," in *Proceedings of the IEEE International Conference on System of Systems Engineering*, Los Angeles, April 2006.
[53] DiMario, M. J., "System of systems interoperability types and characteristics in joint command and control," in *Proceedings of the IEEE International Conference on System of Systems Engineering*, Los Angeles, April 2006.
[54] Meilich, A., "System of systems (SoS) engineering and architecture challenges in a net centric environment," in *Proceedings of the IEEE International Conference on System of Systems Engineering*, Los Angeles, April 2006.
[55] Abbott, R., "Open at the top; open at the bottom; and continually (but slowly) evolving," in *Proceedings of the IEEE International Conference on System of Systems Engineering*, Los Angeles, April 2006.
[56] Jamshidi, M. (Ed.), *System of Systems—Innovations for the 21st Century*, Wiley & Sons, New York, 2008.
[57] Jamshidi, M. (Ed.), *System of Systems Engineering—Principles and Applications*, CRC Press, Boca Raton, FL, 2008.
[58] Pearlman, J., "Creation of a system of systems on a global scale: The evolution of GEOSS," Keynote speech, *IEEE International Conference on System of Systems Engineering*, Los Angeles, CA, USA, April 24–26, 2006.
[59] Keating, C., "Critical challenges in the maturation of the system of systems engineering paradigm," Keynote speech, *IEEE International Conference on System of Systems Engineering*, Los Angeles, CA, USA, April 24–26, 2006.
[60] Jamshidi, M., "Theme of the IEEE SMC 2005, Waikoloa, Hawaii, USA," http://ieeesmc2005.unm.edu/.
[61] Sage, A. P., and Cuppan, C. D., "On the systems engineering and management of systems of systems and federations of systems," *Information, Knowledge, Systems Management* 2(4), 325–334, 2001.
[62] Kotov, V., "Systems of systems as communicating structures," Hewlett Packard Computer Systems Laboratory Paper HPL-97-124, pp. 1–15, 1997.
[63] Carlock, P. G., and Fenton, R. E., "System of systems (SoS) enterprise systems for information-intensive organizations," *Systems Engineering* 4(4), 242–261, 2001.
[64] Pei, R. S., "Systems of systems integration (SoSI)—A smart way of acquiring Army C4I2WS Systems," in *Proceedings of the Summer Computer Simulation Conference*, pp. 134–139, 2000.
[65] Luskasik, S. J., "Systems, systems of systems, and the education of engineers," *Artificial Intelligence for Engineering Design, Analysis, and Manufacturing* 12(1), 11–60, 1998.
[66] Manthorpe, W. H., "The emerging joint system of systems: A systems engineering challenge and opportunity for APL," *John Hopkins APL Technical Digest* 17(3), 305–310, 1996.

[67] Sahin, F., Jamshidi, M., and Sridhar, P., "A discrete event XML based simulation framework for system of systems architectures," in *Proceedings of the IEEE International Conference on System of Systems Engineering*, April 2007.
[68] Sahin, F., Sridhar, P., Horan, B., Raghavan, V., and Jamshidi, M., "System of systems approach to threat detection and integration of heterogeneous independently operable systems," in *Proceedings of IEEE International Conference on Systems, Man and Cybernetics*, Montreal, 2007.
[69] Lee, W. H., and Sanderson, A. C., "Dynamics and distributed control of modular robotic systems," in *Proceedings of the 26th Annual Conference of the IEEE Industrial Electronics Society*, pp 2479–2484, 2000.
[70] Murata, S., Yoshida, E., Tomita, K., Kurokawa, H., Kamimura, A., and Kokaji, S., "Hardware design of modular robotic system," in *Proceedings of the 1999 IEEE/RSJ International Conference on Intelligent Robots and Systems*, pp. 1579–1585, Korea, October 1999.
[71] Yoshida, E., Kokaji, S., Murata, S., Kurokawa, H., and Tomita, K., "Miniaturized self-reconfigurable system using shape memory alloy," in *Proceedings of the 1999 IEEE/RSJ International Conference on Intelligent Robots and Systems*, pp. 1579–1585, October 1999.
[72] Fryer, J. A., and McKee, G. T., "Resource modeling and combination in modular robotics systems," in *Proceedings of the 1998 IEEE International Conference on Robotics and Automation*, pp. 3167–3172, May 1998.
[73] Denneberg, V., and Fromm, P., "OSCAR: An open software concept for autonomous robots," in *Proceedings of the 24th Annual Conference of the IEEE Industrial Electronics Society*, pp. 1192–1197, September 1998.
[74] Opp, W. J., and Sahin, F., "An artificial immune system approach to mobile sensor networks and mine detection," *Proceedings of 2004 IEEE International Conference on Systems, Man and Cybernetics* 1, 947–952, 2004.
[75] Kauler, B., *Flow Design for Embedded Systems*, R&D Books, Lawrence, KS, 1999.
[76] Santo, B., "Embedded battle royale," *IEEE Spectrum* 28, 36–41, 2001.
[77] Labrosse, J. J., *MicroC/OS-II: The Real Time Kernel*, CMP Books, Lawrence, KS, 1999.
[78] Stevens, D. S., and Ammar, M. H., "Evaluation of slot allocation strategies for TDMA protocols in packet radio networks," *IEEE Military Communications Conference* 2, 835–839, 1990.
[79] Papadimitriou, G. I., and Pomportsis, A. S., "Self-adaptive TDMA protocols for WDM star networks: A learning-automata-based approach," *IEEE Photonics Technology Letters* 11, 1322–1324, 1999.
[80] Kanzaki, A., Uernukai, T., Hara, T., and Nishio, S., "Dynamic TDMA slot assignment in ad hoc networks," *International Conference on Advanced Information Networking and Applications*, pp. 330–335, March 2003.
[81] Young, C. D., "USAP: A unifying dynamic distributed multichannel TDMA slot assignment protocol," *IEEE Military Communications Conference*, vol. 1, October 1996.
[82] Burr, A. G., Tozer, T. C., and Baines, S. J., "Capacity of an adaptive TDMA cellular system: Comparison with conventional access schemes," *IEEE International Symposium on Personal, Indoor, and Mobile Radio Communications* 1, 242–246, 1994.
[83] Ali, F. N., Appani, P. K., Hammond, J.L., Mehta, V. V., Noneaker, D. L., and Russell, H. B., "Distributed and adaptive TDMA algorithms for multiple-hop mobile networks," *IEEE Military Communications Conference* 1, 546–551, 2002.

[84] Colomi, A., Dorigo, M., and Maniezzo, V., "Distributed optimization by ant colonies," European Conference on Artificial Life, pp. 134–142, 1991.
[85] Dorigo, M., and Gambardella, L. M., "Ant colony system: A cooperative learning approach to the traveling salesman problem," *IEEE Transactions on Evolutionary Computation* 1, 53–66, 1997.
[86] Wang, Y., and Xie, J., "Ant colony optimization for multicast routing," IEEE Asia-Pacific Conference on Circuits and Systems, pp. 54–57, December 2000.
[87] Hoshyar, R., Jamali, S. H., and Locus, C., "Ant colony algorithm for finding good interleaving pattern in turbo codes," in *IEEE Proceedings on Communications* 147, 257–262, 2000.
[88] Parpinelli, R. S., Lopes, H. S., and Freitas, A. A., "Data mining with an ant colony optimization algorithm," *IEEE Transactions on Evolutionary Computation* 6, 321–332, 2002.
[89] Tsai, C. F., Tsai, C. W., and Tseng, C. C., "A new approach to solving large traveling salesman problems," in *Proceedings of the International Joint Conference on Neural Networks*, vol. 2, pp. 1540–1545, May 2002.
[90] Bonabeau, E., Dorigo, M., and Theraulaz, G., *Swarm Intelligence—From Natural to Artificial Systems*, Oxford University Press, Oxford, 1999.
[91] Mercer, J. L., and Lenoir, A., "Individual flexibility and choice of foraging strategy in *Polyrhachis laboriosa* F. Smith (Hymenoptera, Formicidae)," *Insectes Sociaux*, 1999.
[92] Collet, M., Collet, T. S., Bisch, S., and Wehnar, R., "Local and global vectors in desert ant navigation," *Letters to Nature* 394, 269–272, 1998.
[93] Holldobler, B., *Chemical Ecology: The Chemistry of Biotic Interaction*, National Academy Press, Washington, DC, pp. 41–49, 1995.
[94] Andersen, S. L., "Landmine detection research pushes forward, despite challenges," *IEEE Intelligent Systems* 17(2), 4–7, 2002.
[95] GAO, "Landmine Detection, DoD's Research Program Need a Comprehensive Evaluation Strategy," United States General Accounting Office, April 2001.
[96] Hewish, M., and Ness, L., "Mine-detection technologies," *International Defense Review* 28, 40–45, 1995.
[97] Arkin, C., "Cooperation without communication: Multiagent schema based robotic navigation," *Journal of Robotic Systems* 9(3), 351–364, 1992.
[98] Nicoud, D., "Vehicles and robots for humanitarian demining," *The Industrial Robot* 24(2), 1997.
[99] Pynadath, V., "Multi-agent teamwork: Analyzing the optimality and complexity of key theories and models," in *Proceedings of the International Joint Conference on Autonomous Agents and Multi-Agent Systems*, 2002.
[100] Gallistel, C. R., *The Organization of Learning*, MIT Press, Cambridge, MA, 1990.

Index

A

A posteriori estimate 84
A priori estimate 84
ACE Center, 249
Action-value function, 266, 267, 270
Actuation layer, 373
Ad hoc networks, 378, 414
Adaptive sampling interval for wireless NCS, 257
Adaptive TDMA (ATDMA), 373
  Development, 378
  Protocol for swarm robots, 380
Agent, 263, 382
Agent-in-the-loop simulation, 189, 222–225, 227
Airport as an SoS, 38
Alpha-Cut Fuzzy Sets, 109
Alpha-Cut Sets, 110, 147
Analysis of Fuzzy Control Systems, 132–135
Angular momentum, 16
Animal–robot interaction, 347
Animal–robot–human integrated system, 297
Ant colony, 54
Ant colony optimization (ACO), 363
  Based mine detection algorithm, 363, 382–383, 388, 397
  Cognitive maps, 393–394, 397
  Long range recruitment (LRR), 363
  Foraging, 363, 386–388, 391–397
  Freezing, 387, 391
  Model, 387
  Recruitment of nests, 384
  Reduction factor, 395–396
  Scent following, 389, 397–398, 400
  Short range recruitment (SRR), 366
  Threshold time, 388, 392
  Visual cues, 349, 383
  Waiting, 397, 400–401

Antecedents (IF part), 125
Antipersonnel landmines, 297, 298, 354
Approximate reasoning, 123
Aral Sea Basin, 54
Architecture, 202
Artificial immune systems, 54
Artificial life, 41
Asymptotically stable, 35, 143, 144
ATDMA, 373, 378, 380–382, 409

B

Balancing problem, Cart-pole, 67
Banner of fuzzy logic, 100
Biosensors, 303
Blending parameter, 84
Boeing 747 airplane, 38
Boeing 787 Dream Liner, 46
Building, consensus, 254

C

C4ISR, 59
Canada, 54
Cartesian products, 117
Cart-pole balancing problem, 67
Center of gravity defuzzifier, 129
Centralized strategy, 254
Centrifugal governor, 9
Centrifugal matrix, 20
Challenging problems in SoS, 39
Characteristics, 38
Classes of fuzzy control stability problems, 136
Classical set complement, 103
Classical set intersection, 102
Classical set operations, 102–103
Classical set union, 102
Classical sets, 101–104
Cluttered environment, 298, 302, 340, 346, 353
Code division multiple access (CDMA), 378

Command, controls, computers, communications and information (C41), 38
Communication in space, 50–52
Communication protocol, 245, 252, 366, 374, 376
Communication submodule, 368, 373
Consensus, 254
Complex system, 37
Conjunction, 114, 115
Consensus building, 254
Consensus-based control, 252–253
Consequent (THEN part), 125
Control, 373
Command, 15
    Input, 72
    System, 2
Control, feedback, 14
Control, networked, 255–259
Control of robot manipulators, 16
Control of system of systems, 233
Control surface, 125
Controlled variables, 10
Controller 11, 13
Controllers to actuators, 258
Cooperative control, 252, 255
Coriolois matrix, 20
Corrective motor commands, 282
Cost of controlling, 72
Covariance, error, 85
Critic 268–269, 293
Crystal oscillator, 333

D

Data aggregation, 189, 190, 198
Data terminal, 283–285
Decentralized control
Decentralized control of SoS, 240
    Law, 244–245
    Navigation control, 243
Decrescent functions, 35
Deductive inferences, 119–121
Deepwater Coastguard Program, 59, 60
Defense example of SoS, 61
Definite, positive, 21
Definition of SoS, 38
Defuzzification, 125, 129
Defuzzifier, 124

Degrees of freedom, 20
Delays, 258
    Actuators, 256–258
Department of Defense, 45
Derivative gain, 13, 30
Design, control, 99, 124–146
Differential velocity control, 304
Discrete action space, 266
Discrete event system specification (DEVS), 196–197
Discrete Kalman filter, 83
Disturbance, 11, 31, 271
DEVS, 197
    Activities, 223
    Atomic models, 195, 199, 224
    Components, 209
    Coupled models, 195, 224
    Entity, 210
    Formalism, 195, 224, 228
    Modeling, 188, 223
    Model, 203
DEVS-HLA, 196
DEVSJAVA, 196, 198–200, 202
    Environment, 199–200, 209
    Plot Module, 199, 200, 210
DEVS-XML, 42, 198
    Format, 198, 202
    Simulation framework, 43
    SoS simulation, 215
Disjunction, 114, 115
Distributed systems, 38
Distribution, posterior, 95
Dynamic task uploading, 375
Dynamics, Newtonian, 11

E

E-enabling, 46–47
Efficient locomotion, 300
Eigen values, 72, 77
Electrical power systems, 52, 53
Elements of a control system, 2
Embodiment, 263
Energy, potential, 17
Emergence, 1, 266
Emergence in SoS, 43
Emergent behavior, 38, 40
Engine, inference, 125, 126–128
Engineering of SoS, 40

Index

Enterprise SoSE, 38
Entity-centric abstraction model, 57
Equality/Equivalence, 114, 115
Equation, recursive, 85
Equilibrium points, 31
Error, measurement of, 75
Error covariance, 85
Estimated parameters, 23
Estimation error, 76
Evolution of the social infrastructure, 38
Expected future rewards, 266–268, 273
Exploration vs. exploitation, 271
Extensible Markup Language (XML), 41
Extension principle, 110–111

F

FCM system, 61
Feedback control, 14
Feedback correction, 29
Feedback gain vector, 73, 77
Feedback gains, 79
FDMA, 378
Forest environment, 298, 302, 340, 344, 348, 352
Formation of the system of rovers, 250
Framework, of simulation, 195–202
Frequency division multiple access (FDMA), 378
Frequency-domain methods, 138
Front agents, 250–251
Function, reward, 264
Future combat missions, 60–61
Fuzzification, 125
Fuzzifier, 125
Fuzzy chips, 99
    Composition, 113
    Control design, 99, 124–146
    Control system, 124, 133
    Logic, 99, 100, 121–122, 147
    Proposition, 121
    Relations, 112
    Rule-based system (FRBS), 124–125
Fuzzy sets, 99, 100, 104–108
    Complement, 108
    Intersection, 108
    Union, 106–108
Fuzzy Systems, 99

G

Geographic distribution, 38
Global Earth Observation SoS (GEOSS), 58–59, 60
    Positioning satellite (GPS), 51
Goldcare, 47
Graphical user interface (GUI), 398, 403
    Comm menu, 404
    Robot status panel, 405
Gravity matrix, 20
Great Lakes, 54
Ground-penetrating radar, 301
GroundScout, 6, 215, 217, 222–228, 231, 364, 365, 368–369, 371–379, 381–382, 397
Group innovation process, 281

H

Hardware in the loop simulation, 252
Hardware modularity, 369
Health Care Systems, 57–58
Heterogeneous systems, 37
Hierarchical control of SoS, 233
    Four-system SoS, 237
    Reformulation of SoS control problem, 239
    Structure, 234
High-impedance obstacle, 345
Hot temperature as a fuzzy variable, 101
Human-induced disasters, 59

I

I-button, 283, 384
ICSoS, 40, 57
Identification, 4
IIT-Kanpur, 249
Implication, 114, 116
Importance sampling, 95
Industrial application of SoS, 5
Inertia matrix, 20
Inference engine, 125, 126–128
Infrastructures, 47

Integral parameter, 15
Integration, 192, 195, 343, 347, 366
Intelligence, 1
Interactions among energy
    and atmospheric, 55
Interdependencies among systems, 43
Internal models, 264–265, 268
Internal states, 2
International consortium of system of
    systems engineering, 40
Interoperability, 191, 363
Interrupt pin, 339
Inverse dynamics, 261
Interval matrix, 143
Intervals, sampling, 25
Inverted pendulum, 130

J

Joint Planning and Development
    Office, 56

K

Kalman filter, 67, 79, 83
Kinetic energy, 17
Kinetic potential, 18

L

Lagrange equation, 17
Large-scale integrated system, 38
    Complex, dynamical system, 52
Law of contradiction, 109
    Excluded middle, 109
Learning, supervised, 261, 292
Legged field robot, 297, 302
Linear regulator, 72
Linguistic label, 344
Locally positive definite functions, 34
Locomotion, 364, 368, 370
LQG problems, 258
Lyapunov stability, 138–142
    Requirement, 34

M

Machine Intelligence Quotient
    (MIQ), 99

Machine operating skills, 287
Magnetic quadrupole resonance, 301
Mamdani implication, 128
Manipulated variables, 10
Markov decision process, 265
Markov state, 265, 268–269, 272
Mars, surface of, 252
Master agent, 248–251
Matrix, interval, 143
Maximum difference matrix, 143
Mean, 79
Measurand, 337
Measurement error, 75
Measurement noise, 3
Mechanical filter, 345
Membership, 3
Membership function, 105
Memory primitives, 264
Meta-architecture generation for
    financial markets, 42
Metal detectors, 298–300
Microcontroller, 225, 283, 333, 336
Micromodular swarm robots, 6
Mine detection, 300, 363
    Problem, 364, 382, 385
    With ant colony-based swarm
        intelligence, 382
Mobile swarm agent, 210, 212, 215, 224
    Communication layer, 225
    Control layer, 224
Model, 210
Mobile threat, 212
Model, 212, 223
Model based controller, 27
Model of the dynamics, 4
Model, sensor, 94
Modern control systems, 9
Modularity, hardware, 369
Modular robotics, 365, 368
Modular robotic systems (MSR), 368
Modus ponens, 118
Modus tollens, 118
Momentum, 12
Momentum, angular, 16
Motion coordination, 245–259
Multiagent rovers, 249
Multi-robot approaches, 353
Multisensor systems, 300

Index

N

NASA, 50
NASA's Space Constellation Project, 50
National Healthcare Information Network (NHIN), 57
National security, 61–62
Navigation in space, 50–52
Negation, 114, 115
Net-centricity, 4
Netcentric community-of-interest SoS ecology, 44
Network control system (NSC), 259
Network-centric operations, 47
Networked control, 255–256
Networks, sensor, 48
Neutral equilibrium point, 34
Newtonian dynamics, 11
Noise, 79
Noise, measurement, 3
Noise, process, 3
Nonlinear dynamics, 3

O

Observation, 2
Observer, 79
Observer design, 67
Observer-Observable pattern, 224
Observing and filtering, 73
Obstacle avoidance behavior, 262
Olfactory capability, 298
Open software concept for autonomous robots (OSCAR), 369
Open system approach to SoSE, 39
Orchestration, 180, 294
Ordinary differential equation, 25
Oscillator, crystal, 333

P

Parameter, blending, 84
Parameter uncertainty, 24
Particle filter, 94
Particle Swarm Optimization (PSO), 364
   Algorithm, 364
Passive dynamics, 303

PD controller, 12, 15
Performance, tracking, 27, 30
Petroleum-based transportation SoS, 54
Parameters, estimated, 23
Phase portrait, 79
Pheromone, 363, 382–384, 397, 399
Photo resistor, 338–339
Plant, 9
Policy, 256, 264
Ponens, modus, 118
Positive definite, 21
Positive definite functions, 34
Posterior distribution, 95
Portrait, phase, 79
Potential energy, 17
Power systems, 52
Predicate logic, 114–121
Principle, extension, 110–111
Process noise, 3
Properties of classical set, 103–104
Properties of fuzzy sets, 108–111
Proportional gain, 13
Pulse-width modulation, 333, 371

Q

Q-learning, 267
Q network, 274, 276
Qualitative feedback, 293
Quality of performance (QoP), 258
Quality of service (QoS), 258

R

Recursive equation, 85
Regulator, voltage, 335
Reinforcement based learning, 5, 265, 293
Renewable energy, 52–53
Resistor, photo, 340
Reward, 266
Reward based behavior adaptation, 263
Reward function, 264
Riccati Equation, 236, 238
Rigid body manipulator, 17
Robot manipulator, 21

Index

Robotic,
  Swarms, 54–55, 56, 363–367
  Colony, 365
Robots finding start of a fire, 43
Robust threat detection, 189, 197, 202
  Simulation, 205
Runge-Kutta integration, 24

S

Sagittal diagram, 112, 116
Sampling, importance, 95
Sampling intervals, 25
Seismoacoustic methods, 301
Sensing and perception, 336
Sensor, 210, 212, 372
Sensor model, 94
Sensor networks, 48
Sensors to controllers delays, 258
Service industry, 47, 48
Signal sensor, 338
Simulation, 41–42, 191
Simulation framework, 189, 195, 200
  XML hierarchy, 209
Situatedness, 263
Skill innovation, 281, 285, 291
Smart grids as SoS, 52
Software modularity, 368, 374
Sonar sensor, 262, 336, 340
SoS aircraft design via SoSE, 46–47
SoS, 38
  Architecting, 41
  Architecture, 189, 193
  Characteristics, 189, 363
  Concepts, 189
  Definitions, 190
  Engineering, 37
  Integration, 38, 42–43
  Management, 44–45
  Message architecture, 198, 202
SoSE, 37
Space exploration, 50
Stability, 31
Stability of fuzzy control systems, 135–136
Stability via interval matrix method, 143–146
Stable equilibrium point, 31

Standard deviation, 79
Standards of SoS, 40
State, 70
State space, 70
State space methods, 67
State space model, 70
State variables, 70
Strictly increasing function, 35
Sub-systems, 1
Super system, 37
Supervised learning, 261, 292
Surface, control, 125
Surface of mars, 252
Sustainable Environmental Management from SoSE Perspective, 53–54
System, transportation, 55–56
  Intelligence, 365
  Robot, 200, 202
  Robotics, 363–364
  Robots, 200, 206, 220
Synchronization, 223–224
System, 10
Systems, distributed, 38
System of systems, 37, 191, 215

T

Takagi-Sugano rules, 127, 147
TD error, 268
TDMA (time division multiple access), 55, 373, 378, 381, 404
Temporal difference based learning, 267, 270
Threat, 189, 190, 202
  Detection scenario, 197, 202, 223
  Detection scenario with several swarm robots, 208
Time division multiple access (TDMA), 373
Time-domain methods, 136–138
Tollens, modus, 118
Totally stable system, 141
Trace explosive detection systems, 301
Tracking performance, 27, 30
Trained mongoose, 348–349
Transducer, 301, 338
Transportation system, 55–56

Turbine, 11
Two-point boundary value (TPBV) problem, 235

U

Uncertainty, parameter, 24
Unifying slot assignment protocol (USAP), 378
  Multiple access (USAP-MA), 378
University of Texas, San Antonio, 249
Unstable, 14
UTSA, 249

V

Value function, 266, 268, 270
Variables, 10
Vector Q learning, 272

Vector reward function, 264, 275
VideoRays underwater rovers, 253
Voltage regulator, 333, 336

W

Water resource management, 59
Wireless networked control SoS (WNCSOS), 256

X

XmlStrEntity, 210–211
XML and DEVS, 198, 218
XML-based SoS simulation, 205
XML hierarchical architecture, 215
XML-like common language, 367
XML SoS real-time simulation framework, 222, 232